TELEPRESENCE

BT Telecommunications Series

The BT Telecommunications Series covers the broad spectrum of telecommunications technology. Volumes are the result of research and development carried out, or funded by, BT, and represent the latest advances in the field.

The series includes volumes on underlying technologies as well as telecommunications. These books will be essential reading for those in research and development in telecommunications, in electronics and in computer science.

1. *Neural Networks for Vision, Speech and Natural Language*
   Edited by R Linggard, D J Myers and C Nightingale
2. *Audiovisual Telecommunications*
   Edited by N D Kenyon and C Nightingale
3. *Digital Signal Processing in Telecommunications*
   Edited by F A Westall and S F A Ip
4. *Telecommunications Local Networks*
   Edited by W K Ritchie and J R Stern
5. *Optical Network Technology*
   Edited by D W Smith
6. *Object Oriented Techniques in Telecommunications*
   Edited by E L Cusack and F S Cordingley
7. *Modelling Future Telecommunications Systems*
   Edited by P Cochrane and D J T Heatley
8. *Computer Aided Decision Support in Telecommunications*
   Edited by P G Flavin and K A F Toffon
9. *Multimedia Telecommunications*
   Edited by W S Whyte
10. *Network Intelligence*
    Edited by I G Dufour
11. *Speech Technology for Telecommunications*
    Edited by F A Westall, R D Johnston and A V Lewis
12. *Mobile Systems*
    Edited by I Groves
13. *ATM for Service Providers*
    Edited by J Adams
14. *Computing Systems for Global Telecommunications*
    Edited by S West, M Norris and S Stockman
15. *The Internet and Beyond*
    Edited by S Sim and J Davies

TELEPRESENCE

Edited by

P. J. Sheppard and G. R. Walker
B.T. Laboratories,
U.K.

SPRINGER SCIENCE+BUSINESS MEDIA, B.V.

A C.I.P. Catalogue record for this book is available from the Library of Congress.

ISBN 978-1-4613-7414-5 ISBN 978-1-4615-5291-8 (eBook)
DOI 10.1007/978-1-4615-5291-8

Printed on acid-free paper

Contents

Contributors

S D Benford	Reader in Computer Science, Nottingham University
P A Bowman	Media Environments, BT Laboratories
J M Bowskill	Wearable Computing and Communications, BT Laboratories
A Bullock	Department of Computer Science, Nottingham University
D Burraston	Multi-modal Perception, BT Laboratories
I B Cockburn	Wholesale Finance, Syntegra
M Collins	Telematics Applications, BT Laboratories
P Cordell	Multimedia Applications, BT Laboratories
G Cosier	Applied Research and Technology, BT Laboratories
J M Courtenay	Value Added Service Development, BT Laboratories
K J Fisher	On-line Futures, BT Laboratories
C J H Fowler	Education and Training Research, BT Laboratories
P Garner	Telematics Applications, BT Laboratories
D L Gibson	Speech Platform Technology, BT Laboratories
C M Greenhalgh	Department of Computer Science, Nottingham University
N Haque	Systems Integration Sales, Syntegra
D J T Heatley	Mobile Systems, BT Laboratories
M R Hinds	Shared Spaces, BT Laboratories
M P Hollier	Multi-modal Perception, BT Laboratories
P J Lawrence	Media Environments, BT Laboratories
A V Lewis	Speech Coding and Adaptive Audio Systems, BT Laboratories
F T Lyne	Mobile Systems, BT Laboratories

D Machin	Natural Communication Systems, BT Laboratories
J T Mayes	Centre for Learning and Teaching Innovation, Glasgow Caledonian University
S McConnell	Natural Communication Systems, BT Laboratories
J Morphett	Shared Spaces, BT Laboratories
D G Morrison	Video Coding and Standards, BT Laboratories
A N Mortlock	Natural Communication Systems, BT Laboratories
P J Mulroy	Distributed Information Systems, BT Laboratories
I Neild	Mobile Systems, BT Laboratories
I Parke	Broadcast and Conferencing Services, BT Laboratories
D Pauley	Applied Research and Technology, BT Laboratories
R W Picard	Associate Professor of Media Technology, MIT
S J Powers	Shared Spaces, BT Laboratories
A N Rimell	Multi-modal Perception, BT Laboratories
D A D Rose	Telematics Applications, BT Laboratories
S Rudkin	Distributed Systems, BT Laboratories
M Russ	Audiographic Conferencing, BT Laboratories
P J Sheppard	Natural Communication Systems, BT Laboratories
C K Sidhu	Human Factors, BT Laboratories
D M Traill	Media Environments, BT Laboratories
J Tromp	Department of Computer Science, Nottingham University
R M Voelcker	Multi-modal Perception, BT Laboratories
G R Walker	Telepresence Campaign Manager, BT Laboratories
S M Webster	Telematics Applications, BT Laboratories
M W Whybray	Video and Audio Coding, BT Laboratories
A K Williamson	Systems Integration, BT Business Sales
L Willis	Applied Research and Technology, BT Laboratories

Preface

More than a century ago, when Alexander Graham Bell and his assistant made the first telephone connection, there was much debate and speculation about whether anyone would find a practical use for such a contraption. Who could have envisaged that Bell was laying the groundwork for something as dynamic and complex as the telecommunications infrastructure that exists today?

Yet no matter how sophisticated our telecommunications system becomes, the greatest use is still the ordinary phone call. Telephony is the ultimate example of how successful technology adapts to the very human need to communicate; and, whether it is a call home while travelling on business or an interactive journey through the Internet, bringing people together remains the foundation of the global telecommunications industry.

Telephony is a basic form of telepresence — enabling human interaction at a distance, creating a sense of being present at a remote location. As we move from a physical to an information economy, demand will grow for services which transport bits not atoms and which support the output of the human mind. In the next millennium we are certain to be increasingly 'telepresent' both at work and at play, with an expanding market for telepresence services.

Building on a strong base of established audiovisual conferencing and interactive multimedia services, BT Laboratories, together with its research partners, is leading the way in the multi-disciplinary exploration of the social opportunities and technical challenges of future telepresence services. The content of this book has been selected to provide a snapshot of these ground-breaking activities — conveying the complexity, uncertainty and enthusiasm involved in this challenging and exciting work.

The opening chapter overviews the status and prospects for telepresence providing a context for the remainder of the book. Telepresence services will be built on a diversity of technologies, and a wide-ranging coverage is provided of the future of audio, data, video and virtual environments. These chapters highlight the remarkable functionality that will become available through continued advances in processing power and global network connectivity.

We are ultimately concerned with telepresence as a service, not as isolated technologies. The starting point for all such communications services is a customer need, a particular desire for contact. The importance of the human experience in achieving that contact through telepresence, including issues of usability, quality of service and emotional interaction, should not be underestimated. We have therefore included a number of chapters that address the human factors of user experience and assessment of service quality.

In looking forward to prospective telepresence services we can anticipate a future of greatly increased diversity of media. There will be a much wider choice in matching an appropriate degree of presence with the initial desire for human contact. Making that match, with a cost-effective and easy-to-use solution, will be the key to commercial telepresence. With this in mind we have included not only some chapters which combine the various technologies to meet generic needs and also a few early examples of the application of telepresence to specific markets, namely education, medicine and finance.

The telephone has set high standards for telepresence services that stress 'people as content'. The future promises a remarkable diversity of multimedia telepresence — we invite you to enjoy the various contributions presented here as pointers to just some of the exciting possibilities.

Graham Walker and Phil Sheppard
BT Laboratories

1

TELEPRESENCE — THE FUTURE OF TELEPHONY

G R Walker and P J Sheppard

1.1 INTRODUCTION — THE TELEPHONE HERITAGE

Telepresence is the enabling of human interaction at a distance, creating a sense of being present at a remote location. As a well-established telepresence service, the telephone extends human speech and hearing. Whatever your metric, it is an outstandingly successful service — 600 million lines world-wide, 100% interoperability, comprehensive directory services, simple interface, consistent quality of service, low cost, and universal accessibility.

Moreover, the telephone has moulded both users and society to its capabilities and limitations. There is established telephone etiquette for both business and personal calls, and 'user training' is an assumed part of modern childhood. As we plot a course into a future of ubiquitous, multimedia telepresence we must be conscious of this heritage. The telephone is the 'existence theorem' for such universal services, but has also set a high level of customer and market expectations.

In the next millennium we are certain to be increasingly 'telepresent' both at work and at play, with an expanding market for telepresence services. As we move from a physical to an information economy, demand will grow for services which transport bits not atoms and which support the output of the human mind. Moreover, technology continues to advance at an unprecedented rate, and it can be suggested that we are reaching key breakpoints in client processing power and global network interconnectivity. Against a background of these favourable long-term trends, this chapter overviews the status and prospects for telepresence, providing a context for the remainder of the book.

We start with a review of key telepresence technologies, the building blocks of future services, addressing issues in audio, data, video and virtual environments. This segmentation of media is a reflection of traditionally distinct technical disciplines, but we move on to consider the reality of telepresence as a complex multimedia service, where developments in component technologies must be complemented by advances in integration, interface and service. We conclude with a commentary on current vertical market applications and more generic future telepresence solutions.

1.2 TELEPRESENCE TECHNOLOGIES

In setting the scene for later chapters on current and future work, Chapter 2 also serves as enlightening historical background. Its review charts the development of BT's conferencing services from proprietary analogue systems in fixed studio locations, through gradual evolution and occasional revolution to standards-based digital services with widespread availability.

As we speculate on a potential explosion in telepresence services enabled by decreasing costs and increasing mobility in networks and consumer hardware, it is salutary to note that the chapter chronicles several decades of remarkable technical advances but only a handful of commercial services. Indeed, pundits world-wide have been repeatedly over-optimistic in their predictions of market growth. Spectacular growth in PC ownership (180 million in 1995) and Internet hosts (40 million in 1996) confirms the potential for revolutionary developments, but in aspiring to truly mass-market telepresence services, it should be remembered that 50% of the world's population have yet to make their first phone call. Achieving technical interoperability is certainly a prerequisite, but we will return later to the importance of parallel advances in aspects of interface and service surround in pushing forward a market with a chequered history for matching technical capabilities to customer needs.

1.2.1 Audio

In real-time communications, audio is the core of the media experience. When receiving information it can be argued that 'a picture is worth a thousand words', but, lacking a direct visual output capability, human-to-human interaction is dominated by speech.

There are well-established and evolving standards for the delivery of 'telephone quality' monaural speech over ever-decreasing bandwidths. Chapter 3 discusses service advances which promise to make multi-party audio-conferencing almost as accessible and routine as one-to-one telephony. Audio-conferencing is currently a closely supervised, premium service, ideally suited to pre-arranged business meetings but inappropriate for informal or social discussions. The chapter goes on to describe developments in unattended conferencing which put the customer in control and offer flexible, cost-effective and scalable conferences.

However, monaural speech falls well short of our rich audio experience in the physical world, and bridging that gap would be a significant boost for telepresence. As coding techniques and processing power advance and network costs decline, we can anticipate higher audio bandwidths and multi-channel sound formats. Chapter 4 reviews the principles and promise of spatial audio, a technology that can recreate a complete audio environment with directional sound sources and location-specific acoustics. The advantages of spatial audio have been proven in complex control environments such as aircraft cockpits, and compelling three-dimensional sound for computer games is currently the focus of much commercial activity — benefits in more routine applications are readily envisaged.

Further opportunities are afforded by interpretative management of the audio information. A combination of speech processing and natural language techniques promises spoken language systems supporting direct man/machine dialogue. Once again, the pre-eminence of speech as the human 'output channel' suggests that this is certain to be a dominant element in future service interfaces. Westall et al [1] provide an introduction to this area.

Looking at wider technology trends, PCs, and computers in general, have stressed the importance of the visual display, and in application development audio is typically a poor relation in the allocation of system resources such as CPU cycles. Practical issues such as audio interference in open-plan offices are one reason, but as we move to a future of interactive, mobile devices, with screen space at a premium, there will be increasing emphasis on 'auditory displays'. The creation and management of personal audio workspaces is a critical telepresence enabler, with miniaturized, wireless transducers competing with free-space solutions. Delivering flexible audio landscapes into hostile settings, such as open-plan and reverberant offices, presents substantial technical challenges, but with potential for commensurate rewards.

1.2.2 Data

Remote interaction and presentation of data is a key element of many telepresence services. Even asynchronous exchanges by e-mail or file transfer can be viewed as telepresence, extending our ability to communicate at a distance. Indeed, a tighter integration of real-time conferencing with established desktop groupware will be an important market development. Directories and interfaces will be required to support a seamless transition from asynchronous to real-time services, mirroring the reality of multi-faceted human communication around a single transaction or relationship. Even a one-off conference involves pre-meeting set-up and post-meeting 'wash-up' phases, which a customer would prefer to see integrated into a single service.

Within a real-time service, data-conferencing is typically used in parallel with other media for the direct human/human interaction. Chapter 5 reviews progress in collaborative data sharing, and describes a trial audio-graphic conferencing service developed by BT. Shared access to files within a workgroup is already commonplace, and the chapter points to a future of collaborative applications on a shared desktop.

In extending such services beyond the local team one aspect of data-conferencing requiring particular attention is security. For media such as audio and video, security equates primarily to the protection of privacy in real-time streams, whereas shared data may imply access to persistent file storage, typically protected within a corporate firewall. This issue is addressed in Chapter 6. At one level, it is appropriate for us to address telepresence services in isolation from developments in underlying network protocols, trusting that gateways will smooth over problems of interworking between a diversity of switched and packet-based networks. However, services must ultimately be 'network-aware', recognizing and adapting to the data transport capabilities, and these wider aspects of Internet conferencing, including discussion of emerging standards, are also considered in Chapter 6.

Finally under the heading of data, we consider 'text chat', an alternative or supplement to speech, which is prevalent in current Internet games and chat rooms. Text dialogue, while tending to be regarded as a short-term expediency for overcoming limitations in network performance, still has a long-term appeal where cost savings, persistence and pseudo-real-time nature are of direct benefit. For example, in applications that do not demand 100% attention, a text chat user can sustain more than one simultaneous session — a limited form of concurrent telepresence with multiple identities [2].

1.2.3 Video

In many respects, video is an obvious and desirable extension to audio in enhancing any telepresence experience. We have already referred to vision as the most powerful human 'input channel', and although sometimes subtle and overlooked, visual output in the form of gesture and body language is a key element of rich dialogue in a physical encounter. However, transmission and display of high-quality moving images is enormously costly compared with audio, and current options for intelligent, content-based processing and control are even more restricted than with speech signals.

Videoconferencing services have met with only limited commercial success, with image quality a frequent concern. It can be argued that television has established a benchmark for customer expectations, and reaction to the 'pixelated jerk-vision' achievable over traditional audio bandwidths is often disparaging. There are also many types of meeting where visual contact is relatively unimportant compared with audio and data interaction, and anecdotal evidence is peppered with references to minimized video windows and disconnected cameras in desktop applications. The greatest penetration of videoconferencing is in high-end 'board room' systems, which deliver acceptable image quality over high-bandwidth links, and where discursive interaction is strongly supported by visual cues.

Uptake of multipoint videoconferencing services is similarly limited, with specific issues of scalability in the distribution and display of multiple images. 'Continuous presence', where all locations are tiled into a single image, is generally preferred to voice switching or manual control, but there is no sense in which participants achieve a feeling of co-presence in a shared meeting space. Chapter 7 discusses the uptake of videoconferencing, focusing on social and interface challenges and seeking inspiration in the history and techniques of film and television.

Such service issues notwithstanding, there has been tremendous progress in technical standards and interoperability for video, and this is reviewed in Chapter 8. The chapter also points towards future developments in content-based coding and processing which will open opportunities for more sophisticated yet cost-effective services.

Moving beyond general conferencing, video is an essential element of many specific telepresence services. In remote surveillance, medical or engineering applications, there may be objective requirements for image quality, and trade-offs can be made in meeting the particular needs. For example, a mixture of poor-quality moving and high-quality still images may be appropriate, or the quantifiable benefits may justify more expensive equipment and network connections. We will return to such 'vertical market' applications later.

This section concludes with an assertion that in the long term video **will** be an assumed component of most telepresence services. Although we have outlined a number of factors which often make the cost-benefit equation unfavourable in the short term, some degree of benefit is invariably present, and will ultimately be realized in the face of declining costs.

1.2.4 Virtual environments

Our final telepresence technology is the all-encompassing area of virtual environments. At one level, virtual environments promise a faithful recreation of physical reality with complete sensory telepresence. Beyond that, they need not be constrained by physical laws and could even amplify or modify conventional human presence. More realistically in the immediate future, we are concerned with computer-generated three-dimensional graphical environments, which may include other media components such as real-time streamed audio or video. This flexible mixing of media — audio, video and data, real-time and persistent — in a single consistent environment is the unique appeal of virtual environments.

In conferencing applications the focus is on multi-user or 'inhabited' virtual environments, which serve as shared spaces for communication and interaction. A key concept in such spaces is the 'avatar', a graphical embodiment through which the user interacts with their surroundings and other users. Depending on the technology and application, the avatar may be more or less realistic, with non-human forms common in multi-player games. Chapter 9 describes a relatively sophisticated virtual environment ('Inhabiting the Web') which was used for a series of meetings between BT Laboratories (BTL) and five UK universities over the SuperJanet network. Virtual environments can potentially deliver highly scalable telepresence services, and the chapter discusses some of the performance issues and architectural options.

Of particular interest in non-business applications, avatars afford the option of anonymity, which provides equality between users and promotes open dialogue in a setting of safe intimacy. In exploring service concepts in this area, a recent publication describes 'The Mirror', a multi-user Internet virtual environment, based on the emerging VRML standard [3]. The Mirror mixed professional content with social chat and interaction in a vision of 'Inhabited TV', which was delivered to two thousand viewers of the BBC series 'The Net' (see Fig. 1.1).

Fig. 1.1 Avatars 'inhabiting' a bouncy castle in 'The Mirror'.

Although there are a wealth of options and opportunities in virtual environments for telepresence, it is clear from these and other trials that we remain in the early stages of experimentation. There are few standards for technical interoperability, and studies of interface and performance are in their infancy. Chapter 10 reports on the state of the art that can be achieved in a research setting. While virtual environments can portray complex behaviours, this is of limited value when user interaction is constrained by the mouse and keyboard, and the chapter describes the use of gesture tracking to facilitate the transition from real to virtual world. Further advances in such intuitive, multimodal interfaces will be important developments.

Chapter 11 considers a complementary approach, which seeks to mix real and virtual worlds in order to maximize their respective strengths. The chapter focuses on collaborative working, and highlights a number of applications where 'enhanced reality' telepresence services offer more appropriate functionality than their fully virtual relations. As with video, virtual environments will find early niche applications in telepresence services which deliver benefit for a specific remote interaction as opposed to supporting general-purpose conferencing. Opportunities include remote manipulation in hostile locations and also control of miniature instruments, where virtual environments can deliver a translation of physical scale. Such applications may engage senses other than vision and hearing, with force feedback offering clear benefits.

In the face of so many options in applications and implementation — desktop versus room versus immersive, and the physical form of future computing 'appliances' are just two areas that we have not considered — our brief review can only hint at the issues and potential of virtual environments for telepresence. The 'dream' of immersive telepresence is the driver for several long-term network and computing research programmes, such as the US Internet2 initiative [4], and many challenges remain. Not the least of these is identifying the most appropriate combination of technologies to serve specific market needs, and then packaging them into an attractive commercial service. Contrasting this technological complexity with our baseline telepresence service, the telephone, highlights the importance of **service** above **technology** and leads on to the next section of our discussions.

1.3 TELEPRESENCE SERVICES

In the preceding section we considered specific telepresence technologies, able to extend selected human senses and interactions at a distance. However, we are ultimately concerned with telepresence as a service, not as isolated technologies. Our interest is motivated by specific customer communications requirements, which can be satisfied by an appropriate combination of media and control.

The starting point for all communications is a customer need, a particular desire for contact. The nature of that need is then reconciled with the available technology in selecting an appropriate service. At present, the choice of technology for many people may be restricted to a phone call, fax, post or a physical meeting, with e-mail an increasingly prevalent option. However, we can anticipate a future of greatly increased diversity of media, where there will be a much wider choice in matching an appropriate degree of presence with that initial desire for contact.

Making that match, with a cost-effective and easy to use solution, will be the key to commercial telepresence. Interoperability is perhaps the leading challenge for component technologies, and we have referred to the varying states of standards development in the areas of audio, data, video, and virtual environments. In this section we move on to consider the wider user experience, and to discuss specific service opportunities.

1.3.1 User interface

Returning once again to our baseline telepresence service, we speak of 'picking up the phone' to establish a telephony connection — a powerful expression of simplicity in a ubiquitous interface. However, even minor variations in national tones and dialling procedures can confuse the uninitiated, and greater complexity in mobile handsets, office PABX telephones and network services is fuelling an increasing weight of manuals and user guides. In multimedia communications we are far from 'picking up the camera' or even 'clicking the mouse' to initiate a dialogue. Day-one demand for world-wide services and global manufacturing ambitions will help to minimize international disparities, but will do nothing to halt this spiral of complexity and ultimately customer confusion.

While this burgeoning complexity can be handled by the expert user in the short term, we expect developments in agent technology which will assist in the set-up and management of future services. In the opening of this section, we characterized a telepresence service as a reconciling of customer need for remote presence with appropriate media and control. As the options multiply, there will be increasing need for transparent automation in this matching of needs with available technologies. Current conferencing protocols can exchange details of terminal 'capabilities', such as coding schemes and camera control. Future agent-based solutions will extend to the negotiation of scheduling and media selection, using personal profiles, history logs and diary information.

In addition to simplifying service set-up, it is also important to provide familiar default interfaces which mimic established services in their interaction with both the environment and the technology. As new technology becomes widely accepted, this requirement for conventional interface metaphors will diminish. For example, the computer 'desktop' is a comfortable and flexible spring-board, but does not extend efficiently into immersive virtual environments, which literally open new dimensions in interface design. Chapters 9 and 12 give particular consideration to interaction in virtual environments.

Chapter 13 also explores a new dimension of user interfaces, in its overview of 'affective computing'. The role of emotional intelligence in enabling human decisions and communications is highlighted, while a future is anticipated in which computers can support and mediate such interactions. As with the promise of virtual environments, increased processing power and innovative hardware will progressively overcome practical limitations which have traditionally moulded interfaces to the strengths and weaknesses of computers, not those of their human users.

1.3.2 Service assessment

In parallel with advances in individual technologies and interface design, we must progress our ability to assess and monitor complex multimedia communications, both during development and in service. Once again, the telephone provides instructive pointers, with a wealth of understanding on objective measures and also subjective performance assessment.

Subjective tests have been used for many years as an effective means of evaluating the transmission quality of telephony networks. In a typical test, subjects are asked to rate intelligibility of dialogue under a carefully controlled set of network and environmental conditions, e.g. different levels of loss and noise. However, for live network testing, there are both practical and cost advantages in using test equipment to provide objective assessment of systems. Chapter 14 describes the development of such an objective system for audio quality assessment, and the chapter goes on to present an extension of that work to multi-modal assessment, for services which involve more than one sense. Multi-modal assessment presents a major challenge in understanding the psychological factors of importance and mapping these on to appropriate measurement technologies, but it is essential if we are to be able to assess objectively the perceived quality of future telepresence services.

1.3.3 Vertical markets

As we resolve both technical and service issues, the future of telecommunications will see increasingly flexible and ubiquitous multimedia telepresence progressively displacing 'bread'n'butter telephony'. However, the commercial starting point in the face of such complexity must be a selection of 'vertical markets' with clear telepresence needs and quantifiable short-term benefits. These highly focused applications must be able to exploit the capabilities of early generations of specialized hardware, such as wearable computers, and we will also see a growing role for kiosks and other forms of public computing in providing wider access.

A number of chapters point to early opportunities for telepresence:

- multi-player gaming is currently the leading market for residential multimedia 'conferencing', perceived demand for these services driving developments in both hardware and software, which aim to deliver compelling desktop virtual environments over relatively modest bandwidths — Chapter 15 addresses long-term challenges in the scalability and performance of such applications, motivated initially by the latency and persistence requirements of games, but of equal relevance to other inhabited virtual environments;
- tele-medicine offers immediate benefits in cost and quality of care, and sets relatively clear goals for telepresence services — Chapter 16 reports on a range of early trials which stress the importance of focused design and integration of the overall service;
- tele-education is another important market, where telepresence can have an immediate impact — Chapter 17 presents a vision of learning as a social and collaborative process, ripe for the application of telepresence principles, reporting practical lessons from a range of robust prototypes which have been used in real educational settings;
- the world of finance is our final example of a vertical market for telepresence services — Chapter 18 describes an integrated desktop environment, which provides a tailored mix of information and communications services to dealers; the skills and interactions supported are typical of many industries in an information economy, and similar developments can be envisaged for other business sectors.

1.3.4 Immersive conferencing

Finally, and looking further forward into the future, we outline a broader vision of another two generic telepresence scenarios with potentially widespread application. Our first vision is of 'immersive conferencing', a telepresence conferencing service which is 'better than being there'. Figure 1.2 illustrates the concept for two people seated at a desk. Large video images extend the desk 'through the screen' to participants at remote locations and a consistent, spatial audio environment gives a true sense of co-presence. Opportunities to go beyond a real life meeting arise through the use of intelligent processing in the interface, perhaps to provide access to on-line information services, an automatic secretary or even language translation.

We can conceive of a personal immersive conferencing workstation. In the 'SmartSpace' concept demonstrator, described in Chapter 11, a wrap-around visual display and personal audio 'sound bubble' provide the sense of immersion,

Fig. 1.2 Desktop immersive conferencing.

while the tablet is a flexible local desktop. Figure 1.3 extends the concept to a purpose-built room. Wearable computing is used to identify participants, record and customize their interactions, and track their location. A virtual 'greeting agent' handles security and access to the main conferencing area, where tracking

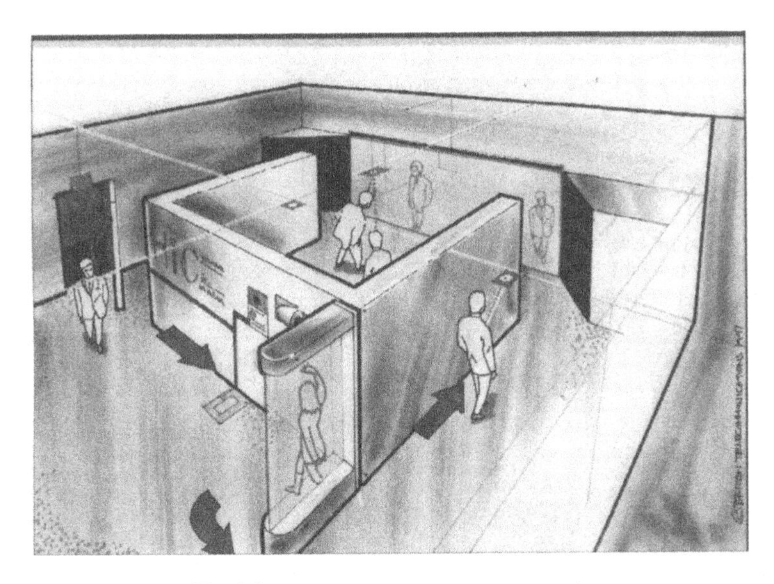

Fig. 1.3 Room-based immersive conferencing.

information can be used to optimize both the hands-free audio and video environments.

1.3.5 Persistent communities

'Persistent communities' is a telepresence service which extends beyond real time conferencing to meet the ongoing communications needs of a geographically distributed customer community. We referred earlier to the vision of 'Inhabited TV', a service in which professional broadcast material is mixed with social chat, and communities form around content and celebrities. We can envisage communities forming around a sports team, soap opera or a pastime such as gardening. User involvement and participation is essential in building communities where interactive communication complements traditional passive broadcasting.

Another expression of this vision is the company shared space, which supports a community built around a brand or product. Such a service, or 'corporate portal', would provide elements of a high-street shop, a telephone support line, physical brochures and on-line information. Portal communities would aim to personalize marketing and build customer relationships analogous to the proliferation of physical world club and loyalty schemes. They would also exploit the power of 'referral marketing' by promoting direct contact between existing and potential customers. Finally, they would target the cost saving achievable through 'self help' of the type currently practised on electronic newsgroups and bulletin boards.

Figure 1.4 illustrates 'Going Spaces', a travel community, where members can experience possible destinations in the company of both experts and fellow travellers.

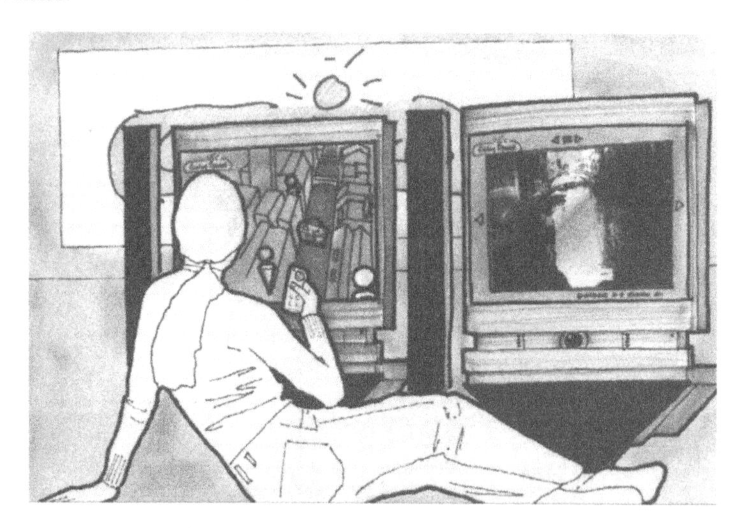

Fig. 1.4 'Going Spaces' travel persistent community.

1.4 CONCLUSIONS

Multimedia telepresence is the future of telephony — diversity of media, matched to customers' needs, and capabilities in delivering an appropriate degree of remote presence. However, we must not be blinkered in merely replacing and updating current communications channels. Our closing scenarios hinted at opportunities for telepresence to revolutionize existing processes. Much of the success of telephony has been achieved by meeting previously unrecognized needs, helping to solve problems that people did not know they had. Similar imagination and flexibility will see the emergence of novel telepresence services, which will again transform commercial and social life.

In outlining our vision we have highlighted a range of opportunities and challenges, many of which will be revisited in greater depth in the following chapters. The telephone has set high standards for telepresence services, and an ongoing depth and breadth of research will be essential in meeting customer expectations.

REFERENCES

1. Westall F A, Johnston R D and Lewis A V (Eds): 'Speech technology for telecommunications', Chapman & Hall (1997).

2. Turkle S: 'Life on the screen: identity in the age of the Internet', Simon and Schuster (November 1995).

3. Walker G R: 'The Mirror — reflections on Inhabited TV', British Telecommunications Engineering J, 16, pp 29-38 (April 1997) and http://vb.labs.bt.com/msss/IBTE_Mirror/

4. http://www.internet2.edu/

2

THE EVOLUTION OF CONFERENCING

I Parke

2.1 A BRIEF HISTORY

Visual communication has a long history in the UK with the first public service established nearly 25 years ago, called 'Confravision', which linked London, Manchester, Birmingham, Bristol and Glasgow [1]. Soon after this, advances were made in video coding techniques in which it was possible to compress the video signal and transmit it over digital circuits. This work was ground breaking not just because 100 to 1 compression was achieved on the video pictures, but also because for the first time companies collaborated together to define the standards, thus recognizing that people ultimately needed to be able to communicate outside closed user groups [2].

The first standard was the European H.120 standard for the transmission of audio and video over 2 Mbit/s links. In the mid-1980s it was recognized that it was now technically feasible to utilize the integrated services digital network (ISDN) for the transmission of audio and video. The ISDN provided a digital end-to-end connection, image compression techniques showed acceptable picture quality at 128 kbit/s and close to VHS quality at 2 Mbit/s, and field programmable devices were appearing on the market which meant development costs were coming down. Therefore, at that time, the International Telecommunications Union (ITU), formerly the CCITT, began work on the H.320 standard which is the cornerstone of videoconferencing today [3].

The first commercial implementation of H.320 was developed at BT and was known as the VC2100 codec. This codec was able to operate at all data rates from 64 kbit/s to 2 Mbit/s implementing many of the optional operating modes of H.320 and became a benchmark for the industry.

These H.320 systems were still room based in the early 1990s, but the bit rates at which they operated came down from 2 Mbit/s to 384 kbit/s as users became aware that the video quality from the H.261 video codec was perfectly acceptable at these lower bit rates. The cost of H.320 room-based systems started to come down in price as very large scale integration (VLSI) became available for the computationally intensive parts of the system. With such VLSI the physical size of codecs was reducing all the time and BT embarked on a personal computer (PC-) based H.320 system. This led to the familiar PC videophone, the VC8000, which was launched in January 1994.

The VC8000 revolutionized the H.320 conferencing market; H.320 became more affordable with an entry price of £3500 in 1994, but more importantly it harnessed the power of the PC to bring data collaboration.

Users of PC videophones were able to pass files from one PC to another PC during the call, were able to view documents together and annotate them, and were able to share applications in such a way that a person who did not have the application on their PC could view the remote application and control it. This was all achievable through the use of another set of important standards from the ITU, designated T.120.

The next revolution was the astonishing pace with which PC performance was growing year on year. In the space of the three years that the VC8000 has been on the market, the PC has moved from the 386 to the 486, through the Pentium and on to the Pentium Pro, and now the Pentium II. The central processor unit (CPU) performance has been doubling almost every 18 months. Such processor power soon made it possible to harness the CPU to perform the codec functions of H.320.

It was also the era in which the growth of networked computers, i.e. the Internet but more importantly intranets, exploded. With new protocols such as RSVP (the resource reservation set-up protocol), it was soon going to be possible to use local area networks (LANs) as a means of transporting real-time audio and video traffic. As a consequence, the ITU set about defining a standard by which audio/videoconferencing could take place over LANs and this became known as H.323. H.323 has become an important standard, with significant industry backing from heavy-weights such as Intel and Microsoft.

Image and voice compression techniques have also not stood still over time (see Chapter 8 for more detail). In 1990 H.261 was state of the art in video compression algorithms for low bit rates delivering 10 to 15 pictures per second at ISDN BRI rates with 30 pictures per second at 384 kbit/s plus. The coding was a motion-compensated interframe prediction loop of which the prediction error was subsequently transformed using the discrete cosine transform (DCT), quantized and variable length coded. The VC2100 was the first codec to implement motion compensation in real time [4]. It used a 'full search' algorithm which meant that for every 16 × 16 pixel video block there was an exhaustive search of 961 possible positions in the previous coded frame in which the best match could be found. The computational power required was equivalent to 5800 million operations per second.

Improvements have subsequently been made to the motion compensation, interpolated frames have been added to improve frame rate at low bit rates, and there have been a number of other advances to the coding which have brought about an approximate 2:1 enhancement in performance overall; this has been incorporated into the H.263 standard. The audio has also been improved, with the latest generation of codecs transmitting near toll-quality speech at 6.3 kbit/s (G.723.1) instead of 16 kbit/s (G.728) as before.

Such improvements have meant it is now possible to videoconference over analogue telephone lines using the latest generation of modems (V.34) and the improved audio and video codecs mentioned above. The ITU have formally approved a standard for this (H.324) and there are now products on the market.

2.2 BENEFITS

The ability to communicate over distance brings about major benefits to companies such as the speed of decision making by reducing the need for physical meetings and hence the travel time associated with them. Costs can be reduced and key personnel can make better use of their time. This has been the power of the global voice network today, keeping people and businesses in touch. There are, however, many situations where voice-only communication is insufficient and there is a need to be able to see each other, e.g. in contract negotiations body language is just as important as the spoken word. In other situations there may also be the need to share information, such as when a mortgage expert assists an applicant to fill in mortgage forms remotely.

The benefits have been well understood over time but the videoconferencing market is still niche and the expected uptake of desktop conferencing has not yet materialized. Why should this be? The answer is that realizing the benefits detailed above places quality thresholds on the audio-, video- and data-conferencing in the particular application for which it is used.

2.2.1 The important factors

In a videoconference the video transmission quality is determined by the resolution and the number of pictures per second. We as users compare the quality against the images we view daily from our television set or video cassette recorder (VCR) at home. With limited bandwidth from the public switched telephone network (PSTN), ISDN or LAN, it is not feasible to deliver such quality end to end. For a typical head and shoulders scene, pictures are delivered over ISDN2 (128 kbit/s) with a spatial resolution approximately a quarter that of VHS and at a picture rate approaching 15 pictures/sec; over PSTN (28.8 kbit/s) the picture rate would be approximately 8 pictures/sec for the same image resolution. Higher bandwidths (384 kbit/s and above) would be required to reach close to VHS quality, with the obvious added cost penalty of increased call charges. This means that applications have to pick carefully the appropriate image quality.

As important is audio quality, which includes lip synchronization and end-to-end delay. If video quality is degraded, the call is not necessarily interrupted; on the other hand, without stable and acceptable audio quality, a videoconference call can be a very irritating experience! The longer the call, the greater the quality required.

For a data-conference, the most important aspect is the integrity of the information exchanged. For example, with a multiple file transfer it is critical that the file be sent without error to each participant. Speed of transmission is also important — annotations on a shared whiteboard need to be immediately displayed on each person's screen as they happen. Here too, delays or interference detract from the efficiency of the conference.

Today, choosing the right videoconference or data-conference solution for professional usage means determining which technology and network delivers the optimum trade-off between reliability, speed and cost.

2.3 TODAY'S CONFERENCING

Businesses are supported with a full range of conferencing solutions today, ranging from custom studios on dedicated corporate networks to desktop devices either incorporated into personal computers or as stand-alone.

All the solutions provide a rich environment for communication. Conference room systems offer high-quality audio and video at bit rates from 384 kbit/s to 2 Mbit/s plus control of a number of peripherals, both at the local and far ends of the link. The desktop-PC solution has the added advantage of providing sophisticated data-sharing capabilities across the link alongside the audio and video, though this is soon to come to the room systems as well.

2.3.1 Desktop systems

This is a system which offers a range of data-conference features, such as a shared 'whiteboard', application sharing or multiple file transfer, together with audio- and video-conferencing on a PC. These generic data features can be tailored to suit particular applications.

The whiteboard is an on-screen workspace viewed by everyone in the conference and is similar to the whiteboard you find in a meeting room. Images can be loaded on to this screen, such as presentations or drawings from other applications, which can subsequently be annotated by all in the conference. The original material that is imported into the whiteboard is not changed, it simply forms one of the presentation layers; each conference end-point is allocated a layer in which their annotations are drawn.

Application sharing is subtly different to whiteboarding, as described above, in that the data that is shared is permanently altered by any annotations carried out. Only one conference participant needs to have the application and work material, the other conference participants are given the ability to share the PC resources on which the application is running. The resources shared are the keyboard strokes, mouse movements and graphics. This technology is platform-specific at present as it involves intercepting commands and information at the operating system level, hence its restriction to common operating systems in a conference.

Multiple file transfer enables a file to be sent simultaneously to two or more participants at remote sites. For instance, a spreadsheet, which has been discussed and changed using application sharing, as described above, in a conference, can be copied to all or a selection of participants in the conference at the end.

2.3.2 Group systems

Group systems are more scalable, offering a range of options in terms of available data rates and the numbers of cameras and screens. The conference user has much greater control of the conference with the ability to control both near and far ends. Group systems also harness the flexibility of a PC for such things as providing the data-conferencing applications and the ability to control the conference through a graphical user interface.

2.3.3 Multipoint conferences

The discussion so far has centred around systems that are primarily designed to operate on a point-to-point basis, facilitating communications between two end users or a user and a network service. Increasingly there is a need to communicate between three or more locations, which is supported through a device known as a multipoint control unit (MCU) [5].

Multiple sites are supported by a star network which joins each location to a central node. At this hub, information can be processed, selected and forwarded to all other locations in the star. Such a configuration does not require extra equipment at the location end-points and can be expanded by cascading MCUs together to form a dumb-bell type arrangement, as shown in Fig. 2.1. Such an arrangement can be useful to concentrate traffic from several sites before transmission across international links for instance.

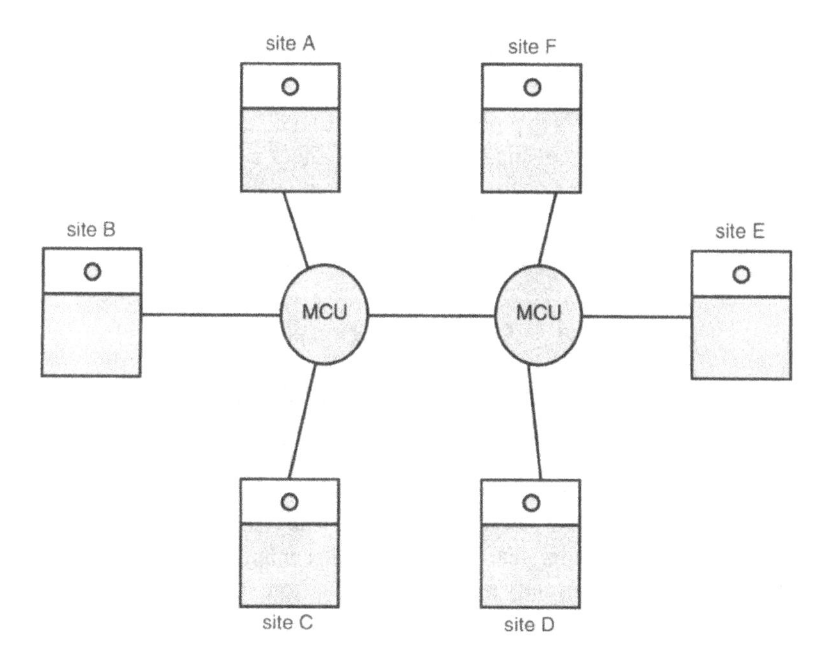

Fig. 2.1 Typical dumb-bell network with three locations.

2.3.3.1 Switched multipoint

These are conferences in which the video information is selected by some means and distributed to all other connected locations. The video can be switched automatically such as by voice activation or it could be manually switched under the control of a chairperson. The audio from each site is mixed together centrally and distributed to all sites.

This is a non-ideal environment for multipoint conferences because each location only sees the location that is speaking, and as a consequence visual cues of other participants are lost in the discussion. If the conversation continually moves between locations, the video switches frequently. The video could also switch on a false trigger, such as a cough or door closing, unless under chairperson control. A more natural environment would be if all locations could see each other during the conference; this is known as continuous presence multipoint.

2.3.3.2 Continuous presence

It is now possible for MCUs to process the video from up to four locations and to distribute to all others, providing the ability to hold a continuous presence conference with up to four locations. This significantly improves the quality of a multipoint videoconference; the screen is not now continually changing in front of you and the visual cues from others in the conference are not now lost.

2.3.4 Conferencing services

Audioconferencing is the backbone of any conferencing service provider as depicted in Fig. 2.2 and probably accounts for 75% of the market today. It is a low-cost solution with no barriers to entry, access is via the ubiquitous telephone, and a high quality of service can be offered.

Videoconferencing has a much higher cost of entry and is more difficult to use and hence has a smaller share of the overall conferencing market. These types of call require significantly more intervention from the service provider to ensure the calls are successful due, in a large part, to the immaturity of the standards and equipment in this market.

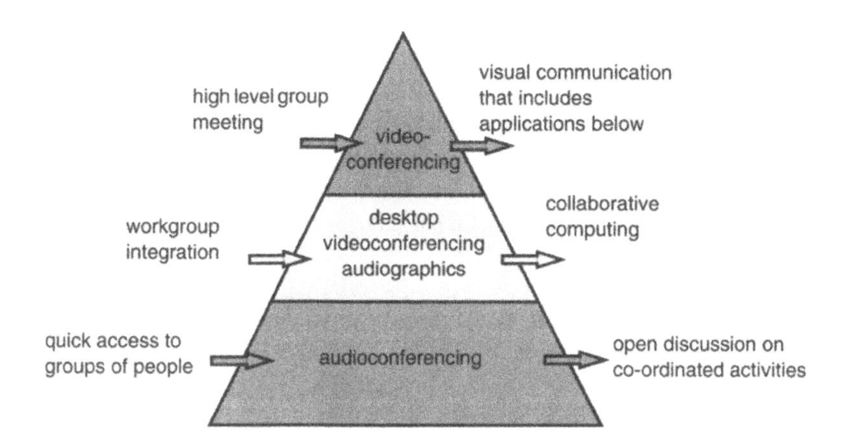

Fig. 2.2 Conferencing needs and tools.

Between pure audioconferencing and videoconferencing lies audiographic conferencing. Audiographic conferencing is still in its infancy with an early set of standards, T.120/T.130, that need to 'bed down'. The cost of entry will be low as it is a software application run on a PC and, once there is a population of conference-'enabled' PCs, this end of the market will grow. Two connections are required, one for voice and one for data. For a teleworker this could be two PSTN lines or one ISDN2 line.

For someone working from their office and whose PC is already connected to the corporate LAN, their existing LAN connection could be used for the data and their phone for the voice.

There are also a growing number of one-line solutions — voice/data modems (V.70), the use of H.324 or H.323 clients, and voice/data over Internet protocol (IP).

With V.70 and H.324 having only appeared in late 1996 and with a number of voice-over-IP implementations, it is difficult to determine which solution will prevail at the end of the day.

2.4 VCnet

One of the largest managed videoconferencing networks is VCnet within BT, with over 55 sites in the UK (see Fig. 2.3) and other sites at country offices around the globe. Each studio room has been specially created to conform with specialist guide-lines on size, lighting, furnishing, air-conditioning and sound-proofing. The criteria for location of facilities is based on having a room within 30 minutes travelling time of 80% of managers.

Fig. 2.3 Map of VCnet sites in the UK.

The equipment is all H.320 standard-compliant and at each site there is a standard build for each facility of a VC5000 double unit accommodating up to six participants per site and an ELMO graphics unit or a display station for document transmission. Some sites are equipped with a VCR and a PC interface. Soon the VS3 will be introduced into the studios which will bring the added benefit of data collaboration.

All sites are operated over the IBTN, a Timeplex Link network, and use 384-kbit/s channels connected locally by direct-cabled RS449 interface, or a 384-kbit/s KiloStream circuit. Some sites are equipped with ISDN6 dialling capability to allow direct dialling to sites outside BT.

The network is centrally managed out of Pontefract and there are five MCUs within the network, allowing multisite conferences.

All conferences are pre-booked via a telephone reservation desk and the IBTN network is configured dynamically to set up the relevant 384-kbit/s connections. All connections are established prior to the conference start time ensuring that when participants arrive at the studios the connections are already up and running and the meeting can begin at once.

2.5 ON THE HORIZON

2.5.1 Internet conferencing

The Internet of the mid-1990s suffers from the chronic shortage of bandwidth and the total lack of quality-of-service levels which makes it incompatible with intensive professional use for teleconferencing applications. However, the Internet is here to stay and, with continued investment and the deployment of protocols in the network that support quality of service (QoS), the Internet will offer significant flexibility and an unprecedented platform for creativity for conferencing in the future.

The first Internet conferencing applications, based on the H.323 standard, are emerging. They integrate standard collaborative working applications such as whiteboards and application sharing, utilizing T.120, as well as Internet telephony services [6]. Significantly they are all PC-based solutions and the majority are software-only solutions.

Given the levels of investment being sunk into Internet solutions by both IT and telecommunications companies, it is likely that within 18 months the Internet will become an attractive alternative for data-conferences and that Internet-based videoconferencing will be feasible within two years. Not only will PC processing power be even greater, codec technologies will also have been optimized, and bandwidth-on-demand protocols for the Internet will have been standardized and deployed, guaranteeing users a certain quality of service.

Where the real revolution is happening is in corporate networks, or intranets as they have become known, which are now ubiquitous in large businesses. H.323 offers large and small corporations the ability to route telephony, data and real-time multimedia services to the desktop over a single connection — the office LAN. It allows provision of service from basic telephony right through to sophisticated computer/telephony integrated call management. With corporations being able to reserve IP network bandwidth, H.323 will be the inter-site communication connection of choice.

H.323 has been developed to extend the ITU's H.320 set of videoconferencing standards to operate over non-guaranteed quality of service LANs (such as Ethernet) and the Internet.

H.323 systems will interwork with H.320 systems on the ISDN and H.324 systems on the PSTN by way of gateways (see Fig. 2.4). In addition to the terminals that sit on the LAN side, there are also gatekeepers. These are used for bandwidth-reservation (or perhaps more accurately bandwidth-restriction) and address-translation services. Terminals, gateways, and gatekeepers are therefore the three main functional units described in H.323. It should be noted, however, that, in practice, the gatekeeper functionality might be implemented as part of a particular vendor's gateway unit, or, indeed, be implemented in one of the terminals.

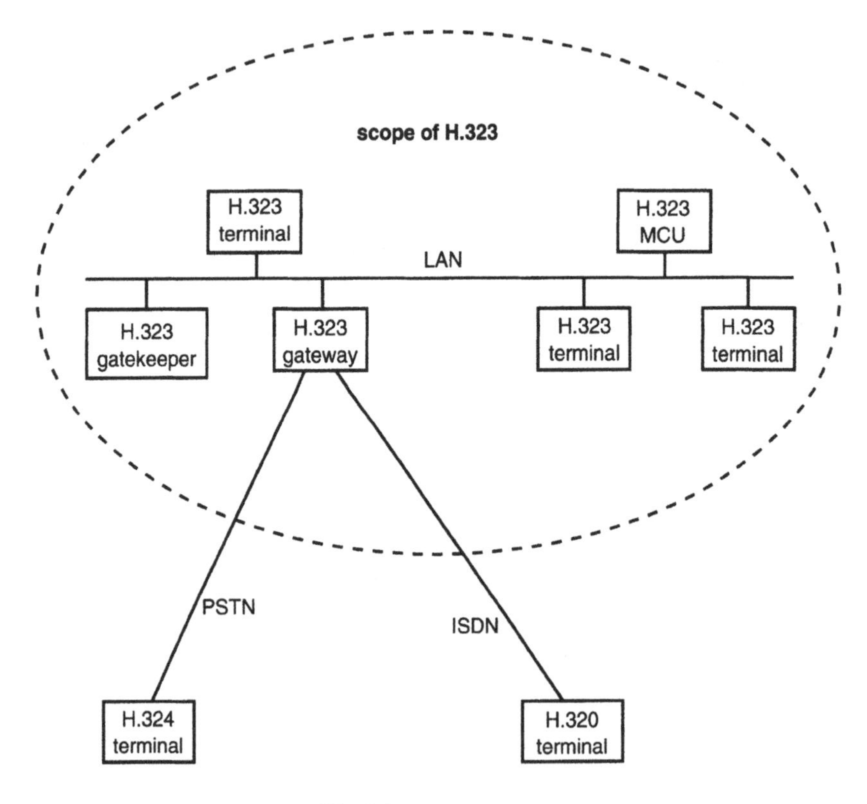

Fig. 2.4 Scope of H.323.

2.5.2 Conference control

The conferencing environment will become richer in time; the T.120 protocols, which provide the infrastructure for data-conferencing, are about to be enhanced with a further set of protocols that allow the manipulation of audio and video media streams in a conference. This set of protocols is currently being defined in the ITU and is known as T.130. The services provide support for user-controlled distribution of real-time streams and of automated stream processing, and distribution services such as video switching, audio mixing, and continuous presence. Other services include stream identification, floor control and 'on-air' indications. In addition, peer-to-peer functions such as source selection, privacy and remote device control are supported. A forerunner to this is a BT development called Passepartout (see Chapter 5), which has set out to develop a richer environment for audioconferencing by linking audio and data-conferencing together in order to generate and prove some of the more complex protocols mentioned above.

2.5.3 Broadband

In time the availability of bandwidth will increase and the current restrictions in the local access loop will be overcome to bring bit rates in excess of 2 Mbit/s to homes through the use of technologies such as ADSL and VDSL [7]. In the corporate LAN, improved switching and segmentation of the networks will bring 100 Mbit/s plus to the desktop.

Increased bandwidth offers the opportunity to deliver high-quality conferencing with TV-quality images and stereo audio, making the user experience of conferencing much closer to the real thing.

2.6 CONCLUSIONS

Teleconferencing is entering a new era in which new conferencing protocols are able to harness the flexibility of PCs and computer networks to bring a richer environment to users. The advent of conference-enabled personal computers by the year 2000 will stimulate a whole host of new services, some of which have been touched on in this chapter.

REFERENCES

1. Haworth J E: 'Confravision', POEEJ, 64, p 220 (January 1972).

2. Nicol R C and Duffy T S: 'A codec system for worldwide videoconferencing', Professional Video (November 1983).

3. Carr M D: 'Video codec hardware to realize a new world standard', BT Technol J, 8, No 3, pp 28-35 (July 1990).

4. Parke I: 'A hardware motion compensator for a videoconferencing codec', IEE Colloquium E4 and E14 (July 1990).

5. Clark W J, Lee B, Lewis D E and Mason T: 'Multipoint audiovisual telecommunications', BT Technol J, 8, No 3, pp 36-42 (July 1990).

6. Babbage R *et al*: 'Internet phone — changing the telephony paradigm?', in Sim S P and Davies N J (Eds): 'The Internet and beyond', Chapman & Hall, pp 231-254 (1998).

7. Young G, Foster K T and Cook J W: 'Broadband multimedia delivery over copper', in Whyte W S (Ed): 'Multimedia Telecommunications', Chapman & Hall, pp 139-163 (1997).

3

UNATTENDED AUDIOCONFERENCING

D L Gibson, D Pauley and L Willis

3.1 BACKGROUND AND INTRODUCTION

It is well known that the patent for the telephone instrument was filed by Alexander Graham Bell in February 1876. The potential of the telephone was rapidly seized on by the commercial world in both New England and London, and within two years the first telephone company in the UK was established. The first London exchange opened just three years later with a manual switchboard, and by 1927 it was possible for one 'subscriber' to dial another on the same exchange without any manual intervention.

However, while in these very early days of telephony the operator had the ability to connect several parties together by telephone such that they could all converse, a commercial public service to provide this same capability was not introduced in the UK until the early 1980s with the introduction of BT's 'Rendezvous' service. This service was, and its successors still are, operator based and conference calls generally have to be booked in advance.

The operator establishing a conference call carries out many vital processes, perhaps invisibly, in the creation of the call, e.g. explaining to called parties that an audioconference is being set up and why, that if lines are busy or engaged further attempts can be made to include them in the conference, and that, if the wrong person answers the telephone, attempts can be made to find the correct party or suitable messages can be left on the wide variety of answering machines that may be encountered. Any automatic means of establishing conferences needs to be able to address such issues, and it is only within the last few years, with the growing availability of interactive voice response technology, that provision of such an unattended service could be realistically implemented.

This chapter looks at one such service, Conference Call 'Instant' (CCI), which has been recently introduced by BT. Its implementation can be viewed as a series of three technology layers — the algorithmic layer which provides the raw conferencing capability, the platform layer which provides the environment for the algorithm and connectivity, and the application layer which provides the interaction with its human user. Each of these will now be described in turn.

3.2 BASIC CONFERENCING ALGORITHMS

The transmission path from the telephone instrument to the conferencing bridge is shown diagrammatically in Fig. 3.1. The path from a telephone instrument to an exchange is generally an analogue connection. The line card within the exchange splits the incoming speech path from the outgoing speech path and will provide any amplification required and filtering prior to converting the speech signal to a digital form ('A'-law PCM in the UK) for onward transmission. Thus a network-based conferencing system in the UK must be capable of digitally combining and distributing multiple streams of 'A'-law coded speech samples. Any practical implementation of such an algorithm must achieve a satisfactory compromise between digital signal processor (DSP) utilization, flexibility and subjective speech quality.

A common conferencing algorithm employs the principle of 'principal speaker' selection. This simply monitors the average speech power at each input to the conferencing 'bridge' to determine the principal speaker, and then connects the actual signal from that input to all of the other outputs from the bridge, except the output to the port of the speaker. This technique is very efficient since the input signal is copied directly in 'A'-law form to the output without any conversion. The subjective speech quality of this technique, however, is not ideal, because speakers hear no feedback from the conference participants while talking. They cannot therefore hear the acknowledgement signals sent back to them in a normal conversation, such as, 'mmm', 'yes',

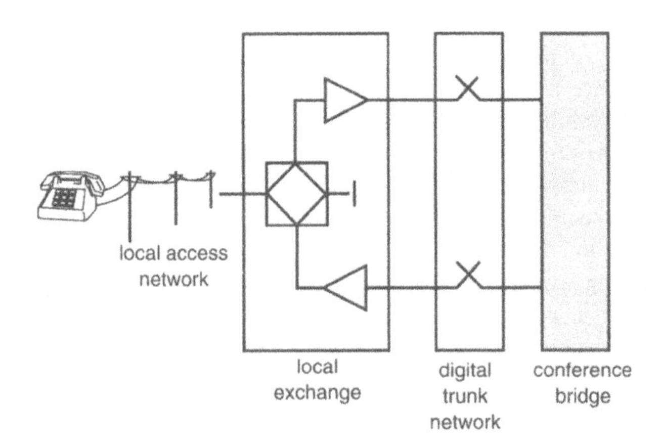

Fig. 3.1 Transmission path.

'That's right'. This has the effect of speakers feeling they are talking to an 'empty room', and is generally unsettling to unfamiliar users, and may discourage people from using telephone conferencing.

The algorithm employed in the BT Conference Call Instant service employs simple linear mixing of the inputs to the 'bridge', so that each of the participants hears the sum of all of the other inputs except their own input. This means that every participant can hear what anybody else utters at any time. This supplies the 'acknowledgement signals' that provide the normal conversation environment, to which people are accustomed, and which they find less discomforting than the single speaker approach.

The linear mixing method is less suitable for large conferences when the summation of the background noise sources can become unacceptable, but the CCI service is not intended to cater for larger conferences. In other applications, such as Passepartout (see Chapter 5), the same algorithm can be used but with the addition of voice activity detectors on each input to control the muting of non-speaking inputs. This eliminates the background noise problem while still allowing acknowledgement feedback.

The DSPs used in the speech platform can process 30 speech channels and it would be straightforward to create a 30-channel conferencing bridge in a single DSP, but, if a conference only has five participants, this would achieve very poor utilization of the valuable DSP resource. The DSP code is therefore arranged to provide five bridges each having six ports. This means that small conferences can be efficiently accommodated, and the bridges can be cascaded when required to accommodate larger conferences.

The six DSPs on a single BT Laboratories (BTL) speech processing card used in the speech platform are sufficient to support any mix of conference sizes from a 30-channel line system.

3.3 CONFERENCE BRIDGE PLATFORM

The particular platform chosen to host the CCI service is the Millennium-CT (MCT). The system is PC-based and includes two main types of cards. The digital line interface card (DLIC) provides a 30- or a 60-channel connectivity to the telephone network. It contains a non-blocking digital switch to enable digital speech paths to be extended over an industry-standard speech bus (multi-vendor interface protocol) to the signal processing cards. These cards each contain six DSPs along with a controlling Motorola 68000 family processor.

The system also contains a (data) network interface card and an alarm card. Finally, the system is controlled by a PC processor and is accommodated in a single 19-inch rack-mountable chassis. A photograph of the CCI hardware is shown in Fig. 3.2.

Fig. 3.2 Conference Call Instant hardware. The Millennium-CT shelf is the black unit in the right hand rack with its associated monitor above.

The MCT's operating environment is Unix, and management/control of the system is vested in a single process. This MCT manager has the capability to run different speech applications on one or more telephone lines, either immediately or on receipt of an incoming call. The manager is also responsible for resource allocation — the assignment of DSPs and lines to particular applications — which it can achieve on a static or dynamic basis.

Once launched by the MCT manager, an application proceeds by requesting the manager to carry out a particular operation such as replaying a message or detecting DTMF tones. At the completion of each operation, the manager will respond to the application with a message indicating the success of the operation along with any ancillary information. Based on the response, the application may then make another request, and then another, and so on, until eventually the application terminates.

The CCI service is implemented as one such application launched in response to an incoming phone call. The World Wide Web (WWW) training facility is launched in a similar way when the customer calls a different number. Support for WWW-requested conferences is provided by an application on the same platform, which the MCT manager launches immediately.

3.4 CONFERENCE CALL 'INSTANT'

The conference calls which people make can probably be split into two types. Firstly, there are those which take place on a regular basis — this might be a weekly sales progress meeting or a monthly management meeting. Secondly, there are those conferences which take place on an *ad hoc* basis, perhaps in response to some event which could not be predicted in advance, such as an act of sabotage. Conference Call Instant can be successfully used in both circumstances. However, it should be remembered that as the size of a conference grows, so also does the amount of effort required to bring together all the various parties, and for a large conference this can become a significant overhead. In such circumstances use of an attended service is to be preferred as the work of collecting together all the conferees can be handled by one or more operators. For this reason the maximum conference size available through Conference Call Instant is limited to ten (which covers more than 90% of all conferences) and distinguishes it from the attended service which can handle more participants and at the same time offer a greater range of facilities, e.g. voting.

3.4.1 Telephone access

When customers register with the CCI service they will be issued with a customer number (CN), a personal identification number (PIN) and given a telephone number and WWW uniform resource locator (URL) by which access to the service can be gained.

A customer wishing to establish an audioconference using a telephone will ring the given telephone number using a normal DTMF telephone. The call will be routed to an MCT platform which will subsequently act as the conference bridge. The MCT will answer the call, greet the caller and initiate a dialogue to elicit the caller's identity and conference parties to be contacted. The principal design objective for the MCT application program, which provides the CCI telephone service, was that it should be straightforward to use, so that it should not be necessary for the user to constantly refer to a written user guide. This requirement had to be balanced against demands to also provide 'world-beating' sophistication. These aims have been met by allowing users to establish straightforward conferences easily by using only the most basic facilities, and to move on to the more sophisticated features as they become more familiar and more confident in using the system. The application provides full voice guidance at every stage of the conference set-up procedure.

On answering the telephone, the CCI application will greet the caller and request them to enter their customer number, followed by their PIN. It will invite new or infrequent users of the system to listen to a spoken description of how to use the service by simply pressing a # when asked for their customer number. After successfully identifying themselves to the system, the user will be asked to enter the number of the person they wish to join their conference. The system will call the given number, and, upon answer, will connect the conference instigator directly to the called party. This is to allow instigators to establish that the wanted person has indeed answered the call, rather than a colleague or an answering service, and to introduce themselves and invite the called party to join the conference. If the correct person has been contacted and is willing and able to join the conference the instigator can instruct the system by pressing a key on the keypad, to accept the called party, and connect them to music-on-hold until all of the other members of the proposed conference have been contacted. At this point instigators will instruct the system to connect all the participants into the conference, to begin their discussions. The conference is terminated by the instigator by simply hanging up the telephone.

3.4.2 WWW access

The World Wide Web is an alternative means by which conference calls can be established and, for the office user with Internet access, may be a more convenient mechanism. To run the service customers use their browser to access the CCI home page (http://www.bt-instant.com). The WWW pages have a simple format and can be displayed with all popular browsers. The home page provides links to other pages providing further information about the CCI service and the attended service, the registration process, down-loadable user guides and training facilities, as well as to screens allowing conference calls and personal conference lists (PCL) to be established.

Just as the CCI telephone service catered for both the novice and the experienced user, so also does the WWW system by proving a simple conference set-up as well as an advanced set-up facility. In either case, the essential information to be entered is the same — a customer number, a PIN and the telephone numbers of the conference participants. Assuming a valid CN/PIN combination, the first number in the list is the telephone number which the customer supplied at registration. This number can be edited on the screen as appropriate if, for example, the customer is using the facility while away from his normal place of work. Other numbers are added in like manner until the list is complete. Clicking on the 'submit' button will result in the conference request being dispatched and a response being displayed on the customer's screen as to whether the request can be currently serviced (see Fig. 3.3). The WWW screens play no further part in the conference establishment.

The first stage of conference establishment is that the bridge will ring the chairman who will be expecting the call. If the telephone is engaged or unanswered, the CCI system will subsequently make a number of further

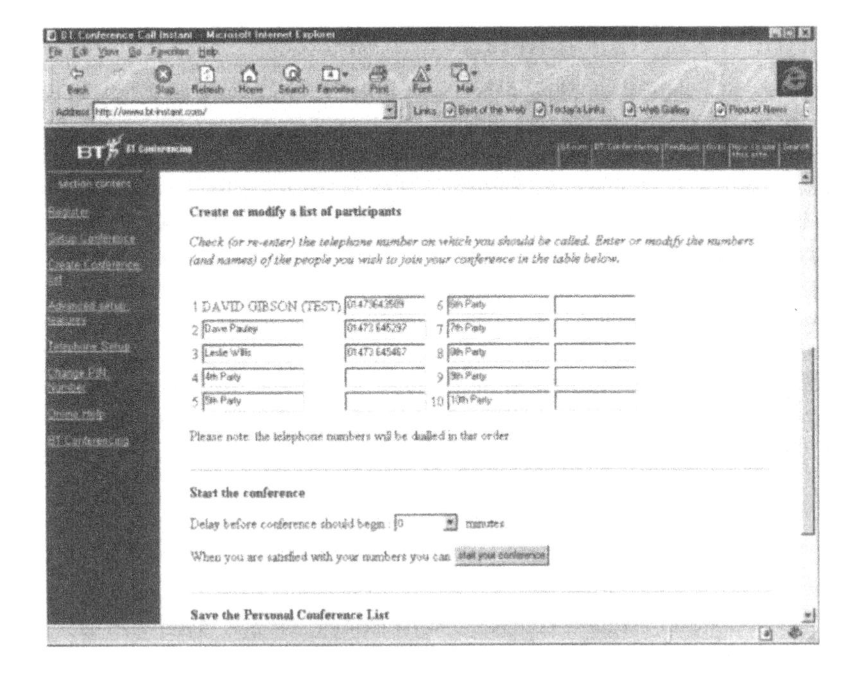

Fig. 3.3 CCI WWW conference set-up page.

attempts and if these are unsuccessful the conference request will be discarded. After announcing the service, the first stage in the dialogue is to ask the chairman to press a key on the telephone keypad as confirmation that the call has not been answered by an answering machine. Upon acknowledgement, the system will then attempt to contact each of the conference parties listed, in the same way as the telephone-based service, with the chairman being asked to accept or reject lines in the usual way with accepted parties being given music-on-hold while the conference set-up progresses. Once all numbers in the list have been contacted the conference is automatically started with the number of other participants being announced. The chairman can subsequently control the conference composition by accessing a special 'options menu'. The conference terminates when the chairman replaces his handset and a customer data record (CDR) is generated for billing purposes.

3.4.3 Personal conference lists (PCLs)

One of the more sophisticated features of the application is the provision of personal conference lists. It can be quite arduous to key in more than two or three full-length telephone numbers to set up a conference. Also the user may feel under pressure once they have some parties connected to their conference on hold waiting for the conference to begin, and consequently more likely to mis-key the number entries. The provision of PCLs allows users to prepare, in advance, lists of telephone numbers of parties they wish to join in a conference, and then specify the list they wish the system to use by simply entering a single digit. It is also often the case that customers regularly wish to establish conference calls to the same groups of people and the use of PCLs can speed this process. These lists can be prepared 'off-line', i.e. outside of an actual conference set-up using the telephone or WWW browser. Users can check and correct each number entry at their own pace and thereby avoid any pressure which they may have otherwise experienced if they had had to enter each number during the conference set-up. If a user sets up a conference by keying in on a telephone the numbers during the conference set-up, the user can, at the conclusion of the conference, instruct the system to store in a PCL the numbers of the parties which were connected. Users can then simply instruct the system to implement that PCL the next time they wish to confer with the same parties.

3.4.4 WWW training

While a design aim has been to make the CCI as simple and as intuitive to use as possible, some aspects of the advanced WWW conference set-up may, perhaps, not be completely apparent to people who have had little exposure to the WWW. To meet this need a WWW training facility has been introduced to guide the customer through PCL and conference creation. To invoke the facility the customer should 'click' on the appropriate link on the CCI home page. The customer will then be advised of a telephone number to ring. Once the call is connected, the MCT instructs the training application on which pages to display depending on the user's DTMF response to audio instructions. Audio explanation of the operation of the various WWW pages is given over the telephone and relevant areas of each page are highlighted. The training is terminated by the user selecting a particular option on the telephone or by simply hanging up. It should be noted that this facility calls for the use of 'server push', a feature which is not supported by all browsers.

3.4.5 Service administration

3.4.5.1 Registration and customer support

Before using CCI a customer must first contact BT to register and thereby gain access to the service. Such contact may come by referral through operator services or as a result of accessing the CCI WWW home page, itself indirectly linked to BT's public home page. In either case the customer will be asked to provide contact information for billing purposes. The customer is also asked to provide his normal telephone number to be used when establishing conference calls using the WWW facility. Once satisfied on these points all billing details are entered on to the main database and a CN/PIN generated. A 'welcome' pack is then sent advising the customer of the CN/PIN, the URL, the telephone number for the service, and the contact number of an operator for further assistance.

A number of special WWW screens are available to assist the service operator to administer the service. These screens are necessarily username/ password protected and the system manager can control the number of administrators able to access customer information. Using these screens an administrator can access the database to identify and update CN and PIN information and thereby access other customer data, such as PCL content. The use of WWW screens enables the systems administration not to be tied to a particular type of equipment or location, giving the service operator greater freedom in how the service is administered.

3.4.5.2 Billing

At the conclusion of a conference, a CDR is produced by the Millennium platform and held locally. These CDRs provide information concerning the CN, bridge, channels used, method of conference establishment, telephone numbers contacted and call duration. These CDRs are periodically transferred to the billing system where they are used for bill production. This system uses a complex algorithm and charging tables, to establish the price of a given conference call. The charge is based on the cost of the outgoing calls together with a conferencing charge for use of the MCT.

3.4.5.3 MCT operation and configuration

The telephony aspect of the CCI service is provided by multiple MCTs connected to the PSTN via ISDN-30 links. Each MCT has a maximum capacity of 60 channels. A single telephone number is associated with channels on these trunks in such a way as to distribute inbound traffic equally across all MCTs, exercising all available time-slots. The MCT will associate calls on each of these time-slots with the CCI telephone access application or the training application, neither of which is invoked until call arrival. Other time-slots in a trunk are associated by the MCT with the WWW application which is invoked on system start-up. The relationship of application to channel is held in a single table within the MCT and can be readily changed by the MCT administrator, but naturally has to align with the channel/telephone number association provided at the public exchange. The number of channels associated with each application should be determined from traffic estimates/measurements; initially, from each 30-channel group, five will be available for WWW conference creation and two for WWW training with the balance available for conferences to be set up by telephone.

In the event of catastrophic MCT failure, a hardware relay will operate which can be used to raise an alarm. Such a situation might arise due to mains power or trunk loss. In the case of less severe errors, where there is redundancy (such as an individual DSP failure or entire signal processing card failure), the MCT system attempts to reconfigure itself to ensure continuity of service, though perhaps at reduced capacity, and keeps a log of all errors for subsequent inspection. Thus the internal configuration of resources is dynamic, but can be controlled and influenced by the MCT administrator.

MCT administration may be either via a local console or remote via the data network. Generally, there is little need for local access to the system, and system maintenance and support can be provided at a distance.

3.4.6 Service enhancements

The above CCI service meets the essential requirement of establishing conference calls on demand. However, there are a number of additional features, listed below, which are under consideration for introduction to widen the appeal and usefulness of the service.

- Dial-in conference joining

  These are many reasons why the need for this arises. Firstly, it may be that a chairman has left a message for a would-be participant who was unavailable at the time the conference call was being established. If a conference dial-in facility existed, the participant could then join the conference when available. Secondly, a conference participant using a mobile telephone might lose telephone contact and with such a facility could rejoin the conference when conditions prevail. A third reason might be that the meeting had been prearranged but some participants could not specify in advance a number on which they could be contacted. With a conference dial-in facility they could join the conference at the previously agreed time. Finally, a fourth reason might be simply to spread the cost of the conference call!

- Conference recording

  A conference call will usually be made with the intent of discussing particular matters, and it may be desirable that a transcript of the discussion be later produced. Thus a facility whereby the chairman can record some or all of the conference discussion for subsequent replay is desirable.

- Use of speech recognition

  While selection of PCLs within CCI is easy via the WWW because they can have names, its use on the telephone is dependent on the user being able to remember the number of the particular PCL that should be used. This problem could be overcome if speech recognition technology were used instead — users would simply have to speak the name of the conference group they wished to use, e.g. 'sales meeting', 'project review meeting'. Speech recognition could probably also be used successfully as an alternative method of responding to certain prompts.

In providing any or all of these enhancements great care needs to be taken to retain the ease of use of the system while ensuring the security and reliability of the service.

3.5 AUDIO- AND DATA-CONFERENCING

The purpose of a conference call is generally to discuss some particular issue, and in conducting the discussion participants will often have relevant information at their disposal, perhaps circulated beforehand. At present, this information will generally be in paper form and may be personal to each participant. The usefulness of the meeting will often be significantly improved were all participants able to view data pertinent to the discussion, and may be improved even further if this data could also be manipulated.

Conference Call Instant goes some way towards making this possible. Once a conference call is established, participants with WWW access could view a common page. If one participant (perhaps the chairman) could publish to this known URL, others could also view it and see updates, provided they used their browser's refresh option periodically. While such an approach to combined audio/data conferencing may be acceptable in certain circumstances, better data conferencing tools, based on the ITU T.120 standard which allows a richer interaction, are now becoming commercially available. To meet this need a useful extension to CCI would be the ability to optionally initiate a T.120-based data-conference to enable such sharing and viewing of electronic documents.

This approach supposes that the audio interaction takes place before the data interaction. However, with the advent of developments such as 'shared spaces' on the Internet, this supposition is no longer true. Thus, if the primary contact with other participants is in a data world, the question arises as to whether the unattended audio bridge would take the same form as in a service such as CCI. The answer is that if the control of the audio/data interaction is vested in the data platform the audio bridge must change to act as a server device, performing its operation in response to requests made by the controlling data platform.

To enable the data platform to successfully establish and control audioconferences, a messaging protocol needs to be established between the two platforms. Such a system has been implemented based on using RFC1006 protocol to provide an X.224-like transport service to be established across the TCP connection between the two platforms. Using this protocol, messages can be exchanged between the two systems to control the operation of the audio-conferencing platform. These messages fall into five general categories:

- registration functions — these allow the two platforms to pass the necessary set-up information between themselves to register platforms, conferences and lines;
- telephony functions — these allow a data platform to query a line status, clear calls and make calls;
- bridging functions — these allow mixing, speaker and microphone muting, the mixing functions allowing the complete conference to be mixed, groups within a conference to be mixed together as if they were a conference in their own right, and individual mixing functions for each person;
- support functions — these allow requests for conference and line status;
- inquiry functions — these allow the data platform to request that the audio platform ask a question of the user, the query being a text string which is converted to speech by the audio bridge; any response required is given in the form of DTMF tones which the audio bridge passes on to the data platform.

Two forms of system software allowing such control from a data platform have been implemented using an MCT platform. The first form uses the same MCT manager as for CCI and as such its implementation currently precludes some operations such as sub-conferencing. The second implementation takes the form of a modified MCT manager and potentially allows a more complete implementation.

This form of the bridge has so far been used in conjunction with data platforms to provide two different services. One such system is RISE/MERLIN (see Chapter 17) which is used to assist in the teaching of English as a foreign language. In this system the data platform is a WWW server/database. The other system is 'Passepartout' (see Chapter 5) which is a T.120-based data-conferencing platform which has an iconic representation of conference participants.

3.6 FUTURE DIRECTIONS

There are many possible additional features to the audio bridges to further enhance the overall conferencing capability. These include such items as using voice activity detectors within the bridge not only to perform a microphone muting operation but also to indicate to an attached data platform who is speaking. In data-led applications there is also a perceived need for faster audioconference establishment by parallel dialling on the bridge instead of the present sequential dialling process. A further enhancement is a facility for participants to signal to the bridge using DTMF tones — this might be used, for example, to flag attention or perhaps vote on an issue.

Linking the conference creation capability to other tools on the corporate desktop, such as a meeting scheduler or messaging facility, would also be a valuable enhancement. An integration with 'Groupware' products such as Lotus Notes is another opportunity.

In the further future, conference establishment and management might be established by a 'personal assistant' rather than by the individual. The assistant might appear as a talking head and accept natural language verbal requests for telephone call or conference establishment, garnering, displaying and manipulating data as commanded.

3.7 CONCLUSIONS

The last decade has seen an explosive growth in the audioconferencing business and it is now one of the fastest growing parts of BT's business. The opportunities for continued growth brought about by combining the strengths of telephony, interactive voice response technology and the Internet are tremendous, and the next few years will witness many changes which will affect the way in which people work. Services such as CCI and RISE are only just the beginning; new and broader services will continually emerge. It promises to be an exciting new Millennium!

4

SPATIAL AUDIO TECHNOLOGY FOR TELEPRESENCE

M P Hollier, A N Rimell and D Burraston

4.1 INTRODUCTION

A key requirement to enhance the usability of a range of telepresence applications is the exploitation of all the sensory modalities, thus maximizing naturalness. Spatial audio has a vital role to play in enhancing speech communication performance and naturalness in synthetic spaces by giving directionality to individual sound sources and providing feedback about the virtual space in which the user is located. When groups of people meet in synthetic spaces either for business or recreation it is possible to provide services and features which are not available in a face-to-face meeting. In particular, interaction with intelligent agents, manipulation of virtual 3-D objects, data sharing, and automatic note taking may all seem quite natural in the enhanced meeting spaces of the future.

4.1.1 Audio landscape

The sensory modality which dominates perception is normally assumed to be vision and this has been reflected in the attention paid to the visual components of 'virtual world' developments to date. However, for communication it can be readily argued that hearing is the most indispensable modality — imagine a videoconference with no audio! More generally, hearing may provide very compelling cues which offer powerful support for the naturalness of a synthetic environment. For example, navigation cues in a synthetic space are particularly important to avoid disorientation and to facilitate efficient location of persons or items of interest. The required cues can be provided by associating an audio landscape with the world; the location of other talkers and objects of interest in close proximity can then be readily identified by the direction and character of their audio signal (compare this with the poor usability of an auxiliary visual display).

4.1.2 3-D spatial sound

The work reported here concerns high-quality audio systems capable of reproducing a virtual audio source for either a 2-D (full 360° plane around the user), or 3-D (any direction in three dimensions) space. This approach must be distinguished from other pseudo-3-D audio and surround-sound systems found in existing distributed virtual worlds and home cinema systems. For example, a typical home cinema surround-sound system such as Dolby ProLogic does not usually produce 2-D spatial sound (a rear image can be created using advanced steering techniques [1]). DTS and Dolby-AC3 systems provide 5.1 discrete channels (5 full bandwidth channels and a sub-woofer) and therefore could more readily produce a full range of virtual source positions if source mastering and loudspeaker placement for replay were appropriate. In general, of course, it is not required, or desirable, to place distinct virtual sources behind the listener for home cinema since there would be nothing there to see if the user were to turn around and look. However, in a synthetic environment it is entirely possible that an object of interest might be located behind the user who would be able to 'turn around' and look at it within the virtual space.

Research into the exploitation of spatial sound at BT Laboratories (BTL) has been conducted within a team responsible for more general research into perception and the modelling of the human senses for performance assessment. This knowledge may be exploited to design attention-grabbing alarms and to avoid startling or disorienting the individual.

Several of the key technologies for practical spatial audio reproduction have been known for many years [2]. However, their integration into real-time interactive systems has required many advances to be made particularly in system architecture and control. BT has worked with a number of collaborative partners to develop the new techniques reported and relevant partners are acknowledged at appropriate points in the chapter. This chapter discusses the technologies used to create perceptually compelling spatial audio for synthetic worlds.

4.1.3 Unification of visual and audio worlds

The bias towards the visual aspects of virtual environments is particularly evident in the files and structures used to define most virtual world software products. The emphasis in the geometry definition, motion description, and information stored about the physical properties of the world's structural materials are all oriented towards the requirements of video rendering. This is quite reasonable as a starting point since numerous developments in graphical handling and processing were required to produce the visual virtual environments which are currently available. Unfortunately the geometric and materials definitions which have evolved for visual virtual worlds are not always ideal for an acoustic model which requires different geometric resolution and additional information about the structure of the world.

Models of virtual acoustic environments do exist and find practical application. In particular, a number of modelling packages are available for auralizing buildings such as concert halls while these are still in the planning stage. Given the capital value of such projects it is a testament to modern auralization packages that they are deemed sufficiently accurate for this kind of application.

To progress the application of advanced spatial audio techniques within unified audiovisual worlds, a partial integration of existing visual and audio models has been developed at BTL. In one system the visual world software was modified to extract the spatial co-ordinates of listener and source positions. This information is then coded as TCP/IP packets and transmitted to an audio server which runs a separate world model with the required acoustic properties. The audio server then spatializes the audio according to the geometry information received and introduces the appropriate room acoustics such as reflections and reverberation according to the room model loaded. It is possible to move from room to room within the virtual world which allows the different room models to exhibit widely varying characteristics, such as a dead recording studio through to a large reverberent church. The unification of the visual and audio parts of virtual worlds remains the subject of on-going research.

4.2 SPATIAL AUDIO REQUIREMENTS

The emphasis of the spatial audio research reported here has been on high-quality, fully spatialized systems. Lower performance/surround-sound systems are extensively discussed elsewhere and are, in general, unable to position a virtual sound source at all points around the user.

Spatial audio requirements can be categorized by the number of users, restrictions on sound reproduction techniques, and the type of spatial audio percept required. Some examples are listed in the following sections.

4.2.1 Single users with headphones

A common situation is expected to be a single user in front of a VDU, who is likely to look continually at the screen. In this case the headphones need not be motion tracked since the user will face forwards. An advantage of headphones is that they do not typically produce sound which is intrusive to others. When one or more users are equipped with headphones and have freedom to move about, head-tracking is necessary to keep the audio landscape static; for example, as a user turns round, a sound source should appear to stay still, coming from behind the user when he turns his back towards it.

4.2.2 Single users with loudspeakers

A typical configuration is for the user to have a pair of loudspeakers either side of a VDU. Transaural reproduction has the advantage that it requires two conventionally located loudspeakers and can provide very wide angle frontal spatialization. However, it cannot currently provide compelling audio images behind the user.

A development of this technology, known as Soundbubble, with improved rear spatial images, is under development between BTL and the University of Essex.

4.2.3 Multiple users

When multiple users require a consistent spatial audio percept, conventional stereo, transaural and ambisonic all have limitations due to their limited sweetspot. The sweetspot is the ideal listening position, away from which spatial sound reproduction is impaired. Various techniques have been investigated to provide spatial sound to groups of users.

Steered mono offers some advantages for videoconferencing since it provides spatialization which is independent of listener position, can be achieved with a single audio channel and simple control information, and results in an echo-cancellation requirement that is potentially simpler than multi-channel techniques. True 3-D spatial sound implies that all users will hear a virtual sound source constantly following any desired trajectory.

A technique developed at BT to provide spatial sound for a group of users in the VisionDome immersive environment is described in section 4.4.3.

4.3 SPATIAL AUDIO TECHNOLOGIES

A variety of spatial audio techniques have been investigated and developed at BTL to suit a range of applications [3]. A number of these technologies are introduced below.

4.3.1 Binaural

A binaural signal (recorded with a dummy head or synthesized) recreates the soundfield present at the entrance to the ear canal when listening to an actual sound source [4]. The sound waves impinge the outer ear (pinna) at a given angle; the pin then filters the sound with an angle-dependent transfer function (head-related transfer function — HRTF). The combination of HRTFs and time delay enable the listener to locate sounds — HRTFs have the greatest influence at mid and high frequencies, and time delay at low frequencies. Figure 4.1 illustrates the encoding process where HRTFs are encoded into each ear's signal. To prevent the sound being filtered by the pinna twice it is necessary to use headphones to listen to the binaural signal. A dummy head recording is made with an anatomically accurate artificial head with microphones positioned at the entrance to the ear canal whereby the artificial pinnae provide the HRTF filtering. Synthesized binaural encoding requires the signal to each ear to be convolved with the appropriate HRTF — for a moving virtual source location a considerable amount of convolution must take place. The HRTFs are measured with a dummy head or by placing microphones inside a subject's ear canal with the system impulse response being recorded under anechoic conditions.

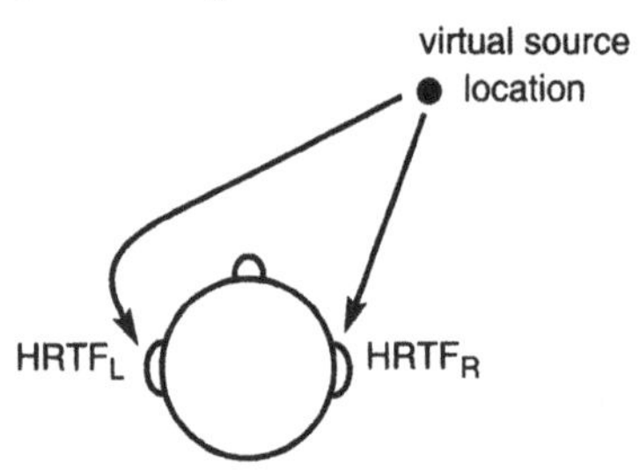

Fig. 4.1 Binaural signal encoding.

Binaural encoding produces very realistic sounding audio, especially when it also encodes the HRTF of each individual reflection. All of the processing is carried out at the encoding stage and the spatialized audio can be reproduced on conventional audio equipment. The principal disadvantage is the necessary use of headphones.

4.3.2 Head-tracked binaural

Because binaural signals are heard with headphones, any movement of the head will result in the same movement of the spatialized audio signal. In telepresence systems it may be desirable to keep the sound location static with respect to the physical environment, resulting in movement as the listener rotates their head. By applying different HRTFs for different head positions, the sound source location can be kept static; a typical practical implementation [5] uses 128 different HRTFs.

The solution comprises a pair of headphones with a rotational position sensor mounted on top and DSP-based hardware which encodes each of the two audio signals with the appropriate HRTF filters. Figure 4.2 shows the effect of head movement both with and without head tracking applied.

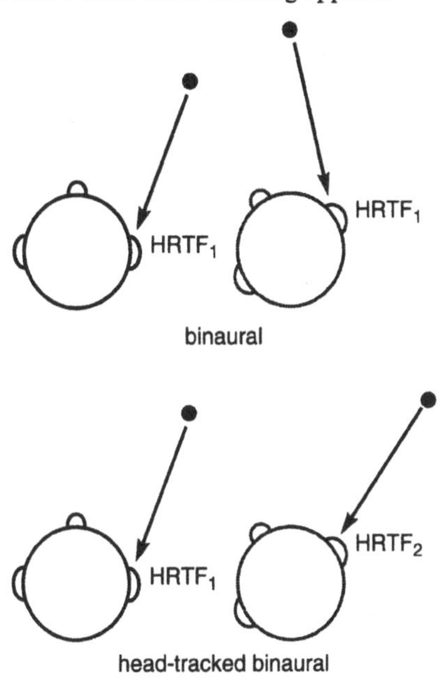

Fig. 4.2 Binaural signal with head tracking.

Unfortunately, head tracking requires real-time digital filtering; however, this is offset by audio image locations remaining static during listener head movement — this is useful in virtual meeting environments as each participant's voice has a constant localization with respect to the video presentation. Like standard binaural systems, headphones are required for sound reproduction; however, they now require a positional sensor to be attached or built in.

4.3.3 Transaural

A disadvantage of binaural encoding is the necessity of headphones for playback. Transaural systems use crosstalk cancellation to enable HRTF processing to be used with loudspeakers. Any number of loudspeakers may be used and they may be located at any angle with respect to the listener. Figure 4.3 describes the principles behind the shuffler crosstalk cancellation network introduced by

Cooper and Bauck [6]. The signal *S* represents the transfer function from the loudspeaker to the nearest ear, and the signal *A* represents the transfer function from the loudspeaker to the opposite ear. The functions *S* and *A* are defined as the HRTFs from each speaker to each ear. In this example, both loudspeakers are at an equal angle from the listener resulting in the same values of *A* and *S* for left-to-right and right-to-left signal paths. Where the loudspeaker arrangement is not symmetrical there will be two values of both *S* and *A*, and if there are *N* loudspeakers there will be *N* values of *S* and *A*.

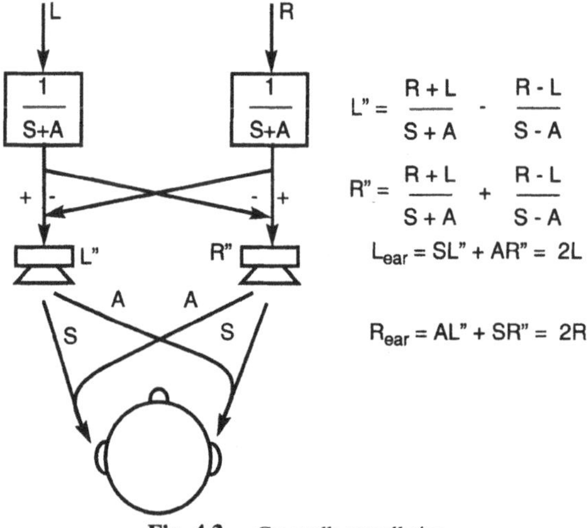

Fig. 4.3 Crosstalk cancellation.

Transaural coding gives good auditory localization and avoids the use of headphones in the playback system. Transaural is more suitable for individual workspace implementation rather than a multi-user environment as the correct combination of signals from the *N* speakers only occurs within a relatively small sweetspot.

4.3.4 Ambisonics

Ambisonic theory presents a solution to the problem of encoding directional information into an audio signal [7-9]. The signal is intended to be replayed over an array of at least four (for a horizontal plane only — pantophonic system) or eight (for a 3-D space-periphonic system) loudspeakers. Like a binaural system, the sound space can be recorded with a specifically designed microphone [10] or synthesized.

The signal, termed B-Format, consists of three components for pantophonic systems (*W* the ambient component, *X* the front-back component and *Y* the left-right component) and four components for periphonic systems *(W, X, Y* and *Z*, the up-down component).

Figure 4.4 shows the geometry used in the B-Format encoder, the equations shown define the *W, X, Y* and *Z* components of the signal, where the 3-D spatialized sound to be encoded is placed within a unit sphere.

During the encoding stage no knowledge of the loudspeaker positioning is necessary — the decoder is programmed with the loudspeaker lay-out. The output at each of *N* speakers in an equally spaced pantophonic array is given by:

$$P_N = \frac{1}{N}W + 2X\cos(\phi) + 2Y\sin(\phi)$$

where *W, X* and *Y* are defined as:

$$W = S \cdot \frac{1}{\sqrt{2}} \qquad X = S.\cos(\phi) \qquad Y = S.\sin(\phi)$$

The equation can be expanded for a periphonic array with a regular eight speaker array or an irregular *N* speaker array. A regular array gives:

$$P_1 = \frac{1}{4}W + \frac{2X}{\sqrt{2}} + \frac{2Y}{\sqrt{2}} \qquad P_2 = \frac{1}{4}W - \frac{2X}{\sqrt{2}} + \frac{2Y}{\sqrt{2}}$$

$$P_3 = \frac{1}{4}W + \frac{2X}{\sqrt{2}} + \frac{2Y}{\sqrt{2}} \qquad P_4 = \frac{1}{4}W - \frac{2X}{\sqrt{2}} + \frac{2Y}{\sqrt{2}}$$

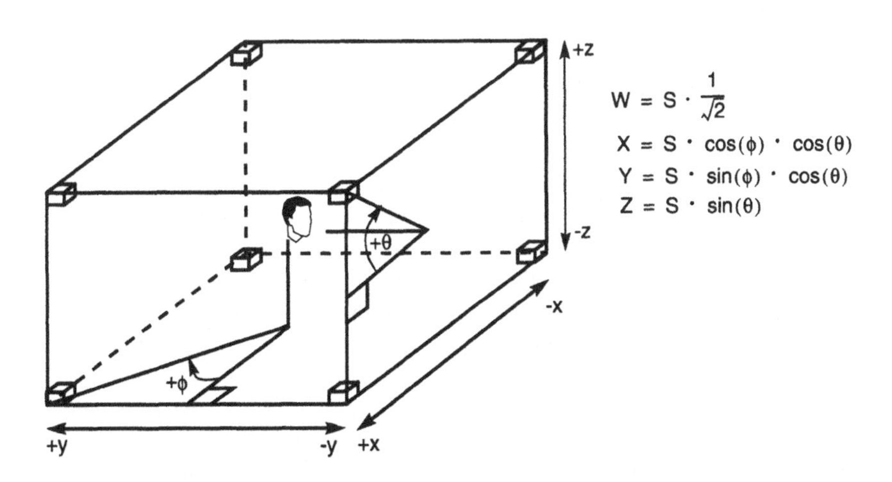

Fig. 4.4 Ambisonic system geometry.

Ambisonic encoding provides a realistic spatial audio percept; however, because the theory is based on wavefront reconstruction, the correct summation of the loudspeaker outputs only occurs within a small sweetspot, as with transaural systems, making an ambisonic system most suitable for personal workspace environments.

While one advantage of ambisonic systems over transaural systems is that the speaker lay-out does not need to be known at the encoding stage, a greater number of loudspeakers are required.

4.3.5 Strategies for multiple users

4.3.5.1 Nonlinear panning

In multi-user environments (such as the ARC Vision-Dome, described below in section 4.4.3), it is necessary to provide spatialized audio for a group of users. The two loudspeaker-based methods described above only fully spatialize the audio within a small sweetspot area. The use of intensity panning across the N loudspeakers gives listeners who are situated near a loudspeaker an improved spatial percept, due to the elimination of the ambisonic ambient sound signal (W). However, the signal will now consist of N loudspeaker position-dependent channels rather than the 2 or 3 of a B-Format signal.

Where the spatialized audio is synchronized to a moving video image projected on a concave screen (as in the ARC VisionDome — a typical example of a multi-user environment), a nonlinear panning is required to keep the perceived sound and video image locations synchronized.

4.3.5.2 B-Format warping

The aim of the warping algorithm is to change from a linear range of x and y values to a nonlinear range. Consider the example where a sound is moving from right to left; in order to remain consistent with the visual display on a large concave screen, the sound needs to move quickly at first, then slowly across the centre and finally quickly across the far left-hand to provide a corrected percept.

B-Format warping takes an ambisonic B-Format recording and corrects for the perceived nonlinear trajectory. The input to the system is the B-Format recording and the output is a warped B-Format recording (B'-Format recording). The B'-Format recording can be decoded with any B-Format decoder allowing the use of existing decoders. An ambisonic system produces a sweetspot in the reproduction area where the soundfield reconstructs correctly and in other areas the listeners will not experience correctly localized sound.

Given the B-Format signal components, it is possible to determine the original values of x and y as the original signal can be reconstructed to give $S' = W\sqrt{2}$ from which values of x and y can be determined:

$$x' = \frac{X}{S'} \quad \text{and} \quad y' = \frac{Y}{S'}$$

so

$$x' = \frac{X}{W\sqrt{2}} \quad \text{and} \quad y' = \frac{Y}{W\sqrt{2}}$$

A suitable warping method is obtained by squaring the value of x and y, or in the more general case by raising it to a power, $2i$:

$$X' = X \cdot (x')^{2i} \quad \text{and} \quad Y' = Y \cdot (y')^{2i}$$

$$X' = X \cdot \left(\frac{X}{W\sqrt{2}}\right)^{2i} \text{ and } Y' = Y \cdot \left(\frac{Y}{W\sqrt{2}}\right)^{2i}$$

$i = 0$ no warping

$i > 0$ nonlinear warping

Because the values of x' and y' may be positive or negative it is necessary to ensure that X' and Y' have the same sign as X and Y respectively. This is achieved by making the multiplier an even number, i.e. by raising

$$\left(\frac{X}{W\sqrt{2}}\right)$$

and

$$\left(\frac{Y}{W\sqrt{2}}\right)$$

to an even power ($2i$).

The resultant signal, X', Y' and W, is referred to as the B'-Format signal, as shown in Fig. 4.5.

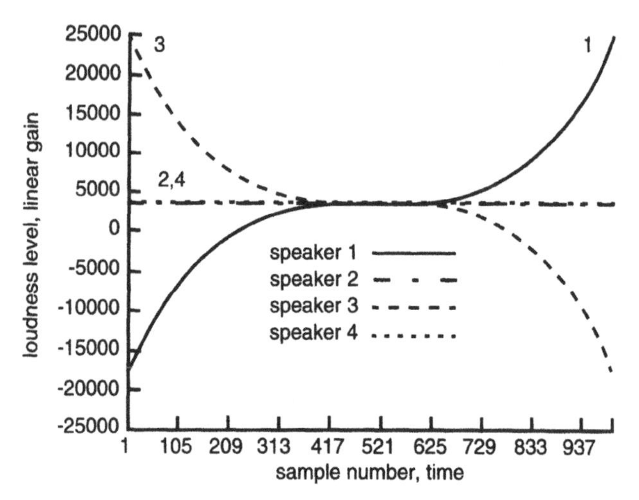

Fig. 4.5 Nonlinear loudspeaker feed profiles after transformation of the B-Format signal to suit a non-ideal listening room.

A further adaptation of the B-Format signal to reduce or remove anti-phase components has also been investigated to avoid misperception of spatial sound by listeners close to ear channel speakers. .

Warping the B-Format signal[1] has the advantage that standard recordings can be manipulated to suit specialist environments such as the VisionDome.

4.4 EXPERIMENTAL SPATIAL AUDIO SYSTEMS AT BTL

This section of the chapter introduces a series of demonstrators which were developed to prove a series of spatial audio concepts:

- Cube World represented an early realization of a real-time interactive 3-D world with high-quality spatial sound;
- BIER-link introduced spatial audio to a distributed shared space for two people;
- spatial audio to suit groups of users and non-ideal acoustic environments were implemented for the BT VisionDome;
- RevRooms introduced the concept of having different acoustics in a number of separate rooms;

[1] Aspects of the manipulation of the B-Format signal developed at BTL are the subject of a patent application.

- NSAS demonstrated the benefit of spatial sound in multi-user distributed environments.

4.4.1 Cube World

The first steps in creating a virtual environment with spatial audio was realized with Cube World. This was an elementary visual representation, consisting of a large room with four cubes inside. The cubes were positioned at different heights towards the four corners of the room. Each cube has a different sound source that emanates from its location in 3-D space. The sound sources used were ambulance, lorry, plane and car horn. This was a fully interactive, real-time, multiple sound source, audio/video environment. A mouse or trackball was used to navigate around the world. The graphics were displayed on to a large back projection video screen. An eight-speaker array was used for the spatial audio playback.

The system consisted of an SGI Indigo 2 workstation, a Lake Huron Digital Audio Workstation and an eight-speaker array. The virtual world was rendered on the Indigo 2, and the control information for user and sound source position was sent to the Huron via TCP/IP sockets in real time. The Huron rendered a B-Format ambisonic soundfield, with Doppler shift, for the eight speakers in real time. This is continuously updated as the user navigates around the world, giving a perceptually compelling 3-D soundfield.

4.4.2 BIER-link

The BIER (Broadband Interactive Enhanced Reality) link was developed from the body tracking used in the ALIVE technology from the MIT Media Lab [11]. The BIER link enables two people to have a virtual conference over ISDN with a multi-speaker ambisonic system. The audio was encoded using a G.722 CODEC giving a 7 kHz bandwidth and transmitted with control information over ISDN. At each end the user is free to move around in an area 4.5 m × 4.5 m, their position being monitored by a video camera. This area is the shared space that both users inhabit for the duration of the conference. The body-tracking software allows for rudimentary gesture input to animate the user's avatar. The position of the user's avatar is continually adjusted in real time to match the user's real location. Figure 4.6 shows the virtual environment inhabited by two users.

With the location of the users known, it is possible to transmit the local audio signal to the remote user and position this in 3-D space. Localization problems arise when the user moves away from the geometric centre of the sound system and hence out of the sweetspot. A novel means to overcome this problem for a single user is to transmit the position co-ordinates to the decoder in real time. These co-ordinates are then used to update the decoder to relocate the sweetspot

Fig. 4.6 Two users rendered in the BIER shared environment.

to the user's current location. This approach, including the performance of a real-time system, is reported in Burraston et al [12].

4.4.3 BT VisionDome

The BT VisionDome is an interactive, digital virtual-reality environment, which enables group interaction around common data or a shared application [13]. The use of a hard plastic dome as the listening room creates many acoustic problems, mainly caused by multiple reflections [14]. The problem was further compounded by the requirement to provide spatial audio for multiple listeners inside the dome. The multiple reflections were reduced by covering the inside of the dome in acoustic tiles.

Reproducing spatial audio for multiple listeners presented another new problem. In this case there is no sweetspot to reconstruct the audio wavefront, because all the listeners are spread out inside the dome. To overcome this problem, a nonlinear panning strategy was developed in collaboration with Lake DSP, and implemented on their Huron system [15]. The system receives the X, Y, Z location of the sound source, the X, Y, Z locations of all the speakers, and the monophonic sound source. The sound is then panned from one position to another using a nonlinear panning characteristic. This system worked well in the dome.

The nonlinear panning technique described above can only be used where the co-ordinates are known for each sound source, e.g. in an interactive computer program. A BTL-developed B-Format warping algorithm allows pre-recorded, standard, B-Format material to be acoustically rendered in the VisionDome to provide compelling spatial sound for a group of listeners.

4.4.4 RevRooms

RevRooms (Reverb Rooms) was the first system developed that used spatial audio rendering coupled with room acoustic simulation, known as auralization [16]. The spatial audio used a transaural playback system to render sound from two loudspeakers. The virtual world was a simple model incorporating four rooms of increasing size from small room to large hall. Each room had a fixed human shaped figure, with a small segment of looped speech playing from the co-ordinates of its head. The system ran in real time and allowed the user to freely travel around the environment. The hardware used to spatialize sounds and acoustically model room parameters was controlled using MIDI (Musical Instrument Digital Interface).

This system effectively demonstrated the ability to interact with a virtual world, while generating an auralization in real time for each room. Users experienced a greater feeling of immersion in the environments due to the inclusion of room acoustics, which accurately matched the visual percept of room size. The ability to accurately model/synthesize room acoustics is essential in creating a compelling sense of presence. Figure 4.7 is a screen-shot of the system, which shows the avatar of another user and a doorway to another room.

Fig. 4.7 View of RevRooms environment.

4.4.5 NSAS (Network Spatial Audio Server)

NSAS is a current project which utilizes several aspects of the technologies described above to provide a networked shared meeting space with high-quality spatial sound. Each participant receives a 3-D audio rendering of the soundfield around them in the virtual environment. This includes the virtual room acoustics of the environment and any sound source considered to be within a predetermined range. The sound sources include the speech of users in the environment and other sounds generated within the virtual environment. The virtual environment in the concept demonstrator consists of a number of rooms where users are allowed to enter and hold meetings. It would be feasible for groups of users to create conferences by providing lectures which would be audible to attendees within a designated room. The number of users is not fixed and individuals can enter or leave the environment at any time.

An overview of NSAS and its associated components is shown in Fig. 4.8. Audio and control messages will be sent across a computer network. The system consists of three sub-systems and *N* number of clients. The three sub-systems are:

- network spatial audio server (NSAS) for rendering the spatial composite and auralization for each client;

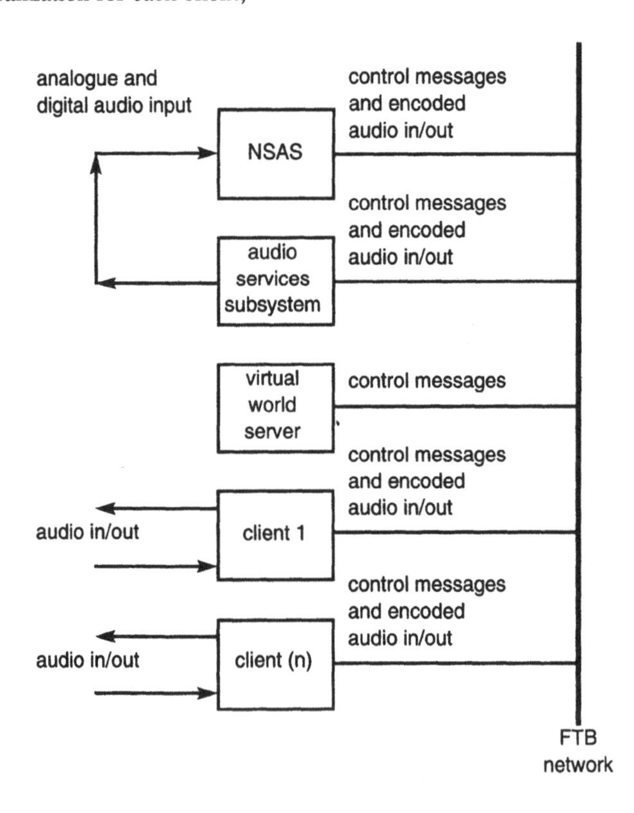

Fig. 4.8 Overview of NSAS and associated components.

- audio services sub-system will provide the system with any data- or event-driven sounds, voice annotation/voice messaging and rudimentary data sonification;
- virtual world server will control the virtual environment for each user and the interface between the user and the other sub-systems.

A prototype system exists as two clients, connected via balanced analogue cables, the NSAS server, the virtual world server and a small audio services sub-system. The virtual environment consists of a corridor with five doorways to enter any of the three meeting rooms or two lecture/conference rooms, each having simulated acoustics including doorway simulation. Each room contains a mannequin reciting a short monologue and one room also contains a moving sound source. This is useful for demonstrations, allowing the system to show its own capabilities.

When a user logs into the environment, their avatar is created and made visible to all other users. The avatar always represents the user's location within the virtual environment, and when they speak, their voice will always be at the same location as their avatar. The result is a natural, intuitive means to communicate within a virtual environment or conference.

4.5 DISCUSSION AND CONCLUSIONS

Much of the leading edge development in virtual reality and multimedia has focused on the video component, with the audio part often being underdeveloped. BT's history as an audio services provider has given a valuable perspective for the development of advanced telepresence applications in which the importance of audio performance is acknowledged. The provision of high-quality spatial audio can greatly enhance the intelligibility of multiple voices and is an invaluable orientation and navigation aid for a variety of applications including virtual spaces. Various spatial audio technologies have been summarized and a series of leading edge demonstrators which employ them have been introduced.

REFERENCES

1. Griesinger D: 'Multichannel matrix surround decoders for two-eared listeners', J Audio Eng Soc (abstracts), 44, No 12, p 1169, Preprint No 4402 (1996).

2. Begault D: '3-D sound for virtual reality and multimedia', Academic Press (1994).

3. Hawksford M O: 'Letter on sound localization using HRTFs', Electronics World and Wireless World (September 1996).

4. Moller H: 'Fundamentals of binaural technology', Applied Acoustics, 36, pp 172-218 (1992).

5. Lake DSP Pty Ltd: 'Huron digital audio convolution workstation user manual', Section 2-21 (1996).

6. Cooper D and Bauck J: 'Prospects for transaural recording', J Audio Eng Soc, 37, No 1/2, pp 3-19 (1989).

7. Gerzon M A: 'General metatheory of auditory localization', Presented at the 92nd Convention of the Audio Eng Soc, Preprint No 3261 (1992).

8. Gerzon M A: 'Surround sound psychoacoustics', Wireless World, 80, pp 483-486 (1974).

9. Malham D and Myatt A: '3-D sound spatialization using ambisonic techniques', Computer Music Journal, 19, No 4, pp 58-70 (1995).

10. Gerzon M A: 'The design of precisely coincident microphone arrays for stereo and surround sound', Presented at the 50th Convention of the Audio Eng Soc (1975).

11. Wren C, Azarbayejani A, Darrell T and Pentland A: 'Pfinder: real-time tracking of the human body', SPIE, 2615, Conference on Integration Issues in Large Commercial Media Delivery Systems (1995).

12. Burraston D M, Hollier M P and Hawksford M O: 'Limitations of dynamically controlling the listening position in a 3-D ambisonic environment', Journal of the Audio Eng Soc (abstracts), 45, No 5, p 413, preprint No 4460 (1997).

13. Walker G, Traill D, Hinds M, Coe A and Polaine M: 'VisionDome: a collaborative virtual environment', British Telecommunications Eng J, 16 (October 1996) (also http://www.labs.bt.com/people/walkergr/IBTE_VisionDome/index.htm).

14. Rimell A and Hollier M: 'Reproduction of spatialised audio in immersive environments with non-ideal acoustic conditions', Presented at the 103rd Convention of the Audio Eng Soc, Preprint No 4543, New York (September 1997).

15. http://www.lakedsp.com/index.htm

16. Kleiner M, DalenBack B and Svensson P: 'Auralization — an overview', J Audio Eng Soc, 41, No 11, pp 861-875 (1993).

5

DESKTOP CONVERSATIONS — THE FUTURE OF MULTI-MEDIA CONFERENCING

M Russ

5.1 INTRODUCTION

The history of human beings documents an ongoing desire to communicate at distance. The close-up interaction of body language works at very short distances, while the spoken word can work to large audiences over longer distances. The spoken word is an attempt to translate thoughts into a code which effectively conveys meaning to a listener, and this 'translation and coding' element forms a common thread through all communications theory. Where the spoken word is too quiet, then shouts can be used over longer distances at the cost of a significant loss of both intelligibility and speed of transmission.

Where audio signals fail, then visual methods can be used. Shouting is frequently accompanied by gesticulation — normally arm-waving and exaggerated body movements. The arm waving works much better if the length of the arms and the position of the hands can be emphasized, which results in the use of flags, and ultimately, in telegraph systems using large mechanical arms mounted with lights. The electrical analogue of these telegraph systems shares the same name, and uses simple electrical signals to convey the information — either via multiple wires and an indication device, or by using a time-based coding scheme like Morse code. Extrapolations of this technology to telegrams and telex merely return to the written word as the coding method, and the physical transport is replaced with an electrical equivalent.

The telephone returns to the spoken word, but uses electrical signals to transmit an audio signal across distances. Facsimile reuses the technology of the telephone to provide the transmission of pictures, and, although this does not seem significantly different to a telegram or telex message to users of languages which use the small number of characters like the Roman or Cyrillic letter forms, for users of pictorial languages with large numbers of characters, like Chinese or

Japanese, the facsimile is a very different communications device indeed. The videophone combines a telephone with, typically, a video picture of the two participants — although it is sometimes also possible to replace the video with the ability to share a common view of a picture or document.

All of the spoken or visual systems mentioned so far allow communication between groupings of people — either 'one-to-many', or 'many-to-many'. In contrast, the electrical systems have been 'one-to-one', where two participants can communicate either by coded words or by audio. Extending this 'communication between two points' to multiple points uses the concept of a gathering together of people to talk — a conference. Teleconferencing is the process of linking together several participants in a multipoint conversation using telecommunications, and it can use audio and video as well as additional data (text or pictures). Just as the telephone allows people to talk even though they are geographically separated, so teleconferencing allows a conference to be held when the participants are in different locations.

To arrive at a desktop teleconferencing system, the way that the telephone and the desktop work needs to be studied, and then the standards infrastructure that is required to make a broadly applicable product, which exploits the merging of telephone and computer, needs to be considered.

5.2 TELEPHONY — TWO-WAY INTERACTION

The telephone appears such a natural device for providing conversational interaction that it is now difficult to envisage a world without it. But the original intention of the telephone was to provide access to audio signals at a distance — which is where the name comes from. The word 'telephone' is made up of two Greek parts — *tele* meaning 'remote' and *phone* meaning 'voice'. Comparisons between the radio and the telephone as a way of listening to music or speech appear unlikely in a modern context, but the use of the telephone to carry out conversations which are disjointed by distance is a serendipitous result of the technology, not its original intended usage.

Although it tends to be used mainly for person-to-person communication, it is possible to use the telephone to connect several people together — this is called audioconferencing. Compared to a face-to-face situation where several people are gathered around a table, audioconferencing provides no visual information about who is speaking, who is shuffling, or who keeps glancing at their watch. Telephone calls to more than one other person, with or without video, tend to be used mostly for business purposes. Adding video pictures to the telephone can provide some of the information that is lost when using just audio connections. Despite this, the videophone has only slowly moved from speculative fiction to real-world usage, whereas the telephone is almost ubiquitous.

5.3 THE DESKTOP — ONE-WAY RETRIEVAL

The desktop is the word used for the graphical user interface (GUI) that is presented on the screen of many personal computers. It is called 'the desktop' because it uses some of the elements of a real desktop to provide a metaphor for controlling the computer. The wastebasket or trash-can is perhaps the most obvious example — to throw something away you put it on top of a small picture or 'icon' that represents the wastebasket, and the item is then deleted by the computer. Documents are represented as icons which have the appearance of documents — although they are still stored as data files inside the computer. Storage locations inside the computer are represented by icons which appear either as pictures of the storage device itself (a floppy or hard disk, perhaps) or as a folder to indicate several documents filed in one place. The contents of an icon or folder can be displayed in an on-screen 'window' area — the equivalent of opening up the folder and looking at a catalogue or directory of the contents. To move documents from one place in the computer to another, they are dragged from one icon, folder or window to another.

Because many of the desks found in IT-aware companies are equipped with a telephone and a networked computer, the desktop is a familiar interface to many business users. Unlike the telephone, the desktop has only been in common usage for a few years, and so its use is still developing, although by adopting the metaphor of a real desktop it carries with it a number of associative characteristics. The difference between resources which are local to the computer, and resources which are provided by a local area network (LAN), is hidden from the user — they appear as icons or windows, and can be manipulated in exactly the same way. This is being extended to wider network resources like the Internet, where future operating systems and GUIs will provide effectively 'transparent' access to the WWW.

The main focus of the desktop is one-way — often just the retrieval or publishing of items of information. Just as sending a letter necessarily involves a time delay while it is in transit and being processed, and a response sent back, so desktop systems tend to have corresponding delay mechanisms. E-mail merely shortens the time in which a reply may be received, but provides no guarantee that it will be any faster than conventional mail. The Internet provides access to huge quantities of information, but the inherent interaction is normally with a remote database rather than a remote person, and so is characterized by the download of large quantities of text and pictorial information, while the upload of control signals normally occupies significantly less bandwidth.

Business users who have a desk equipped with a telephone and networked computer are also likely to be users of conferencing. At the moment, conferencing is synonymous with audio- or videoconferencing, and, while the desktop may be involved in the set-up of the conference, it is rarely used interactively other than as a way of displaying the video picture. When collaborative tools are provided, the use of interactive working with another person is unfamiliar in the context of a desktop, and so is underutilized. The desktop metaphor needs to be extended so that it also becomes associated with interactive bi-directional usage instead of unidirectional retrieval or publishing activities. This will involve a change in the public perception of the telephone and the computer, but it also requires an amalgamation of their different standards and methods of working. Computer/telephony integration (CTI) is one aspect of this merging process [1].

5.4 STANDARDS

The convergence of the once separate worlds of computing, telecommunications and the media is now an accepted part of the fabric of society. The increasing use of digital technology has acted as the catalyst in this ongoing process, and it brings with it a dependence on the use of defined standards so that successful interworking can happen. Integrating the many layers of standards, protocols and interfaces together into a coherent form involves slotting together different paradigms and metaphors.

Some of the organizations that are working on stabilizing this convergence include DAVIC, the ITU, and the International Multimedia Teleconferencing Consortium (IMTC). The convergence itself is driven by both the standards and the technology. In the case of DAVIC, it is the delivery of multimedia content to end users that is the focus, whereas the IMTC works towards facilitating multimedia teleconferencing using standards like T.120 (see Fig. 5.1) and the H.32x series.

One of the underlying enabling standards for conferencing on the desktop is T.120 [2]. The aim of the ITU T.120 Recommendation is to ease the development of a worldwide dial-up multimedia data-conferencing service [3]. It defines a series of open data-conferencing standards that can support integrated communications between users, regardless of the computer platform or network technology that is being used. The intention is to make a 'multimedia' call as simple to carry out as making a telephone call.

The T.120 suite of recommendations describe a series of communications and applications protocols and services that can provide support for multipoint data communications [4]. The facilities and abstraction that it provides should make it easier to develop collaborative applications which use diverse types of network connections. Some possibilities include desktop data-conferencing, several users sharing a single document on one application, and multi-player gaming. The functionality provided means that it is possible for a single conference to incorporate calls made using many different types of data transport mechanism and end-point capabilities — and yet the conference can be controlled as a single unit. It is possible for the user terminals to exchange

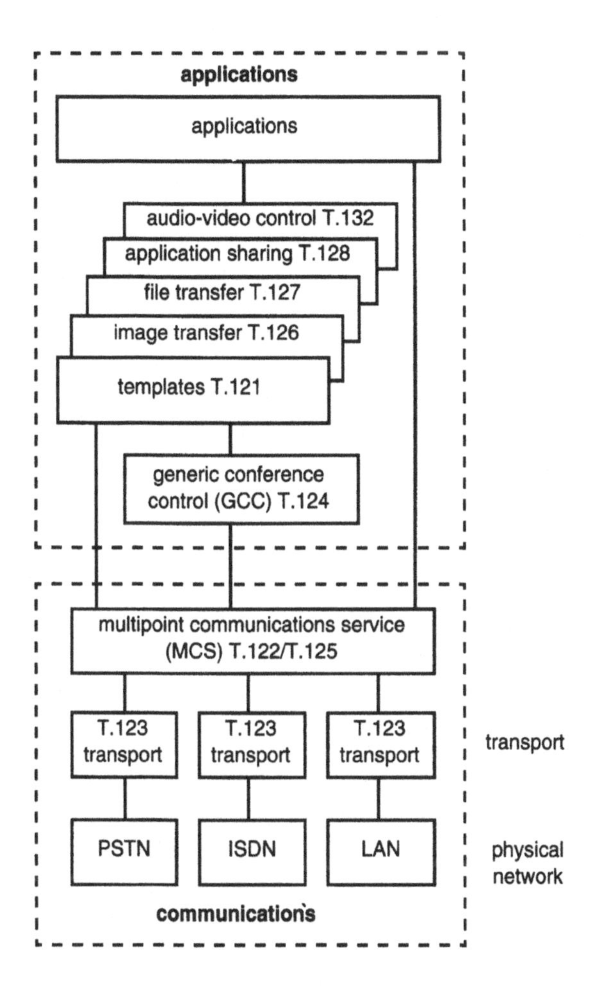

Fig. 5.1 An overview of T.120.

information about their capabilities, and then to take these into account in the management of the conference.

T.120 tries to resolve complex technological issues in a way that is acceptable to both the computing and telecommunications industries. As there are no platform dependencies, and no major limits on the network topology (star, chains, cascades, etc), it offers easy scalability, extendibility, and will co-exist with other ITU standards. It is intended to be the key building block for conferencing — a unifying element.

The T.120 standards suite is designed as a series of layers, with protocols and service definitions linking them together. The lower layers provide a generic mechanism which allows applications access to multipoint communications services. The higher levels define some specific protocols for use by applications to ensure baseline compatibility — 'whiteboards' or file transfer — although it is possible for an application to utilize other non-standard protocols.

At the lowest level is the transport protocol, known as T.123. This interfaces the T.120 data packets, known as protocol data units (PDUs), to the physical networks — PSTN, ISDN, or LANs (TCP/IP or IPX). This provides a uniform guaranteed delivery of data, and can be seen as being the provider of network interfacing. Above these is the Multipoint Communication Service (MCS), described in T.122 and T.125. The MCS is the heart of the T.120 architecture — it connects together multiple channels of information to form what is known as a domain, roughly equivalent to a 'conference'. The MCS controls the information flows between the applications running on the participant computers, and can provide broadcast and private communications. It thus enables the delivery and synchronization of data across a multiple set of end-points — a multipoint conference.

The Generic Conference Control (GCC), T.124, sits above the MCS, and turns the generic connections of the MCS into a support mechanism for conferencing. The GCC maintains a database of the conferences in progress, and so serves as the common point for initiating, managing and controlling conferences. Applications register with the GCC, which then controls the assignment of resources from the MCS and below. The GCC is thus providing the conference administration interface, and it co-ordinates the use of the dynamic data resources below it.

The layers above the GCC can communicate either directly to the GCC and MCS, or can use a defined resource management template which is defined in T.121, the Generic Application Template (GAT). The GAT acts as a standard glue between applications and the lower levels, and so helps to ensure consistency and robustness in their interactions. It provides several services such as an enrolment service for applications, initiating applications on other nodes in the conference, and attaching an application to a domain in the MCS. By providing these facilities the GAT aims to reduce interoperability problems. T.126 is a still image protocol which is intended to facilitate a shared pictorial document, commonly called a 'whiteboard'. It provides virtual workspaces which are shared between the participants, and which can be manipulated by any node. It defines how the images are shared and managed, as well as allowing interactive pointing and annotation by the use of distinct 'planes' of view. T.127 is the file transfer protocol, and enables efficient transmission of files to any or all of the nodes in a conference. With advanced implementations, this allows remote information retrieval from archives to be achieved.

BT has been a major contributor to the T.120 suite of standards, and the knowledge and expertise of the specialists is a valuable asset when designing a conferencing system. The remainder of this chapter will show how conferencing based on these underlying standards can provide an integrated solution to communicating between several people in a rich environment.

By hooking into the shared information in the conference, Databeam's 'neT.120' server provides a simple way of taking the 'whiteboard' content of a T.120-based conference and publishing it on the Internet as a WWW page. It is also possible to use T.120 to provide even closer coupling between applications running on terminals in the conference, by using application sharing. This enables a computer application program on one computer in the conference to be viewed at all of the other end-point terminals by managing the distribution of its graphical output. Arbitration of the control devices, typically the mouse and keyboard, enables the application to be controlled from a different end-point to the one where it is located.

For real-time media information streams like video or audio data, as well as management functions like control and indication, a series of proposed extensions to the basic T.120 recommendations are nearly complete. These are known as the T.130 Audio-Video Control (AVC) standards, and they deal with the way that T.120 handles these types of media streams. The component standards in the T.130 series provide a network-independent control protocol for real-time audio, data and video streams. T.130 uses T.120 data communications services. The movement of data across diverse network boundaries is co-ordinated and managed by T.130, so that end-point terminals with differing capabilities can co-exist in the same conference.

The T.130 suite enables the manipulation of real-time data streams within an established conference topology, but it does not define the mechanisms which are required to set up that topology. The terminal need not implement all of the protocols referenced in T.130 in order to participate in the audio or video parts of a conference, but in this case the control functionality may be limited. The range of control services that is provided includes:

- video switching;
- video processing and transcoding;
- audio mixing and muting;
- control of the 'floor';
- identification and selection of multimedia sources;
- channel management in real time.

BT is also leading the development of this new standard, and the final result will be a comprehensive tool-kit of standards and protocols that will facilitate feature-rich teleconferencing.

5.5 CONFERENCING

The generic word for conferencing which uses telecommunications is teleconferencing — literally, remote conferencing. There are several different types of teleconferencing, and it is normal to describe them by the major method used to deliver information, although audio is normally assumed to be also present in videoconferencing. A video picture on its own would have limited usefulness compared to an 'audio-and-video'conference.

The concept of a teleconference is to provide to the participants the same types of facility that they would expect to find at a real meeting or conference — within the restrictions of the media being used. A desktop conference therefore is limited in the way that it can provide a simulation of a room environment, whereas a videoconferencing suite already has the environment. Immersive virtual reality techniques may alter these divisions in the future, and provide more of the 'sense of being there' that is promised by telepresence. Current conferencing techniques tend to be constrained by pragmatism — which in this case means making the most of available technology. On the office desktop, this probably means using just the telephone and networked personal computer in their basic form.

5.5.1 Audioconferencing

Audioconferencing uses the telephone to provide audio communication between the participants. Three-way calls can be set up directly by the user employing network services, but for more participants special bridging hardware is required — and this is provided by conferencing bureaux services like BT's Conference Call [5]. More detail on the implementation of this service is given in Chapter 3.

5.5.2 Videoconferencing

Videoconferencing adds video pictures to an audioconference. There are two main sub-divisions — room-based systems, and desktop systems. Room-based videoconferencing is normally used by more than one participant, and can be either a permanent installation, or mobile equipment can be moved into a room as required. Because the aim of a room-based system is often to simulate a conventional meeting, the participants normally sit on one side of a table, while the video screens, cameras and loudspeakers are placed on the other side. Desktop videoconferencing is usually designed for single users, and is often associated with a personal computer combined with a microphone and loudspeaker which provides the functionality of a telephone, although stand-alone desktop units are also used. Desktop systems can use either the telephone handset, a headset, or a loud-speaking telephone, and are often designed around the metaphor of a telephone with added video pictures, rather than a meeting room.

Although a videoconference provides the pictures of the participants that are missing in an audioconference, the cost and technical complexity of providing a usable environment that involves audio and video can be prohibitive. Desktop videoconferencing tends to provide additional tools to facilitate the transmission of associated material like documents and pictures, but the link between the video picture on the screen, and documents on the computer is often not easy to comprehend.

5.5.3 Data-conferencing

Data-conferencing typically uses the Internet to provide communications between participants, often using text-only shared 'chat' applications for exchanging written information, or shared picture-based 'whiteboard' applications for annotating diagrams, maps or other pictorial information. Some Internet-based data-conferencing is point-to-point, which restricts the sharing of the data to just those two people. Multipoint data-conferencing enables more than two participants to exchange or share information, and so work collaboratively.

There are many applications for data-conferencing. They include:

- linking sales teams together to share information;
- linking project managers to co-ordinate their schedules and plans;
- enabling marketing teams to work jointly on presentation material;
- enhancing distance learning applications;
- bringing together students from different locations for special lessons or lectures.

5.5.4 Audiographic conferencing

Audiographic conferencing brings together audioconferencing with data-conferencing, and so provides the familiar interaction of a telephone combined with the visual text or pictorial information that is often the focus of a technical or business meeting (see Fig. 5.2).

This produces a dichotomy of user interface — the audio from the telephone is familiar, although loud-speaking telephones perhaps less so, while the presentation of the visual information is probably more unusual, especially in the context of the normal unidirectional retrieval-based use of a computer screen.

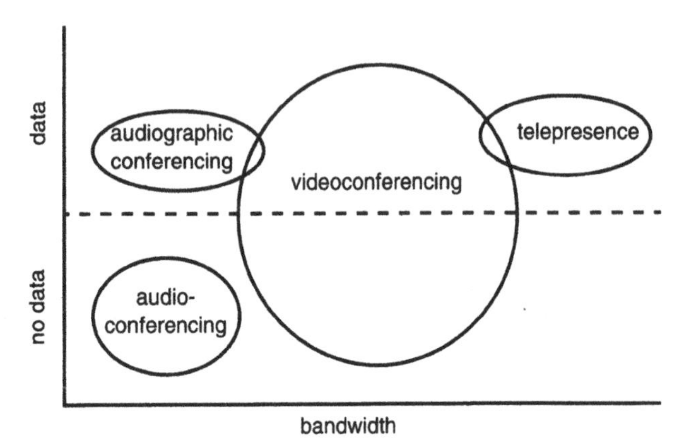

Fig. 5.2 Audiographic conferencing extends audioconferencing by adding in data, while providing a coherent upgrade path towards integrating video and ultimately delivering telepresence features into conferencing environments.

Therefore one of the major challenges of audiographic conferencing is to present the linking between the audio and the data parts of the conference in a way that makes sense to the user. The telephone is strongly associated with audio use, while the desktop uses the metaphor of a desk, and there are no obvious linkages between them. Solving this problem is the key to making desktop conferencing a success.

5.5.5 Desktop conferencing

The term 'desktop conferencing' is used here to emphasize the use of the computer desktop in the role of an interactive, conversational tool. If the normally retrieval-based desktop is to be used by participants in a conference, there needs to be a logical linkage between the audio and the visual elements. This can be observed in audioconferences where it is not possible to know how many participants are present, unless they identify themselves individually. If a new participant joins the conference and does not speak, then it is possible for their presence to go unnoticed. Also, the exchange of supporting visual information in such a conference has to be carried out either beforehand, by post or fax, or afterwards — which may well trigger another meeting because the material does not match with a participant's impression of its content.

A desktop conference needs to bring together the telephone and the computer screen area into one integrated working environment. The actual media forms used merely provide extra information, and do not change this basic requirement. So a desktop conferencing system design must be able to cope with a range of interconnection possibilities — from a simple data-conference, through audioconferencing and audiographic conferencing, to videoconferencing. The provision of a single consistent user environment is the catalyst that makes desktop conferencing a serious communications tool (see Fig. 5.3).

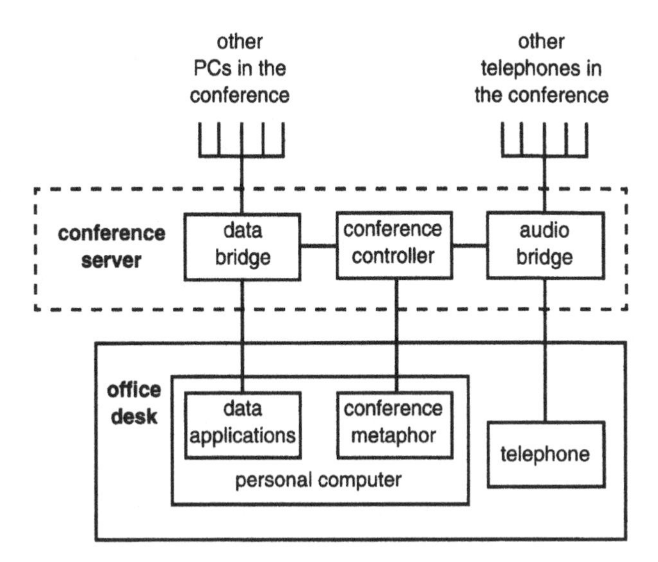

Fig. 5.3 Desktop conferencing. The 'conference metaphor' is the on-screen link between the telephone, the participants and the data sharing applications.

5.5.6 Advantages of teleconferencing

The key advantages to teleconferencing are summarized below:

- cost-effective — can reduce travel costs, and can have low hardware and software costs;
- consistent — all participants receive the same information at the same time;
- enabling — meetings can take place regardless of distance, and can be set up without worries about long journeys;
- exciting — people can exchange ideas from anywhere in the world, to anywhere else;
- faster — travel time can be reduced;
- involving — interaction is immediate, instead of delayed by post, fax or e-mail;
- ubiquity — people can participate from geographically diverse sites by using a variety of transport mechanisms.

5.6 ENVIRONMENTS

Trying to determine the exact requirements of people who wish to interact in a conference environment is not trivial. Audioconferencing can provide audio links to replace the spoken word, but does not provide any visual information about the participants. Videoconferencing enables the participants to see each other and thus receive the non-verbal 'body language' signals that accompany speaking — people still nod their heads and gesticulate with their hands when they are using the telephone. Data-conferencing provides shared access to information that is required in the conference — text documents, diagrams or numbers.

However, the actual working environment of a conference is often very different from the real-world situation. An audioconference, where people gather around a loud-speaking telephone, does not immediately provide information about who else is in the conference — 'disembodied' voices are all that can be heard. Text documents can be discussed via data-conferencing, which enables several people to contribute to a shared document or diagram, but the separation between the contributor and a real person is similar to audioconferencing — having a name, associated with some text that is appearing on the screen, provides only the bare minimum of identification. Videoconferencing provides the missing visual information about the participants in an audioconference, but after the initial introductions, the main focus of many business meetings is not the participants, but the topic that is being discussed.

Other aspects of a real-world conference, that are easily overlooked when a teleconferencing environment is being designed, include the set-up of the meeting, and any subsequent follow-up after the meeting. For all types of meeting, the initial pre-meeting action is normally to circulate an agenda to all of the participants. This is normally dealt with by the organizer of the meeting, and follows the booking of the required resources — time, date, room for 'n' people, overhead projector, whiteboard, video recorder, television, etc. For audioconferences in particular, text or pictorial information that is needed in the meeting is often sent by fax or post to all the participants, so that they will have the information in front of them when the meeting starts.

Audiographic conferencing combines the audioconference with data — the additional text or pictorial information that often forms the focus of the discussion in the meeting. The data-conference can provide several facilities to enable this to happen:

- by allowing documents to be transferred 'in the background' between participants during the meeting, there is no need to send faxes or post printed documents;

- shared 'live' whiteboards allow diagrams or pictorial information to be annotated, revised and discussed *in situ*, as part of the ongoing meeting;

- text documents or spreadsheets can be collaboratively worked on by 'sharing' a suitable application, so that, for example, the exact wording of a paragraph can be determined by consensus, or revisions to the draft of a chapter can be agreed by the co-authors.

However, audiographic conferencing can also bring additional facilities to the teleconferencing environment because it is part of the computer desktop. The initial booking of the meeting or conference requires the organizer to determine a suitable date, and this may involve contacting prospective participants in order to determine their availability. This could utilize the telephone for person-to-person communication, but could alternatively be an examination of their desktop calendar/diary software. The organizer then needs to book or reserve the conference, which can be achieved by a number of methods — human interaction with a booking agent, an interactive voice-prompt system, or by using the desktop, i.e. e-mail, or a WWW form page.

The pre-meeting agenda and any background documents may well be transmitted by using e-mail, and this offers the possibility that it can be integrated into the audiographic conference booking system. The e-mail messages that give details of the agenda can also be used to provide the details of the audiographic conference set-up — so that joining is eased for the participants.

During the meeting, audiographic conferencing provides control facilities which enable participants to mute their microphones or loud-speakers, and 'chair' control features which allow one participant to take control over the conference, thereby requiring people to 'request to speak' and then choosing who can speak. This is made possible because of the linking between the desktop data-conference and the audio bridging equipment which controls the telephone lines.

At the end of the meeting, or perhaps immediately afterwards, the organizer may need to book a follow-up meeting, which might be for a week or a month's time, using the same facilities and the same participants — or a slight variation. Being able to edit the booking from the desktop is more efficient and ensures that the participants have continuity between the meetings. Because of the integration of e-mail into the conference booking, it can be augmented with mailing list news and 'frequently asked questions' (FAQ) distribution for registered users.

5.7 PASSEPARTOUT

Passepartout brings together the experience of telephony, audioconferencing, T.120 expertise, and knowledge of data networks into a single unifying tool for teleconferencing. Passepartout is the BT internal test name for an audiographic conferencing system, developed at BT Laboratories in collaboration with the BT Conference Call team, which is designed to ultimately bring the whole of the teleconferencing environment to the desktop. It integrates many of the wider elements associated with conferencing, which have been described above, into one coherent system, while also providing the strong linking between the audio and data parts of the conference. In the terminology of Fig. 5.3, Passepartout is the 'conference metaphor'. The intention behind Passepartout is to provide a complete tool for enabling productive meetings or conferences with geographically separated participants. The initial meeting or conference set-up is accomplished by using the e-mail message containing the agenda. This also has the details of the teleconference which are required to start the data-conference part of the audiographic conference — these are stored in an attachment to the e-mail. To join the conference, the participant double-clicks on the attachment, which initiates the start-up of the client application. Once the data link is established, the participant's telephone will then ring, and they are connected into the conference. The initial focus of the interaction is thus the desktop e-mail message, followed by the opening of the Passepartout window and the ringing of the telephone. This establishes the desktop as the centre of the participant's attention.

The linking between the audio, the data, the control, and the participants is achieved by using a window which contains a representation of the meeting or conference (Fig. 5.4). In its simplest form, this is a room with a table in it, around which icons appear to indicate the participants. Using this metaphor of a room, or perhaps a videoconference, is a very effective way of providing intuitive information about what is happening in the conference. When a new participant

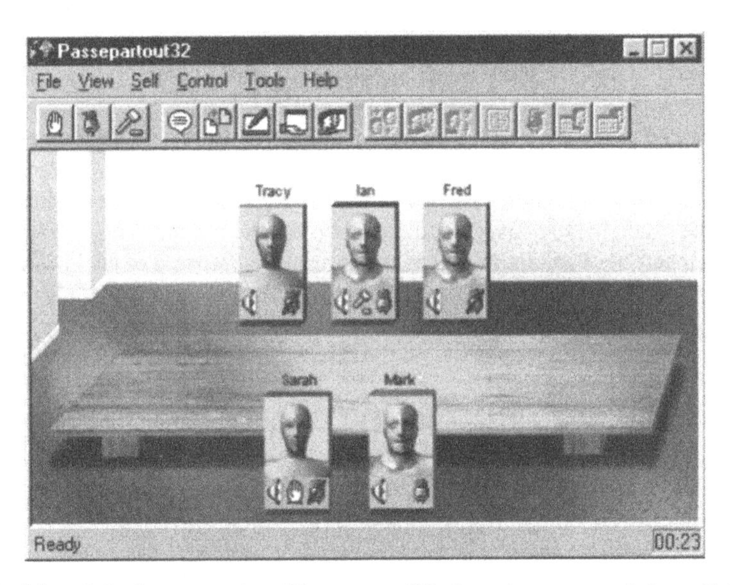

Fig. 5.4 A development version of Passepartout. This shows the on-screen window, with five people in a conference.

joins the conference, a new icon appears. Small glyphs, associated with each participant icon, indicate when they can talk, listen, or when they are requesting to speak. Highlighting of an icon can be used to indicate that they have taken over chairman control, or that they currently 'have the floor'.

Data facilities like file transfer, whiteboarding, and application sharing, are also available from the Passepartout window, so that it is the centre of the participant's interactions with the conference. The telephone is only used to provide audio.

To summarize, the Passepartout audiographic conferencing trial system allows users to:

- monitor and control the conference (see who is in the conference, chairman control, etc);
- use the power of the PC to discuss text, figures and diagrams — tools are provided for remote presentations, sharing documents and spread sheets, etc, the focus for accessing all of these tools being the Passepartout window toolbar and menus;
- easily book the conference — a WWW page is used for booking the conference, although a telephone booking service could also be used;
- easily join the conference — double-click on an attachment in an e-mail invite;
- use their intranet (normal LAN connection) in the office and a modem for dial-up access when away from the desk — T.120 is flexible in its use of networking, and can operate over PSTN telephone lines with a modem, ISDN lines, LAN networks, the Internet, or a modem in a GSM mobile phone.

Passepartout does not require additional hardware in the computer. An office desk with a telephone and a computer (networked or with a modem and an additional telephone line) are the only requirements for audiographic conferencing apart from the software. Because the audio is carried by the telephone, the computer does not need processing power to be used for coding and decoding audio signals carried over the data links, and so a modestly specified computer will give perfectly acceptable performance.

Passepartout exploits BT's comprehensive knowledge of T.120, gained during the years of developing the standard, and so gives a tangible expression to the vision and forethought of the team that have contributed to the standards suite. It represents one aspect of the T.120 view of teleconferencing turned into a desktop controller — in this version enabling audiographic conferencing to happen in a convenient and controllable way.

Future enhancements to the Passepartout concept will continue to build on this in-depth knowledge of the standards, and may well start to incorporate the T.130 standards suite as the audiographic conferencing metaphor is extended to include additional types of multimedia information.

5.8 CONCLUSIONS

Interactive communication using the telephone provides an excellent starting point for a number of enhancements that make the most of emerging technologies. Videoconferencing adds a strong visual element so that the participants can be identified and their body language observed. However, audiographic conferencing links together the audio and data to provide information about who is in the conference without the overhead of video technology. It also allows essential text or pictorial information to be worked on collaboratively when it is required in the meeting — there is no need to fax or e-mail the documentation to all the participants before (or after) the meeting.

The convergence of the underlying technologies and standards is bringing together the fields of conferencing, multimedia, and telephony into a framework which makes the conversational use of the desktop possible. The technology is emerging, and the enabler to the success of teleconferencing on the desktop is the bringing together of a complete system under one unifying metaphor. By easing and simplifying the processes involved in booking, joining, participating, controlling and following up after a meeting or conference, such a comprehensive desktop conferencing system produces a powerful 'on-screen' focus for interactive communication.

The telephone is likely to remain the natural route for conversing with remotely located people, but making the most of a telephone linked to a desktop conferencing system will enable a media-rich conversation instead. Future desktop conferencing systems will see the further integration of data conferencing with directories, calendars, schedulers and other personal productivity and time-management tools. In an increasingly multimedia-based world, desktop conferencing is in an ideal position to enhance and improve business communications.

REFERENCES

1. Hilson G, Hardcastle C and Allington M: 'Callscape — CTI for the small business', British Telecommunications Eng J, 15 (January 1997).

2. Databeam Corporation: 'A primer on the T.120 series standard', (1996).

3. Midwinter T: 'Desktop multimedia', BT Technol J, 13, No 4, pp 113-123 (October 1995).

4. Clark W J and Boucher J: 'Multipoint communications — the key to groupworking', BT Technol J, 12, No 3, pp 72-80 (July 1994).

5. http://www.visual.bt.com/

6

CONFERENCING ON THE INTERNET

P Cordell, M Courtenay, S Rudkin

6.1 INTRODUCTION

Any distributed real-time application operating over an Internet protocol (IP) network has to address the problems of quality of service, packet loss, jitter, and synchronization of different media [1] . However, there are a number of key problems which are of particular significance to IP conferencing applications[1]:

- user or conference location — the discovery of the address of someone who is to be invited to join the conference, or the discovery of the location of a conference that can be joined without invitation;
- latency — the end-to-end delay of control messages and conference media;
- scalability and robustness — the ability to robustly support widely varying requirements in terms of number of users, number of simultaneous conferences, geographical distribution of participants, etc.

This chapter discusses these key requirements and approaches to their solution. It then looks at two different technologies which have been developed to support conferencing over IP. The first is the work of the Internet Engineering Task Force (IETF) whose main goal is scalable conferences. The second is the International Telecommunications Union (ITU-) developed H.323 standard [2], which is aimed at supporting small, tightly controlled conferences. H.323 and IETF conferencing share a common approach to the delivery of media, but are significantly different in terms of how they control conferences. This chapter illustrates these differences with particular reference to the above requirements.

---

[1] The term conferencing application is used in this chapter to mean any multi-party interactive application (such as audio/video/data-conferencing, networked games, shared virtual worlds, and electronic auctions.

6.2 REQUIREMENTS AND APPROACHES

6.2.1 User and conference location

The first step in introducing a new member to a conference is the location of one party by another. This takes one of two forms depending on whether a user is invited into the conference (user location) or whether the user sees a conference announcement and then chooses to join the conference (conference location).

User location involves identifying the current IP address and port at which a given user can be contacted. If users did not move from terminal to terminal, did not have dynamically allocated addresses and were never unavailable, this could be a straightforward domain name service (DNS) look-up. Unfortunately all these complications arise.

Dynamically allocated addresses and the fact that users change location can be handled by requiring users to register with a location server each time their circumstances change (this may be a user-driven or computer-driven action).

When a user is unavailable (intentionally or not), they are likely to want to use some kind of divert facility (e.g. divert all calls, divert on busy or divert on no-reply). Divert all calls can easily be handled at the location server. Divert on busy or divert on no reply are a little more difficult. Some possible solutions include the following:

- a server intervenes in the call set-up path to mimic the private branch exchange (PBX) function — this allows the sequence of search steps to be specified local to the terminal being called, which unfortunately introduces a possible single point of failure;

- the destination system implements a search of alternative systems in a similar way to the above — this removes the single, system-wide point of failure, but it does assume that the system is always switched on;

- the location service returns a divert on busy address and a divert on no-reply address at the same time that it provided the initial address; however, this requires an arbitrarily complex search algorithm (such as ring address 1 three times, then ring addresses 2 and 3 four times and so on) to be encapsulated in some form of protocol — it also means that the calling party must fully implement the intelligence needed to interpret the coded search algorithm;

- after failing to connect with the destination system, the calling system takes the initiative to contact the location service to explicitly ask for a divert on busy or divert on no-reply address;

- multicasting a request for a terminal (perhaps via a location server) is particularly appropriate when it is a workgroup which is to be contacted rather than an individual — for the individual case, it works well for small numbers of users, but does not scale well to many terminals, as each terminal will be interrupted each time an invitation is made to any terminal; however, it may form part of an overall scheme in which a highly mobile set of people or a workgroup are contacted using multicast, and the rest are contacted using one of the above schemes — alternatively, a user location server could try a multicast location after the user location server's search algorithm has failed to find the right person.

Conference location is about determining the IP address and port of a conference typically through a conference announcement. The announcement may be multicast, e-mailed or published (e.g. on a Web page). The conference itself may be multicast or unicast.

6.2.2 Latency

Latency can be a problem both for conference control and media delivery. This section looks at latency of media delivery and considers delays associated with one aspect of conference control, namely, set-up. This area has been selected because it is an essential aspect of every conference.

6.2.2.1 Media delivery delay

A common scenario for current interactive services (e.g. Internet telephony) is where both users are connected to their Internet service providers (ISPs) by 28.8 kbit/s modems (see Fig. 6.1). Often these modems are connected to the local computer via a serial line operating at 38.4 kbit/s. The user datagram protocol (UDP) [3], either unicast or multicast, is typically used for the delivery of media.

Fig. 6.1 A common scenario for today's Internet telephony users.

The consequences for the end-to-end delay are quite horrendous as can be seen in the third column of Table 6.1. Delays which may be pipelined are denoted in Table 6.1 by a bold left-hand boundary line. However, the situation can be improved significantly (as shown in the fourth column of Table 6.1) by adopting the following techniques:

Table 6.1 Sources of delay in Internet telephony.

Source of delay	*Calculation*	*Typical current delay*	*Achievable delay*
Frame size (sampling time)	Determined by codec	30 ms (G.723.1 [7][2])	10 ms (G.729A)
Additional sound card/ API/os latency (over and above sampling time)		~ 30 ms (WAVE API, FIFO based protocol processing)	~ 5 ms (pre-emptible priority scheduling and real-time protocol processing)
Encode/decode	Look ahead + 2* 'frame-size'* required_MIPs/ available_MIPs	< 67.5 ms (G.723.1)	< 25 ms (G.729A)
Packet header	Add 42 bytes	24 bytes becomes 66 bytes	10 bytes becomes 14 bytes (use header compression)
Serial link at transmitter	10/8* packet size/line_speed	20 ms	0 (use internal modem)
Modem tx processing	53 symbols @ 3200 symbols/s	17 ms	17 ms
Modem waiting time	50 ms	50 ms	0 (turn off compression, error checking, transmission of blocks)
Modem transmission	packet/modem_speed	18 ms	4 ms (fewer bytes)
Modem rx processing	103 symbols @ 3200 symbols/s	32 ms	32 ms
Propagation delay	5 ms per 1000 km	25 ms	25 ms
Router queuing delay	~ 10 ms per router	~ 100 ms	~ 10 ms (priority scheduling)
Modem tx processing	53 symbols @ 3200 symbols/s	17 ms	17 ms
Modem waiting time	50 ms	50 ms	0 (turn off compression, error checking, transmission of blocks)
Modem transmission	packet/modem_speed	18 ms	4 ms (fewer bytes)
Modem rx processing	103 symbols @ 3200 symbols/s	32 ms	32 ms
Serial link at receiver	10/8 * packet size/38.4 ms	20 ms	0 (internal modem)
Buffer for network jitter, etc	2*mean variance in queuing delay	~ 200 ms	~ 20 ms (priority scheduling)
Buffer for os latency		30 ms	~ 5 ms pre-emptible priority scheduling
Total		~ 700 ms one way ~ 1.4 s both ways	~ 200 ms one way ~ 400 ms both ways

[2] The speech coding schemes G.723.1 and G.729(A) have the following delay characteristics:

	Look-ahead	Frame
G.729	5 ms	10 ms
G.723.1	7.5 ms	30 ms

- use low-delay codec, e.g. G.729A [4];
- use pre-emptible priority scheduling and use a protocol processing architecture that allows the timing of the protocol processing to be under the control of the application (see, for example, Lee *et al* [5]);
- use header compression [6] to reduce the transmission delays;
- use an internal modem (to avoid communicating with the modem via a serial link);
- turn off modem's error checking, compression, and block transmission;
- use priority scheduling in the network (this minimizes queuing delay and the need to buffer to account for jitter arising from any variation in network queuing delay).

Together these measures can reduce the delay substantially. The resulting end-to-end delay (calculated over 5000 km) would be just about satisfactory for telephony. Further improvements in the end-to-end delay could be achieved either by using more powerful machines to speed up the encode/decode process, or by using a lower latency transmission technology such as ISDN or ADSL, or an even lower latency codec such as G.728. While G.728 is both more complex and uses a higher bandwidth (16 kbit/s) than G.723.1, processing power, and (arguably) bandwidth, will become less of an issue with time, whereas latency will continue to be a problem.

The biggest surprise in Table 6.1 is that for the chosen number of bytes of data, header compression has minimal effect on delay. This is because transmission/serialization delays are masked by the modem processing delays. In the case of an internal modem, the size of the packet makes no difference until the transmission delay for a packet is at least 32 ms, i.e. the packet is at least 115 bytes long. Then each additional byte adds 2*0.28 = 0.56 ms delay one way (@28.8 kbit/s).

In the absence of an internal modem, the size of the packet starts making an impact for packets with a transmission delay of at least 17 ms, i.e. when the packet is at least 61 bytes long. Then every additional byte adds 0.28 ms delay one way. Above 115 bytes every additional byte adds a delay of 3*0.28 ms = 0.86 ms one way. So, when used with 28.8-kbit/s modems as in the chosen scenario, header compression is only useful for any reduction in packet size it has above 61 bytes and is especially useful for the reduction it has above 115 bytes. This is in absolute terms. In relative terms, the longer the packet, the less significant is the saving of ~ 38 bytes. Also, above a certain packet size, the audio sampling time will no longer be hidden by the operating system latency in the API calls to the sound system.

For data interactions, the sound card and encode/decode delays are not relevant. However, as play-out is dependent on the display frame rate, the play-out delay would have to be increased to at least 15 ms. This would give a saving of about 20 ms in each direction resulting in a round-trip delay of around 360 ms.

Unlike audio frames (which describe the audio signal over a period of around 20 ms), each video frame represents the video signal at a particular instant in time — consequently the coding schemes do not impose a minimum delay. However, in practice most software implementations have coding delays of around 100 ms. This is only significant for frame rates above 10 frames per second. Below this rate each frame can be processed before receiving the next frame. With improved PC performance the video coding delay can be expected to decline steadily.

6.2.2.2 Set-up delay

Besides the actual transport of media (audio and/or video), there are key problems that any conferencing system must solve. The first of these (referred to here as call control) is getting remote parties into a conference. The second (media establishment) is deciding which combination of audio and video to use, and telling all the parties involved. Both call control and media establishment contribute to the total set-up delay.

There are two main schemes for setting up conferences — announcement-based and invitation-based. These are shown in Figs 6.2 and 6.3.

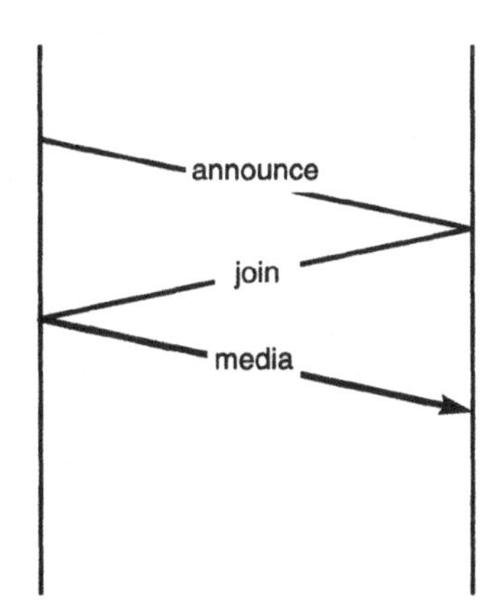

Fig 6.2 Announcement-based conferencing

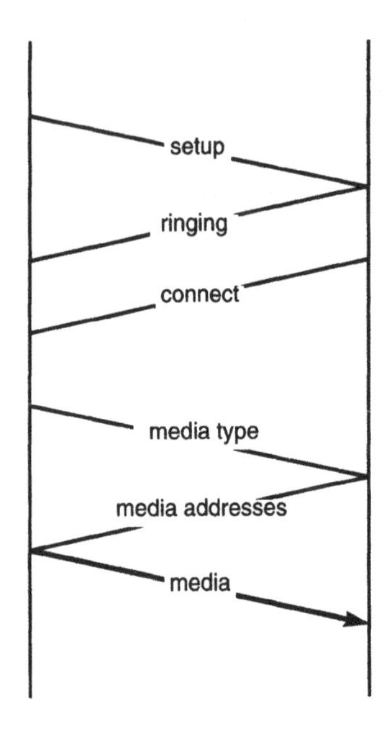

Fig 6.3 Invitation-based conferencing.

Typically when using announcement-based conferencing, call control and media establishment are combined. A party receives an announcement describing the conference including details of the media to be used and the conference address. There is normally no opportunity to negotiate the conference media types. The party simply sends a 'join' message which includes address and port number. Subsequently it receives media from the conference. In multicast conferences, the multicast join latency will depend on the multicast routeing protocol and the tree structure at the time of joining.

Invitation-based conferencing may or may not combine call control and media establishment into one phase. There are three main steps to call control. Firstly, an invitation is sent (this step involves locating the user to be invited). Secondly, the remote party sends an indication that it has received the invitation (this is important for the calling user so that they do not waste time waiting for a call that does not have sufficient resources at the far end for the call to take place). Thirdly, there is an indication that a user is on line[3]. In this chapter, these three phases shall be referred to as *set-up*, *ringing* and *connected.*

---

[3] Connection can be reported by an explicit connection indication or by the arrival of media. An explicit indication has the advantage that it can be used to signal connection to other interested parties such as an intermediate server involved in call set-up (see section 6.2.1). The disadvantage is that it further complicates and slows down the call set-up process.

In general, the choice of media may be imposed by one of the parties or it may be negotiated. The former method is appropriate for announcement-based conferences or for invitations into large conferences. In these cases terminals containing many disparate capabilities may be present, and so performing extensive capability negotiation can be prohibitively expensive.

The situation is different for small invitation-based conferences. The dynamic and unplanned nature of small conferences means it is important to be able to identify and use the best commonly supported codec.

Where media negotiation is required, it may be initiated within the call-control phase or it may follow in a separate media-establishment phase. Figure 6.3 shows the case where media negotiation follows call control and where the called party simply accepts the specified media type[4]. If the media and address messages were removed then it would represent the case where negotiation is initiated or the media is imposed in the initial set-up message. In this case the called party may then choose either to accept the call with the specified media type (as shown) or may offer alternative media types.

Clearly the set-up delay depends on the number of round trips involved; in this respect, specification of the preferred media type in the set-up or announcement message is therefore better than having a separate media establishment phase; it saves one round trip. Furthermore if implicit connection indication is used, negotiated media establishment introduces no extra delay — the invitee simply selects one of the specified media types and starts sending.

The other significant factor in determining delay is whether reliable connections must first be established for the signalling messages or whether soft state is used. For example, when using the transport control protocol (TCP) [3], call set-up can be delayed by retransmissions, so these are only appropriate for very small conferences. Due to the TCP slow start, a packet lost during set-up can cost (in the case of the commonly referenced BSD implementation) 6 s. Additionally, TCP incurs a one round-trip time delay at start-up to synchronize the sequence numbers used for the acknowledgement process. This suggests UDP is preferable for conference control, although issues such as network friendliness, and accessibility to other protocols such as transport layer security (TLS) [8] also need to be considered.

6.2.3 Scalability and robustness

Scalability, when applied to conferencing in Internet environments, can mean a number of things.

---

[4] The address message is only required for unicast media delivery. For multicast media delivery the address can be specified together with the media. However, a separate message is then required for the called party to join the conference.

- Can many small conferences be supported simultaneously as well as a few large ones?
- Can the required conference be located easily as the total number of choices increases?
- Can conferences with large resource and quality of service (QoS) requirements be supported as well as those with modest requirements?
- Can client, 'server' and network software run on equipment with modest resources as well as on equipment with plentiful resources?
- Can conferences allow participants to be widely spread geographically as well as based in the same locality?
- Can conferences support very large numbers of participants (e.g. public conferences) as well as small numbers?

Whether one situation can be supported 'as well as' another depends on a number of factors. Of particular significance are robustness, consistency, performance, reliability of delivery, and cost.

The Internet research and development community has two important mechanisms which can help with the need for scalability in conferencing — multicasting and soft-state control [9], the latter being specifically designed for robustness as well.

Multicasting is a mechanism provided by the network which assures only one copy of a particular packet is transferred across any link in the network, no matter how many recipients there may be. It relieves the sender of data from having to replicate that data to all receivers. As a result, data sources can send to 100s or 1000s of receivers without requiring a commensurate increase in resources (or decrease in the volume of data or its timeliness). Multicasting also provides a convenient abstraction of the set of all receivers of a particular data transmission, which can be shared by all senders. Further, multicast has the benefit that it allows the recipient to control what they receive without specific sender interaction.

Multicasting should not, however, be seen as the panacea for all scalability issues. For example, the efficient support of large numbers of small conferences would depend on the widespread deployment of multicast routeing protocols which do not place too great a load on routers in terms of supporting the state required. Multicast address allocation and multicast inter-service provider policy signalling — such as the border gateway protocol (BGP) — is also immature. There are also issues for dial-up customers. Currently, dial-up users are connected to boxes that contain multiple modems. Because of the way they currently operate, if one of the users connected to the box subscribes to a multicast address, all users connected to the box will receive the stream. This could flood other users' connections. It must be emphasized, however, that these are only short-term problems and are likely to be solved in the next year or two.

Soft-state control is exemplified in a number of Internet protocols (one of the most quoted examples is the resource reservation protocol — RSVP [10]). The basic principle is that control information is constantly refreshed; if a failure occurs which causes refreshes not to take place, the corresponding state will expire. Alternatively, if state is lost due to some failure, it will be recovered the next time the control information is refreshed. Soft-state fits well with multicasting to maintain loose consistency of state in a scalable manner.

It is worth noting that stateless protocols have maximum robustness. The archetypal stateless protocol is the Internet protocol (IP) itself, which was designed specifically for robustness. The stateless hypertext transfer protocol (HTTP) v1.0 [11] has served the World Wide Web (WWW) well over a period of explosive growth, and the durability of IP is unquestionable. Stateless may, however, not be suitable for conferencing, which tends to exist over a defined period of time — hence soft-state is used.[5]

Multicasting can improve the efficiency and timeliness of both the delivery of the 'payload' of the conference (typically audio, video, collaborative data) and the control information associated with the conference. The former aspect has been well progressed by the Internet development community. The latter much less so, especially in the area of invitation-based conferencing. An issue is the lack of consensus about protocols to be used when multicast information must be reliably transferred from senders to receivers. However, in the absence of reliable multicast, soft state provides a very effective alternative.

Another reason for multicasting control information can be inferred from World Wide Web sites which have suddenly collapsed under a siege of overwhelming popularity. Sites supporting unicast control of live events could equally experience similar problems — particularly if multicast or any of the family of 'push' technologies are used to advertise the event. The result could be a 'control implosion' (Fig 6.4), resulting in severely degraded performance (customer dissatisfaction) or even system failure.

A solution to the above dilemma is to design conference control protocols which are 'announcement friendly'. To date, such protocols have not been developed beyond very basic mechanisms, and this is potentially a significant area of future work. It should be stressed that this is still very much a research area, and, as described in section 6.4, an existing system — H.323 — has taken a more traditional and pragmatic approach to these issues. It remains to be seen whether the growth in multicast-based announcement and conferencing will justify confronting the technical challenges posed by distributed control.

[5] Deficiencies in HTTP 1.0 leading to the definition of HTTP 1.1 with persistent connection are attributable to reliance on TCP (transmission control protocol).

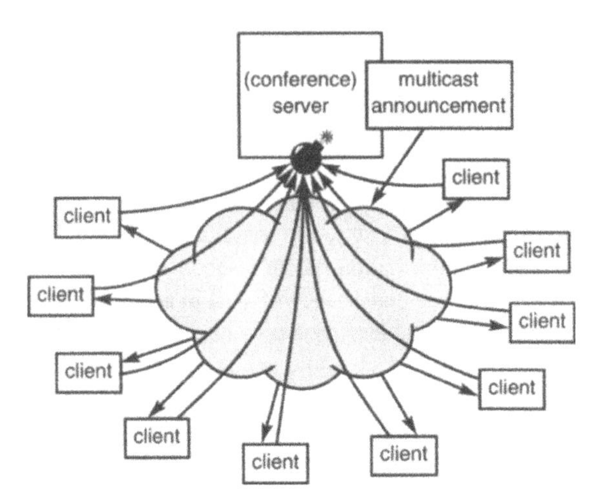

Fig 6.4 Conference control implosion.

6.3 CURRENT WORK IN THE INTERNET ENGINEERING TASK FORCE

The IETF is an open organization dedicated to developing engineering solutions, specifications and standards for the Internet and Internet-related technologies. An area which has been particularly successful is the development of a standard real-time transport protocol (RTP) [12] for use for real-time communications (including, of course, conferencing) over packet networks. This specification has also been adopted by the ITU for the transport of audio and video in H.323 (see section 6.4). The IETF has also developed a number of 'payload formats' for RTP, which state how standard audio and video encodings developed within the ITU and ISO can be carried within the RTP framework.

The IETF has also pioneered the development of multicasting capability for the Internet and the use of soft-state protocols. RTP has been designed to be used in a multicast conference, though it also works well with unicast. A text-book example of a soft-state control protocol is the real-time control protocol (RTCP) [12] which is used as an adjunct to RTP. RTCP provides information about the senders and receivers of real-time streams in a conference, such as counts of packets received and lost. RTCP also provides basic conference membership information, which includes an indication of whether members of a conference have explicitly left the conference (the BYE message) or have stopped reporting due to some failure. RTCP has adopted a simple, but largely effective, approach to congestion control, which is to monitor the amount of bandwidth consumed within a conference and to adjust the retransmission interval accordingly to keep the overall bandwidth within a predetermined limit (5% of the RTP bandwidth). As a result, RTCP is highly scalable, though some concerns have been raised

[13] about its scalability to huge conferences (number of receivers > 10 000). At this point, individual RTCP reports cease to be useful, and can place excessive demands on client resources to track them.

IETF activities on multicasting and conferencing are inextricably linked to the Mbone, an experimental test bed, established in the early 1990s, for multicast on a global scale achieved over the existing unicast-only Internet infrastructure by an overlay technique known as tunnelling. Key designers of RTP and associated control protocols have developed applications which are in wide use on the Mbone for conferencing and broadcasting of events and conferences such as the IETF's own meetings. More recently there has been activity within the IETF to deploy multicast within the wider Internet as an operational service, building on the success of the Mbone.

The IETF has also been addressing call and conference control in the form of the MMUSIC (multi-party multimedia session control) working group. A basic mechanism has been developed for publishing and announcing conferences, particularly for use with multicast. The session description protocol (SDP) [14] provides a general mechanism conveying the information required to participate in a conference, regardless of the transport by which that information is conveyed. The information conveyed specifies the types of media that shall be used, the person who originated the session, what the session is about, and so on. Such transports include e-mail, WWW, invitation or announcement (the latter mechanisms are described below). The session (conference) description includes information on each component media in the conference (see Fig 6.5).

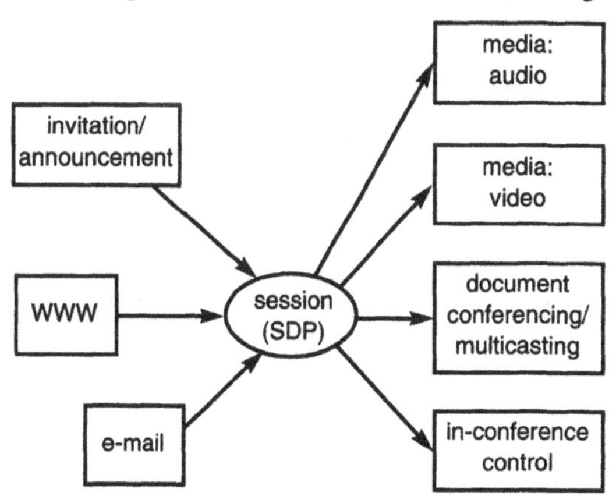

Fig 6.5 Scope of the session description protocol (SDP).

The MMUSIC real-time streaming protocol (RTSP) [15], a protocol to control on-demand real-time streams from multimedia servers, supports the invitation of a server to a conference by another mechanism (such as SIP or H.323), and hence allows the important scenario of introducing stored media to a conference.

Another important area for conferencing over the Internet is that of collaborative applications which do not necessarily have as great a requirement for low-delay delivery but may require reliable transfer of information. Such applications include point-to-multipoint file transfer or shared whiteboards. A number of reliable multicast protocols have been developed, notably SRM (scalable reliable multicast) [16] which is widely used on the Mbone in the form of the 'wb' shared whiteboard.

More recently, an Internet Research Task Force (IRTF) group has been investigating requirements and frameworks for reliable multicast protocols. The T.120 protocol suite (see Chapter 5) is being enhanced to exploit reliable multicast protocols for data delivery, though it currently does not specify the use of a particular protocol to achieve that reliability.

The following subsections describe in more detail the efforts of the IETF in the areas which are covered in this chapter.

6.3.1 User location and invitation

The IETF has initially adopted an alternative mechanism to conventional user location for the establishment of conferences. The basis for this is a format for describing conferences (SDP). Session descriptions could potentially be distributed using e-mail (preferably well in advance), and, assuming that e-mail addresses of potential participants can be reliably obtained from directories and similar means, this could provide a satisfactory mechanism for setting up a conference. For the client, integration between e-mail, conferencing and calendaring/scheduling applications would be highly desirable to make this process more usable.

The IETF has also developed the Session Initiation Protocol (SIP) [17] for invitation. This has some overlap with the call control protocol defined by H.323 (see below). SIP packages a media description (typically an SDP announcement) into a message which is sent using unicast to a remote party and includes acknowledgements so that the sender knows that the message has been received. The transport protocol used can be TCP (reliable) or UDP (datagram); the latter can use repetition of the request to overcome potential packet loss. Once a successful response has been received, it is then possible to proceed with other modes of communication (e.g. audio, video). Figure 6.6 illustrates the use of invitation and announcement to build a multicast conference.

SIP incorproates a number of features for the purpose of user location. The basic assumption is that there may be a number of proxies which may know how to contact the invitee in question, in addition to the invitee's client aplication itself. The proxies can forward the SIP invitations to potential invitee clients or other proxies. Both proxies and clients can also respond to an invite with a re-direction message which will cause a new invitation to be sent to a specified address. In this respect a SIP proxy has many similarities with an H.323 gatekeeper. However, SIP allows for multiple simultaneous invitations to be sent and for multiple responses to be received by the initiating client itself, as shown in Fig. 6.6. This potentially places a greater burden on the client, and may cause problems with adding value via intermediate servers.

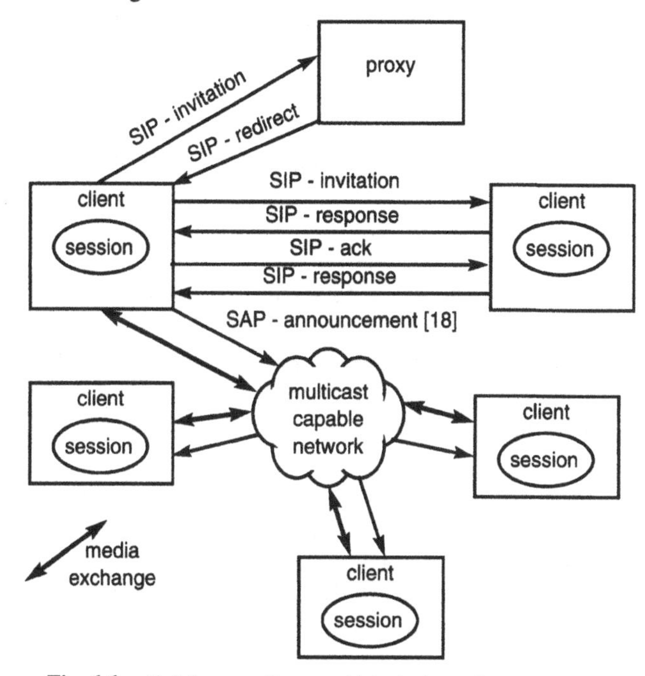

Fig. 6.6 Building a conference with invitation and announcement.

6.3.2 Latency

'Connectionless' protocols may not necessarily provide any gains in terms of latency. For example, an IP packet traversing the network, when no previously cached routeing entries exist, will experience delays as potentially complex routeing decisions are made. If such a packet is multicast, the routeing is clearly more complex; initially, a multicast tree will have to be built, the topology of which depends on a dynamic set of receivers.

Soft-state protocols often incur latency penalties. For example, the multicast membership protocol IGMP (Internet group management protocol) was initially defined to have no explicit leave mechanism, thus relying on time-outs to terminate streams from an unwanted conference. Membership information provided by RTCP will provide delayed indication of a particular participant leaving a conference in the event of the BYE message (the explicit signal of leaving) being lost. In conference control, a request which is multicast to a number of participants may have a delayed response as a result of random back-offs being applied to prevent implosion.

While the use of the datagram protocol UDP (especially in conjunction with multicast) provides for immediate delivery of control information for a conference, packet loss can cause delays corresponding to the retransmission interval used. Therefore there is a need to balance timeliness requirements with available bandwidth and interrupt burden when setting the retransmission frequency.

In summary, the design of protocols which exploit multicast and soft state can lead to many pitfalls in terms of latency. As a result, significant effort is required in the design of these protocols if it is important that latency is minimized.

6.3.3 Scalability

Scalability of conferencing to large numbers of participants has long been a focus in the development of IETF protocols, and indeed inspired the receiver-driven approach to conference membership that is provided by the IP multicast mechanism.

The drivers for increased scalability in RTCP have come from the recognition that the technology being developed can potentially be deployed for the mass market. For conferencing, this could mean large-scale debates, chat lines, etc, as well as networked games and virtual worlds. Anything but the most basic form of control for this sort of environment has, however, not been addressed in any detail in the IETF.

Another scalability issue relates to conference/session 'directories', which collect information about current conferences in progress. Clearly, if the number of conferences occurring simultaneously is large, a single directory (which typically corresponds to a single multicast address and port) will not be suitable for carrying the associated announcements. There are two aspects to this:

- the congestion control back-off will cause an excessive retransmission interval, causing long delays in receiving session announcements;

- the user interface could become excessively cluttered.

The latter problem is less severe, as better user interfaces could be devised.

The issues of scalability of multicast and resource reservation referred to earlier (section 6.2) are actively being addressed in the IETF and have an impact on applications other than conferencing.

6.4 H.323

H.323 [2, 19] has been developed through the ITU. As with most Internet real-time multimedia applications, it uses the IETF's multicast-capable RTP for media transport. Where it differs from other protocols is in the connection set-up mechanisms, and how it decides which media to use.

H.323 version 1 was ratified in June 1996, and version 2 was ratified in February 1998. Much of the discussion below is applicable to both versions, although some of the features discussed are only supported in version 2.

6.4.1 What does H.323 attempt to solve?

H.323 is aimed squarely at small conferences. An H.323 conference containing a few dozen people would be considered large. To support conferences larger than this. H.323 requires support from additional centralized conferencing equipment. As H.323 is intended to handle unplanned, on-demand conferencing, this restriction is not unreasonable. For the largest of conferences though, H.323 on its own is not the right solution. For this reason H.323 conferences can be combined with the IETF SDP protocol described above. This process is described in the ITU H.332 recommendation [20].

Of the many requirements for H.323, one of the first is for interoperability with existing, non-Internet conferencing protocols, which includes H.320 [21] for use on the ISDN and H.324 [22] for use on the PSTN. This is done by gateways. By virtue of interworking with H.320 and H.324, H.323 also interworks with basic voice services on the integrated services digital network (ISDN), and the public switched telephone network (PSTN).

Another requirement is that one should, as a minimum, be able to do with H.323 what one can do with a PBX. A PBX implicitly provides some support for user location. Sophisticated user search strategies, such as 'divert on no-reply', have an impact on the signalling needed, and H.323 has taken this into account.

6.4.2 The H.323 system

An H.323 system contains three different types of main entity. There are also some lesser components, and firewalls must be considered. Some of these entities are shown in Fig. 6.7 which is viewed from the perspective of a corporate intranet.

The first entity type is the terminal which users use to communicate with. Currently these are PC based, though there are indications that stand-alone terminals will appear in the future. These would replace the telephone on the corporate desk and connect directly to the local area network (LAN) via Ethernet.

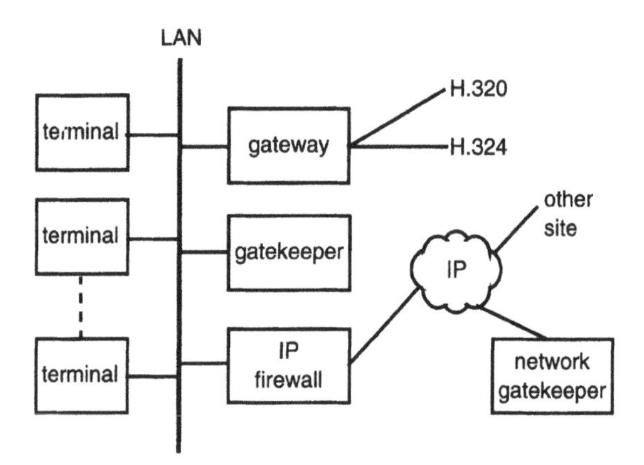

Fig. 6.7 Main components of an H.323 system.

The second entity type is gateways. These provide the bridging between the different networks such as H.323, H.324 and voice. Their use is optional in an H.323 system. Many vendors are in the final stages of developing H.323 gateways.

The third and final main entity is the gatekeeper. This is again optional in an H.323 system. However, it can be most useful as it collects all the features that cannot be easily distributed to the terminals or gateways, such as access control, telephone number to IP number translation, and user location.

Figure 6.7 actually shows two locations where gatekeepers can be used. The first of these is within the corporate environment and performs some or all of the functions listed above. An additional location for a gatekeeper is in a public network, such as the Internet. A gatekeeper in this location allows companies like BT to add value, possibly in the areas of directory services or 0800 number handling.

There are also some secondary entities which are not intended to be stand-alone, but which reside in the other main entities. One of these is the multipoint controller (MC) which will be discussed in section 6.4.5.

6.4.3 User location

H.323 uses the gatekeeper to act as the location server and to intervene in the call set-up path in order to support 'divert-on-no-reply' and 'divert-on-busy'. Providing user location features in a local gatekeeper allows arbitrarily complex user search schemes to be implemented.

User location is based on terminals registering with the gatekeepers. Emulating PBX user location functions, such as 'divert on no-reply and busy', was a key requirement in the design of the H.323 call control. In this commonly used PBX service, the PBX attempts to contact a person at a primary number. If, after a suitable amount of ringing, nobody has answered, then an alternative number is tried.

This sort of scheme requires signalling to go via the gatekeeper. Referring to the set-up, ringing, and connection indications mentioned in section 6.2, the gatekeeper needs to receive a connected indication, so that it can stop the search algorithm. As the RTP media will be routed directly between terminals, and not via the gatekeeper, using the presence of RTP media to indicate that the parties are connected is not sufficient. Hence a connected message is required as part of the group of set-up messages.

When trying to emulate the 'divert on no-reply' service, another indication also becomes apparent. This is an indication that the call has finished. This is used when the gatekeeper has been ringing the first terminal for a period and decides to try another terminal. This indication would tell the first terminal to stop ringing. Once again, this indication cannot rely on, say, the absence of RTP media as no RTP media has been sent at this time. Hence an additional message is required. Consequently the H.323 call control consists of messages associated with set-up, ringing, connected and closed, and does not rely on other events, such as RTP, to signal this.

6.4.4 Latency — H.323 call set-up times

H.323 supports invitation-based call-control and media negotiation. For reasons of interoperability with H.324, H.320, and the PSTN, as well as the desire to reach the market quickly, H.323 version 1 uses separate protocols for the call control and media establishment phases. This removes the possibility of initiating media negotiation within the call control phase, resulting in additional round trips as described in section 6.2. The minimum call control signalling involves an exchange of the set-up and connected indications which results in one round trip.

The H.245 media negotiation procedure involves the following phases:

- specifying to the remote end-point which sets of media can be received (called a capability set),
- from the set of media that could be used, the remote end-point selects the media set that it wishes to use by requesting the opening of specific media channels,
- the request to open media channels is acknowledged by sending to the remote end-point the network addresses to which the media is to be sent.

This represents 1.5 round-trips of signalling and is carried out by both end-points as soon as the H.245 connection is established (Fig. 6.8).

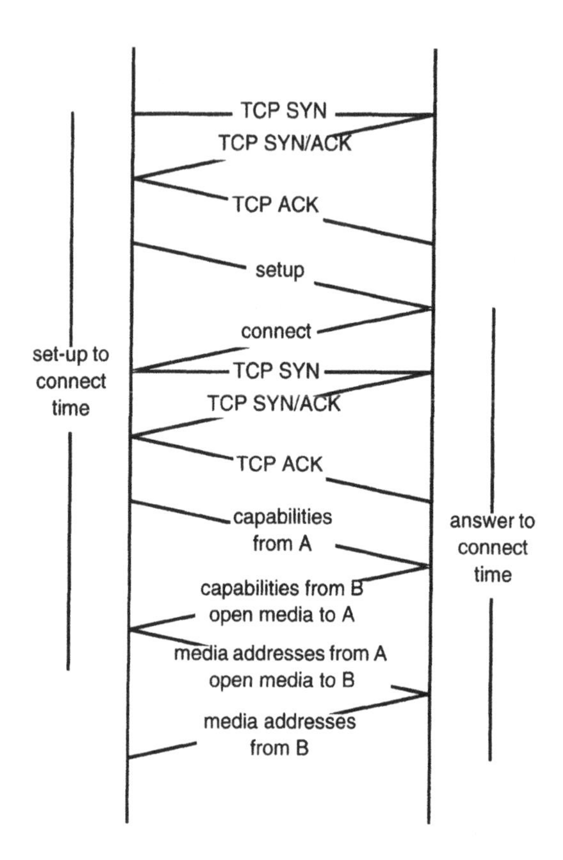

Fig. 6.8 H.323 call set-up.

Another impact on set-up delays is that H.323 uses TCP to carry the call control and H.245 signalling channels. Consequently it suffers two additional round trip delays (one for synchronizing TCP sequence numbers for the call control channel and one for synchronizing the H.245 channel). Also, both channels are subject to the TCP slow start and, in particular, the problems this presents if an initial packet is lost. If the initial packet is lost, a 6-sec delay (in the commonly referenced BSD implementation) is incurred. This figure only drops to about 3 sec after a dozen message exchanges. However, it must be emphasised that this delay is only incurred in the event of packet loss.

The resulting set-up times for H.323 version 1 calls are shown in Table 6.2. The round trip times are calculated from the entries in Table 6.1 associated with signalling delay. The set-up-to-connected time consists of 4.5 round trip times, which consists of one round trip for TCP SYNC for call control, one round trip for set-up and connected messages (assumes immediate answering), one round trip for H.245 TCP SYNC, and 1.5 round trips for media establishment. The answer-to-connected time is similarly calculated, but starts from the point where the connect message is sent.

Table 6.2 Call set-up and answering times for various network delay times. (rtt = round-trip time. Figures for slow and fast dial-up based on the signalling delays from Table 6.1.)

	Slow dial-up 1.2 s rtt	Fast dial-up 360 ms rtt	Intranet 100 ms rtt
Set-up-to-connected with no ringing time	5.4 s	1.6 s	450 ms
Answering to connected with no ringing time	4.2 s	1.3 s	350 ms
Answering to connected with media establishment while ringing (see note 1)	600 ms	180 ms	50 ms

Note 1 Assumes that ringing time is longer than indicated set-up-to-connected time.

It is felt that users who can tolerate an audio round-trip time of 1.4 sec will have the patience to wait five seconds for call set-up. For shorter network delays, connection times are subjectively comparable with the existing switched network set-up times. However, having said that, H.323 version 2 has reduced set-up delays by adding two new features. The first of these is to allow the H.245 media negotiation procedure to be tunnelled within the call control message flow. This removes the need to exchange addresses for the H.245 channel and perform the TCP set-up. This saves two round trips, which results in 2.5 round trips including the TCP synchronization phase. To further shorten the set-up time, a fast start procedure can also be invoked. This involves adding new media capability fields, and the network addresses on which they can be used, to the call control set-up-and-connect messages. This process achieves media establishment in just 1.5 round trips.

6.4.5 Scability and robustness

It has been mentioned that much of the signalling is point-to-point. (Media transport can make use of multicast, and hence is not an issue here.) This presents a problem when communication between more than two parties is required. To cater for this, H.323 uses a multipoint controller, or MC. This acts as a signalling bridge for both the call control and H.245 signalling. The MC is responsible for terminating all of the point-to-point connections from the terminals involved. It receives all the terminals' capabilities and decides which media should be used for this conference, and informs the terminals accordingly. This mechanism is considerably simpler than relying on distributed decision making, and although it may not scale well (beyond, say, 20 nodes), this is consistent with H.323's design goals; also the mechanism is well understood as it is widely used in H.320 and H.324 conferencing. The functionality of the MC can reside in any of the three main entities (terminals, gateways and gatekeepers), and H.245 includes protocol for selecting the most appropriate MC when more than one is present in a call.

The down side of the MC approach is that its functionality should be present for the entire life of a call. This presents problems if the node containing the MC wishes to leave the conference, or the node containing the MC fails. The former situation can be eased by locating the MC in a centralized gatekeeper. This can remain active throughout a call, and handle many such calls simultaneously.

A greater problem is failure. There are two categories of failure mode from which centralized gatekeepers and MCs can suffer — hardware and software. In a client/server world, corporations are becoming accustomed to employing fault-tolerant hardware, such as uninterruptible power supplies and redundant disk drives, on servers. Installing the gatekeeper (with associated MC) on such high MTBF systems would be an ideal choice.

The greater risk to reliability is often software failure. Software failure can be caused by bugs in the gatekeeper/MC code itself, the operating system on which it runs, or other, unrelated, processes causing the system to go unstable. To help with this, version 2 includes extensions that allow gatekeepers to be eliminated as a single point of fialure. As part of the protocol, terminals and gateways are required to register with gatekeepers at regular intervals, such as when they are switched on. As part of this process, the gatekeeper can provide back-up network addresses in case a network card, or some other networking, failure occurs. Indeed, the alternative network addresses given can be for different machines that are being operated in an unspecified redundant configuration that allows a hot-swap when a gatekeeper fails. The gatekeeper can also indicate alternative gatekeepers that are not being run in a redundant configuration, but that can provide sufficient support in the event that a terminal's primary gatekeeper fails.

MC failure should not be ignored as an issue either. However, a number of observations can be made. Firstly, the likelihood of failure is only weakly dependent on the number of terminals in the conference. Therefore, if 10 terminals are taking part in a conference the system is not 10 times more likely to fail. Although complicated, the MC's task is repetitive in nature. Hence freak operating conditions are unlikely, and the various paths through the code are likely to be well exercised. This means that the code is easier to test, and should mature faster in the field.

Also, if a remote terminal is using invalid messaging, only the MC needs to deal with it. The other terminals can be less tolerant to errant conditions and yet still function successfully.

Additionally, although the MC is present throughout the life of a conference, it only has a role to play when terminals enter or leave. If conference membership is stable, an MC failure should not affect the operation of the conference and terminals may well remain oblivious to such an event.

6.5 SUCCESS

SUCCESS (simple universal call/conference establishment sequence) [23] is an attempt to combine the benefits of the IETF and ITU approaches. It is therefore designed to work well for announcement-based conferences and point-to-point calls that perhaps interwork with the ISDN/PSTN. It is fully distributed, and has the ability to use multicast for exchanging control information. The protocol has soft-state features, but also has the ability to 'harden' the state, for example, when interworking with the ISDN/PSTN.

SUCCESS works by announcing that it is in a conference. This is done with a 'Hello' message, which is equivalent to the set-up indication mentioned in section 6.2. The protocol also has a 'Progress' message which allows events like ringing to be conveyed. Where SUCCESS differs from other protocols is that the connected indication uses the same 'Hello' message used for the initial set-up indication. Thus, the calling terminal announces that it is in a conference to the called terminal, and the called terminal answers by announcing that it too is in the conference. In order that control implosions do not occur when in a multicast environment, the 'Hello' message contains fields to indicate which terminal should respond to the message.

SUCCESS also has 'Bye' and 'ByeBye' messages to signal that an end-point is leaving a conference. The 'Bye' message signals that a terminal is leaving the conference. The 'ByeBye' message is an optional acknowledgement of the 'Bye' message. Acknowledged 'Byes' are considered important when ISDN/PSTN interworking is in effect, and charges may be incurred.

Using these messages, SUCCESS can maintain elements of both tight and loose membership information for different terminals in the same conference.

At the present time SUCCESS is a paper design, but it is hoped to get some practical implementations running in the near future.

6.5.1 Scalability and robustness

SUCCESS is designed to be fully distributed. Part of the motivation for this is to achieve scalability and robustness. SUCCESS should easily cope with the largest of conferences as in this mode the protocol collapses back to being a simple announcement protocol like SDP. It has also been designed to work well in a point-to-point mode. These two special cases are fairly easy to model at the design stage. What will be interesting to see is how SUCCESS handles evolving between these two extremes. Although this is part of the design goal, only implementation and exercising the protocol will truly see how well it adapts to conferences of different scale.

Another reason for making SUCCESS fully distributed is to remove dependence on any potential single point of failure. A consequence of being distributed is that SUCCESS uses soft state to signal periodically that it is still in the conference. Compared to always signalling to a central point, this periodic signalling imposes an extra burden on terminals. In a similar vein, it is also important that SUCCESS is network friendly. It is intended that the frequency with which SUCCESS refreshes itself should adapt to the prevailing conditions. This is complicated by the fact that some terminals may be on low-speed or highly asymmetric links. Achieving the desired characteristics while the conference is changing size represents an interesting challenge for the future.

6.5.2 Latency

Set-up latency was not a key design goal of SUCCESS, but low complexity was. Hence SUCCESS can be low latency. As part of this, it combines the call-control and media-establishment signalling within the same protocol. Being designed to exploit multicast, it uses UDP thus avoiding the signalling delays associated with TCP.

SUCCESS can also short cut set-up latency due to the way it supports media negotiation. The 'Hello' message contains two separate fields to indicate the capabilities that a terminal (or an entire conference) can receive and the set of media that the terminal is sending. A short cut can be achieved by including in the capabilities the addresses to which media should be sent if that capability is selected. Thus, for example, different audio codecs can be assigned to different ports, or RTP payload types, and the receiver can use this to select the appropriate decoder when the RTP packets arrive.

This allows the set-up latency to be effectively reduced to half a round-trip time. However, this approach does place additional burden on the receiver which not all platforms will be able to support. In this case, the protocol falls back to the standard sequence of stating what can be received, what is to be sent, and what addresses are to be used.

6.6 CONCLUSIONS

Although H.323 does not represent the ultimate Internet real-time multimedia communications tool, it can be seen that it is most applicable for small conferences, especially those that take place over low-delay networks. Although it does suffer from an MC being a single point of failure, this is common to many other protocols, and there are well-understood mechanisms for handling this. Larger conferences are better serviced using the IETF's SDP protocol.

To provide some support for conferences that require a mixture of loose and tight coupling, the ITU has defined H.332 which describes how H.323 and SDP can be operated together in the same conference.

In the longer term, a single protocol such as SUCCESS [23], which is able to cope with all of the various conference models, may be the way forward.

REFERENCES

1. Rudkin S, Grace A and Whybray M: 'Real-time applications on the Internet', in Sim S P and Davies N J (Eds): 'The Internet and beyond', Chapman & Hall, pp 354-358 (1998).

2. ITU-T Recommendation H.323: 'Visual telephone systems and equipment for local area networks which provide a non-guaranteed quality of service', (May 1996).

3. Braden R: 'Request for Comment 1122, requirements for Internet hosts — communication layers', (October 1989).

4. ITU-T Recommendation G.729 — Annex A: 'Coding of speech at 8 kbit/s using conjugate structure algebraic-code-excited linear-prediction (CS-ACELP) Annex A: Reduced complexity 8 kbit/s CS-ACELP speech codec', (November 1996).

5. Lee C, Yoshida K, Mercer C and Rajkumar R: 'Predictable communication protocol processing in real-time mach', in: 'Proc of Real-time Technology and Applications Conference', IEEE Computer Society Press, pp 220-229 (June 1996).

6. Casner S and Jacobson V: 'Compressing IP/UDP/RTP headers for low-speed serial links', IETF Internet Draft draft-ietf-avt-crtp-04.txt (April 1997).

7. ITU-T Recommendation G.723: 'Dual rate speech coder for multimedia communications transmitting at 5.3 and 6.3 kbit/s', (March 1996).

8. Dierks T and Allen C: 'The TLS protocol', Security Working Group draft-ietf-tls-protocol-05.txt (November 1997).

9. O'Neill A: 'Internetwork futures', in Sim S P and Davies N J (Eds): 'The Internet and beyond', Chapman & Hall, pp 412-440 (1998).

10. Braden R *et al*: 'Resource ReSerVation protocol (RSVP) — Version 1 functional specification', IETF, RSVP Working Group work in progress (draft-ietf-rsvp-spec-14) (November 1996).

11. Berners-Lee T, Fielding R, Irvine U C and Frystyk H: 'Hypertext transfer protocol — HTTP/1.0, Request for Comment 1945', MIT/LCS (May 1996).

12. Schulzrinne H, Casner S, Frederick R and Jacobson V: 'RTP: a transport protocol for real-time applications', IETF, Request for Comments 1889 (January 1996).

13. Aboba B: 'Alternatives for enhancing RTCP scalability', Microsoft Corporation, draft-aboba-rtpscale-02.txt (January 1997).

14. Handley M and Jacobson V: 'SDP: Session Description Protocol', IETF, Request for Comments 2327 (April 1998).

15. Schulzrinne H, Rao A and Lanphier R: 'Real time streaming protocol (RTSP)', IETF, Request for Comments 2326 (April 1998).

16. Floyd S, Jacobsen V, McCanne S, Liu C G and Zhang L: 'A reliable multicast framework for light-weight sessions and application level framing', Proc ACM SIGCOMM, Cambridge, Mass (1995).

17. Handley M, Schulzrinne H and Schooler E: 'SIP: session initiation protocol', IETF MMUSIC Working Group Internet Draft, ietf-mmusic-sip-05 (November 1997).

18. Handley M: 'SAP: Session Announcement Protocol', IETF MMUSIC Working Group Internet Draft, draft-ietf-mmusic-sap-00.txt (November 1996).

19. ITU-T Recommendation H.323 V2: 'Packet-based Multimedia Communications Systems', (February 1998).

20. ITU-T Recommendation H.332: 'Visual Telephone Systems and Equipment for Local Area Networks', Work in progress.

21. ITU-T Recommendation H.320: 'Narrowband visual telephone systems and terminal equipment', (March 1996).

22. ITU-T Recommendation H.324: 'Terminal for low bit rate multimedia communication', (March 1996).

23. Cordell P: 'SUCCESS — simple universal call/conference establishment sequence', internal BT document (Feb 1997).

7

WHITHER VIDEO? PICTORIAL CULTURE AND TELEPRESENCE

A V Lewis and G Cosier

7.1 INTRODUCTION

7.1.1 The magic box

An eight-year-old boy once stood in his grandparent's house, in awe of a cream and black Bakelite radio that smelt hot and mysterious. The heat came from glass bottles that were dangerous to touch, with little fires inside that burned without burning away. That radio could do magic. With a long piece of wire and something called a short-wave band, it could collect voices from far away lands. The boy was intrigued to hear a cold and bitter torrent of words followed at once by a warm and meltingly sweet song from the same place. What kind of contradictory place was this Moscow? His body was standing in the upper Neath valley, but his imagination was somewhere else...

Today, his adult thinking is captivated less by the more advanced technology of videotelephony. Perhaps the passage of time has jaded his imagination. Everyone acknowledges that television is a much more powerful magic box than radio ever was. It is the universal eye that lets us see distant events almost immediately. Nothing escapes the video camera, which can show the tiniest detail of whatever is happening right now. Television is so powerful and pervasive that its values and conventions overwhelmingly shape our attitudes to electronic picture making.

Technological and market convergence is widely forecast for telecommunications, computing and entertainment. Electronic pictures figure prominently in this convergence. Recent developments promise a multimedia mix of telephony with pictures and data, potentially revolutionizing telephone calls and creating new interactive forms of real and virtual environments.

7.1.2 A challenge to perception

But other things have changed, since the days of that Bakelite radio. The American comedian Ernie Kovacs once said that '...television is called a medium because it is neither rare nor particularly well done' [1]. Broadcast television has a paradoxical reputation for the accurate depiction of events combined with a poor evocation of the emotional experience of being somewhere else. The medium today stands accused of creating intrigue without involvement, of portraying events out of context and of reducing everything and everywhere to a uniform banality.

The mirror that reflects the world can distort reality. What promised to be more often seems to be less. When we watch a sporting event, we see what happens without entering into the atmosphere or feeling of the place. We observe but we do not share. Television shows us the world but keeps us, in mind as in body, safely at home.

Is this a deliberate desensitization or simply a result of a dizzy rate of change in technology? What lessons can the evolution of television teach us about the future use of pictures in communications?

7.1.3 Broken dreams

The first public videotelephony service opened in 1936, on a dedicated link between Berlin and Leipzig. In the 1980s, it was forecast that videotelephones would be commonplace, at home and in the office, by the year 2000. Clearly, this was a dream. However, videoconferencing has been successful in niche markets (see Chapter 2). In Greenland, where travel is always expensive and often difficult, videoconferencing is used for legal, medical and educational applications [2]. It is also increasingly popular in the boardroom, where reduced travelling is a major incentive. The experience of the authors is that current products work best when the participants know each other. This makes communication more comfortable, particularly when network delays become noticeable.

Jakob Nielsen, a frequent user of videoconferencing at Sun Microsystems, is less enthusiastic [3]. He complains that it is too easy to ignore the people 'on the box' and that the equipment fails to capture subtle but crucial details. Attempts to launch videotelephony as an enhanced form of mass-market telephony have not been widely successful.

In the early 1970s, AT&T spent an estimated $500 million launching the Picturephone [4]. This service proved to be spectacularly unpopular and became known as the Edsel of the Bell System, despite a better quality of picture than today's PSTN videotelephones.

The reasons proposed to explain this unpopularity were:

- the limited bandwidth of contemporary networks;
- the small size of the pictures;
- the high cost of both the product and the calls.

Noll, who conducted a market survey for AT&T in 1973, goes much further and reports that customers felt no need to see, or even disliked seeing, images of each other on Picturephone [5]. Could there be a fundamental barrier to the acceptability of telephones with moving pictures? Perhaps the 'near and clear but remote and sanitized' nature of broadcast television means that all instant pictures, including those in videotelephony, share this paradoxical set of qualities.

7.1.4 Might more mean less?

Could the very accuracy and immediacy of a video image be a disadvantage? Could an instant electronic picture act to inhibit our imagination and therefore enfeeble our emotions, transforming what is real into a shadow of reality that only increases our awareness of separation? Might a telephone conversation, by speech alone, be an inherently more powerful and subtly expressive kind of human contact than the same conversation by television? Do minds meet more easily when voices are disembodied?

If so, the repercussions for future human telecommunications would be enormous, largely invalidating the idea of telepresence. Videoconferencing would be destined to remain a specialist service, advantageous for formal meetings but unable to convey the subtle and complex aspects of face-to-face interactions. Nothing could make videotelephony popular, if it really is inherently worse than simple telephony. Chapter 13 discusses the potential benefits of adding emotional features to human/computer interactions, but no enhancement to simple telephony has yet proved commercially successful in a wide market.

7.1.5 Signals, sensations and sleepwalking

Both the telephone and television industries started with a simple yet brilliant idea, that defined an interface between technical signals and human sensations. The output voltage of a handset microphone is an excellent way of expressing the content and purpose of a person-to-person communication. Over successive decades, the detailed design of telephone handsets and of television programmes has been refined, enhancing sensations through advances in both design and technology. Telephones that once used carbon microphones and transformers now use electret microphones and silicon chips. The signals in television and telephony have changed from analogue to digital form. But all these developments have not altered what the signals represent.

The signal-to-sensation interface of the telephone handset is so effective that it has become the fundamental and universal metaphor of telecommunications. The industry has been able to sleepwalk into success, blessed with an easy-to-use product that almost everyone liked and wanted. Providing more of the same, at less cost, to more people was an immensely difficult task that occupied generations of engineers. Yet knowing why this should be done was never in question. The past was a good guide to the future for telephony and television.

7.1.6 Where do we go from here?

Internally, both industries are dominated by the objective technology of signal handling, but both products are everything to do with subjective sensation. Neither industry has felt any need to seriously re-examine the link that was established, at birth, between signals and sensations. Attempts to expand this link in telecommunications with video pictures have not been widely successful. The output voltage of a video camera is not necessarily an excellent expression of the content and purpose of each and every kind of person-to-person communication.

The television and telecommunications industries now face questions about what should be done next, because technological options abound. Stephen McClelland, editor of Telecommunications magazine [6], says: '...the communications industry has acquired the reality of a massive, globally dominant business, but, unfortunately, it has also experienced an increased sense of fashion and fad, and marketing hype more reminiscent of showbiz than the respectable (if ponderous) traditions of a community that always relied on universal public service ...'

The important questions are deeper and more fundamental than changes of technology. These are questions of purpose not technique and are about why rather than what. Technology is no longer the problem. The sleepwalking is over.

7.1.7 Flogging a dead horse

Any attempt to develop videotelephony is open to the objection that video pictures, as an extension to ordinary telephony, have already been tried and shown to be unpopular; but the reasons why this happened have a current relevance that stretches far beyond telephony.

Many of today's multimedia products are meant for mildly interactive but contemplative use, as a kind of active paper. The market for such products is like the videotelephony market was twenty-five years ago, in that the idea has not proved as commercially successful as expected.

Further developments in technology promise truly interactive multimedia and 'virtual-world' applications. Engineers, politicians or the public might 'walk through' a proposed building or 'use' a proposed product in a virtual space (see Fig. 7.1), interacting with computer-generated objects and each other. Realizing the potential of this kind of shared-imagination experience means exploiting the subtle power of social communication and group interaction. This is precisely what videotelephony attempts to do and barriers in the signal-to-sensation interface also inhibit these new developments.

Fig. 7.1 Collaborative working in the VisionDome [7].

7.1.8 Meeting in mindspace

The effectiveness of these shared-imagination environments will be much reduced if they are limited to singleton experience, without social interaction. We are unlikely to fully exploit the idea of meeting in mindspace with a computer image if we do not feel comfortable about meeting real people in the same way. In this respect, the future of interactive computer modelling depends less on the number-crunching or display technology and more on the human interface issues of videotelephony. Understanding and assessing these issues, together with their implications for design, matters more than developing bigger and better technology. The appropriate use of video pictures for communication is a design challenge with a remarkably broad and timely relevance.

7.1.9 Outlook

This chapter discusses a selection of engineering, commercial and cultural factors in electronic picture making, to speculate about their future role and development in communications. Little in this chapter is new and much, taken in isolation, might seem obvious. But the authors intend to do more than say the emperor has no clothes. The aim is to draw together disconnected strands of existing knowledge into a more systematic approach, ranking task above technology and challenging preconceptions.

The chapter is meant to stimulate debate and enquiry, with a deliberately provocative outlook that arises from a critique of pictorial culture in broadcast television. That medium has become so universal and familiar that it defines what the public expects video pictures to be able to do. The visual conventions and practices of television have so permeated our unconscious minds that breaking away from those conventions is difficult.

Nevertheless, the business of television is dominated by commercially feasible entertainment, broadcast one-to-many, giving recipients no influence over the pictures they see. Human communication is always interactive and often one-to-one, giving participants marked influence over what they see. A sense of 'being there' can be valuable for entertainment but it is vital in natural presence videotelephony, especially for the many-to-many interactions of meetings and conferences. The technical, cultural and psychological requirements of human telecommunications, in the top left and bottom right areas of Fig. 7.2, are not the same as those of commercial entertainment.

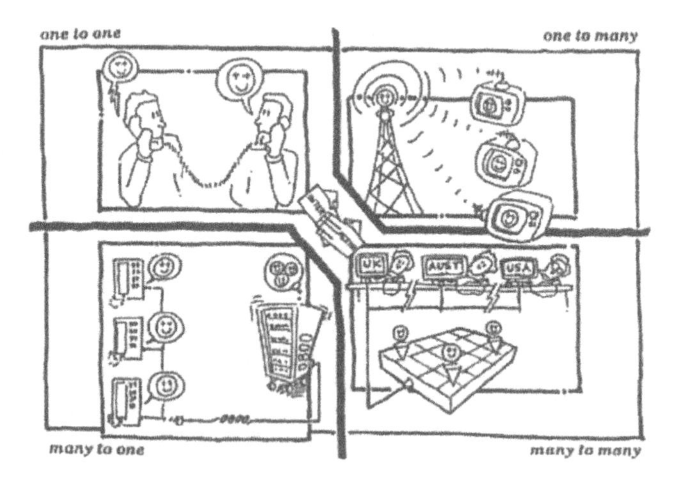

Fig. 7.2 Human communications, grouped by number of originators and recipients.

7.1.10 Structure

The chapter starts with a history of early television technology, which is not the diversion it seems. Next comes a short review of recent developments in videotelephony, followed by a set of proposals for the way forward. Subsequent sections support these proposals by discussing the nature of human communication, debating the use of pictures in entertainment and describing some important aspects of future telepresence systems. Finally, some questions are posed and some issues raised.

7.2 VIDEO HISTORY

7.2.1 Through amateur endeavour

At the end of the 19th century, people in countries all over the world were experimenting with the 'radioscope', 'visual listening' and the 'telephone eye'. Paul Nipkow and Jean Weiller described mechanical devices to scan pictures in the 1880s. In 1907, the Russian Boris Rosing proposed the use of a cathode ray tube for picture reproduction. In 1908, Alan Campbell Swinton was the first to suggest a wholly electronic picture system [8], with magnetically scanned cathode ray tubes for camera and display.

But it was John Logie Baird (see Fig. 7.3) who first publicly demonstrated, in April 1925 at Selfridges's Department store in London, television pictures of naturally lit objects [9]. This system used 16-line mechanical scanning, producing red and black pictures without intermediate shades. Baird demonstrated continuous-tone 30-line television pictures in January 1926, at a meeting of the Royal Society. On 9 February 1928, Baird demonstrated picture transmission across the Atlantic ocean, thirty-four years before the Telstar satellite. Later that same year, the irrepressibly energetic Baird demonstrated colour television with sequential red, green and blue line scans. By 1930, he had experimented with infra-red 'Noctovision' pictures and with stereoscopic television.

With far smaller resources than the silicon revolution deploys today, the television revolution developed much faster. In a staggering burst of inventiveness, television went from crude curiosity to high-definition broadcasting in little over a decade. Amateur enthusiasts, in the UK and America, played a major part in these developments [10].

In America, Ernst Alexanderson experimented with 48-line mechanically scanned television in 1927. Multi-standard receivers, for 24, 36 and 48-line pictures, were commercially available in 1928, the year when Charles Jenkins started regular experimental television transmissions on shortwave radio. In the

Fig. 7.3 John Logie Baird.

same year, Ulysses A Sanabria invented interlaced scanning to reduce picture flicker. The pioneer of television as we know it today was the American inventor and polymath Philo Taylor Farnsworth, who outlined his ideas for capturing images with electrons to his school-teacher at the age of fifteen [11]. He first demonstrated his 'image dissector', a magnetically scanned electronic camera tube, seven years later in 1927 [12, 13].

Vladimir Zworykin, then working for Westinghouse, had applied for a patent of a similar idea in 1923 but had been unable to make a working demonstration. A legal battle ensued and in 1935 the court awarded priority of invention to Farnsworth, in a forty-seven page judgement. Researchers at the Radio Corporation of America tried long and hard to avoid infringing Farnsworth's portfolio of over 100 patents on the fundamental aspects of electronic television. In 1939, RCA conceded defeat and negotiated their first ever agreement to pay third-party royalties. The giant of corporate America bowed to the farm-boy from Idaho. Despite this clear outcome and Zworykin's objections to being called the father of television, record books as eminent as the Encyclopaedia Britannica still attribute the numerous discoveries of Farnsworth to RCA's heavily publicized developer Zworykin.

In 1929 Baird was granted the use of BBC transmitters, after the end of scheduled radio programmes, to radiate his audio-bandwidth video signal. The first outside broadcast was made from the finishing line of the Derby racecourse in 1931. By 1935, several thousand of Baird's 30-line 'Televisors' had been sold to receive a tiny, flickering picture of the half-hour experimental programme that was transmitted around midnight.

The first television transmission by a radio amateur was in January 1933. Reception reports of Baird's test transmissions at 43 MHz, from Alexandra Palace during the early 1930s, were received from Europe, Iceland and Morocco. The first television transmission from an aircraft in flight was made over Hendon aerodrome in 1939. Workers at EMI used a development of the image orthicon camera tube to make an all-electronic 405-line system, which was tested alongside 240-line refinements of Baird's mechanically scanned approach.

This profoundly creative period of experimentation was superseded by professional development when the BBC, from a single transmitter at Alexandra Palace in London, launched the world's first regular television broadcasting service. In America, broadcast television was launched by RCA at the New York World's Fair on 30 April 1939, without full permission from the Federal Communications Commission.

7.2.2 Television develops

Key events in the subsequent history of television in the UK are as follows:

Sept 1939	BBC television ceases abruptly, in the middle of a Mickey Mouse cartoon, shortly before the outbreak of war. Mickey's final words were 'I tink I go home now'.
December 1940	Baird demonstrates the 'Telechrome', a 600-line electronically scanned colour display, but the war effort prevents development.
June 1946	BBC television re-opens and the ending of the Mickey Mouse cartoon is shown.
October 1954	The first test of 405-line colour television, using the modified NTSC system, is transmitted from Alexandra Palace to a single colour receiver.
1955	Independent television is launched.
July 1962	High-definition television pictures are sent across the Atlantic ocean, via the low-orbit 'Telstar' satellite.
1964	625-line television transmissions start, in the London and Midland areas.
July 1967	Colour television is launched, using the PAL system.

7.2.3 A glimpse of the past

Baird had attempted to record his pictures on gramophone discs, applying for a patent for the 'Phonovision' process in 1926, but he never achieved satisfactory playback. By 1930, audio recording equipment was commercially available in the UK, with the 'Silvertone' disc system from Cairns and Morrison Ltd. Some enthusiasts, variously called perceptors, teleseers or audivists, attempted to record Baird's experimental transmissions with such equipment.

Recently, Don McLean [14] has applied complex digital signal processing algorithms to restore a number of these recordings, with excellent results. It is now possible to get a glimpse of what was on television in the 1930s. One privately-made Silvertone recording, of a BBC Baird transmission, is particularly revealing [15]. It shows almost four minutes of the first television revue broadcast, starring the Paramount Astoria Girls, on 21 April 1933.

Despite the technical limitations of the 30-line system and the artistic inhibitions of having to perform in front of a camera the size of an automobile, that rapidly scanned a dazzling spot of light around an otherwise totally dark studio, the performance seems slick and vibrant even today. The production is unlike anything on modern television, since the solo artists enter and leave from the bottom edge of the picture, to minimize the visibility of the coarse vertical scanning.

This recording shows that the pioneers of television, immersed in solving engineering problems, were nonetheless keenly aware that an appropriate choice of artistic culture was not a luxury but a fundamental part of fitting the technology to its purpose. This is still true today, for television and videotelephony. Art and science complement each other at the edge of the known.

7.2.4 Thinking about thinking

If making videotelephony more popular means questions of purpose and not technique, then perhaps we should take the advice of the Irish policeman who, when asked for directions, replied: 'Well if I was you, I wouldn't start from here'. This advice is more insightful than it seems. As Rechtin [16] says: '... step back and think about the best ways to think, especially when creating complex, first-of-a-kind systems.' When designing something quite unlike what has gone before, extrapolation from established principles is unwise.

Most of the development effort in the early days of television depended on synthesis not analysis, and was motivated by dream not logic. This kind of outlook does not fit well with the conventionally taught view of engineering as a deductive, applied science; but engineering has always been an art as well as a science. Concepts like judgement, intuition, creativity, heuristics and metaphorical thinking may seem like the domain of the painter or literary critic, yet they are essential to the engineering of unprecedented things. This is why a knowledge of the early history of television technology, and of the attitudes and motives of its pioneers, can be valuable.

The success of television and the telephone can be a handicap, because of the tangled web of preconceptions that success has created. The authors want to break some of those preconceptions and take a fresh, holistic look at models and metaphors in video telepresence systems.

7.3 COMING SHORTLY ...

7.3.1 Developing standards

For many decades, BT has played a major role in the development of videotelephony and video-related international standards [17]. Table 7.1 lists some of the current ITU Recommendations for videotelephony.

Table 7.1 ITU Recommendations for videotelephony.

Number	*Title*	*Date issued*
H.261	Video codec for audiovisual services at p × 64 kbit/s	March 1993
H.263	Video coding for low bit rate communication	March 1997
H.320	Narrowband visual telephone systems and terminal equipment	March 1996
H.321	Adaptation of H.320 visual telephone terminals to B-ISDN environments	March 1996
H.323	Visual telephone systems and equipment for local area networks which provide a non-guaranteed quality of service	November 1996
H.324	Terminal for low bit rate multimedia communication	March 1996
H.331	Broadcasting type audiovisual multipoint systems and terminal equipment	March 1993

Technical standards are an important foundation for the subsequent work of detailed product design. The development of the VideoPBX system [18], using computer/telephony integration techniques, shows how minor changes of interface design can radically alter users' opinions of the product.

Current BT videotelephony and videoconference products include:

DVS100 PC-based desktop videophone and Internet data-conference terminal

VS1 Self-contained videophone with ten-inch LCD display, for one ISDN line

VS2 Room videoconference system, for ISDN up to 384 kbit/s

VS3 Room videoconference system with auto-tracking, for ISDN up to 2 Mbit/s.

7.3.2 HDTV

Ironically, new standards for videotelephony are much more clearly defined than for broadcast television. The television industry is in turmoil over where to go, what to do and how to do it [19]. After ten years, international agreement on a standard for high definition television (HDTV) is no closer. There is considerable disagreement about whether cable, satellite or terrestrial transmission is the best method for programme delivery. At the convention of the National Association of Broadcasters in April 1997, a consortium of computer companies attempted to define future technical standards for digital television broadcasting, over the heads of the television industry [20-22]. Meanwhile, the industry ponders how much customers will pay for new television sets and how much extra advertising and subscription revenue 500-channel television might generate.

7.3.3 Video for all

The price of television sets, video recorders and camcorders has remained numerically stable, but the price of computer video technology has fallen rapidly — by a factor of almost ten in six years. It is now possible to buy a simple camera for less than $100 and do domestic-quality video editing, albeit slowly, on a high-specification personal computer. On such computers, the quality of videotelephony connections over the Internet is limited by connection bandwidth, but several vendors offer suitable software and hardware [23]. Internet videotelephony is feasible using 28.8 kbit/s PSTN modems, but ISDN connections give better performance.

Future types of general-purpose microprocessor will probably include specific features for digital video manipulation and display. Real-time MPEG encoders, chips and accelerator devices for 3-D rendering are already available for desktop computers, and digital video disks (DVDs) are close to commercial release. New types of video image sensors, based on CMOS technology, promise lower power and lower costs [24]. Organic light-emitting polymers or CMOS digital micro-mirror arrays [25, 26] are just two of the competing technologies in the search for a flat, high-quality and low-cost display. The technology of electronic pictures is becoming increasingly affordable and ubiquitous.

7.3.4 Broad what?

Recent initiatives are exploring a new role for television in society, linked to new forms of cultural experience in the arts and entertainment. Inhabited TV is a multi-user virtual environment, which mixes the ideas of social interaction, audience participation and content production to create a new kind of television programme. The Mirror [7] was a ground-breaking collaborative experiment, created by BT, Sony, Illuminations and the BBC as part of 'The Net' series of BBC2 programmes in February 1997. More than 2000 viewers shared six imaginary worlds, becoming a community involved in both scripted and unscripted encounters. In these three-dimensional worlds, people were represented by avatars or cartoon-like characters, that could move, gesture and converse by text or voice. These experiments were a kind of multipoint conferencing, that offered improved scalability with group size.

Some cynics say that television is evolving from broad*casting* to broad*cluttering*, with a huge choice of channels. If the role of viewer and performer become interwoven or blurred in this way, perhaps an era of broad*catering* lies ahead — combining the diversity of the Internet with the ubiquity of conventional television.

7.3.5 Matching the market

Much debate about videotelephony, including this chapter, concentrates on person-to-person or group-to-group communications. However, there are many applications where a high-quality still picture is more use than a moving picture of lower resolution. The emergency services, the construction industry and many kinds of maintenance services could make good use of instant mobile transmission of electronic photographs. TraumaLink, a BT/Apple Computers trial venture, can transmit pictures, taken by paramedics at the scene of an accident, to trauma specialists at the local hospital within seconds (see Fig. 7.4), allowing additional advice to those at the scene and letting casualty staff anticipate the required treatment. Chapter 16 describes this in more detail.

Fig. 7.4 TraumaLink in operation.

Just as different forms of entertainment use different pictorial conventions, so the purpose of communication changes the technical and human factors of picture capture, transmission and display. The technology of future videotelephony is likely to depend on why the communication is required, whether the purpose is formal or informal contact, through a mobile or fixed link, over a local or international network, for educational, domestic, industrial or business needs.

This variety in matching technologies to markets follows a trend, described by Bonfield [27], where the telecommunications industry departs from its former 'one size fits all' outlook. Anderson of Rank Xerox [28] elaborates on this trend, suggesting a focus on the task not the technology, to extract value from a conjunction of technical and human design features. He suggests that, in recent times, too many attempts to add value have only added technology that seemed to have value.

7.3.6 So what's missing?

Larger pictures with higher definition and better quality at lower prices are therefore probably not the whole solution. Conveying the emotional and behavioural aspects that simple telephony conceals has usually been equated [29] with eye-to-eye contact in 'talking head' pictures, yet non-verbal human communication is subtle and involves more than facial expressions (see Chapter 13). Realizing the full power of videotelephony and creating a sensation of natural presence means more than attaching a high-quality camera and display to a telephone [30].

Further developments in videotelephony, and for multimedia systems in general, depend on a complex and uncertain set of issues. Even asking the right questions will need a multidisciplinary and strongly interdisciplinary mix of skills. In this sense, we have returned to the early days of video pictures, to a world filled with controversy for some and excitement for others, to a world where formerly accepted norms are inadequate and perplexity seems the only sure thing. The reason for this revolution is the technological convergence of entertainment, computing and telecommunications.

7.3.7 Converge or diverge?

Convergence is a fashionable word, but divergence might be a more profitable direction for the use of pictures in telecommunications. Carefully matching the design of a modestly powerful device to a particular application can often give a greater improvement than applying a more powerful, general-purpose product.

In different ages, hand tools were made of wood, stone, bronze or iron. Today, the hand-tool industry has converged on a common metallurgy and much the same production technology. Scalpels, butter knives and meat cleavers are all knives. So why should they evoke such strongly varied emotional associations?

A converged technology does not inevitably lead to a convergence of function or purpose. User requirements can easily create a marked divergence of product design, despite a shared technology.

7.3.8 The way ahead

For the closest approach to natural presence, it is important not to copy pictorial conventions that were developed for a different purpose. If telepresence is to become more popular, the signal-to-sensation interface must be re-examined. This means taking a fresh look at the pictures and their purpose (see Fig. 7.5).

Fig. 7.5 Concept demonstration of a 'smart space' work environment.

Many aspects of the optimum design, construction and operation of videotelephones, and their associated social norms, are poorly defined. Just as telephone handset design evolved to suit its purpose, so should videotelephone design evolve to match the picture to its purpose. This evolution should focus on the following ideas:

- a re-examination of the interface between technical signals and human sensations, in both engineering and social terms[1];

---

[1] Nikias, in the Integrated Media Systems Centre in the University of Southern California, calls the signal-to-sensation interface the 'creator-computer-consumer interface' and seeks its limitations as deficiencies in technology [31]. The authors see these limitations as deficiencies of design as much as technology.

- the development of operational conventions and social skills for comfortable use of still, recorded and live pictures;
- a greater use of videotelephony in education, to foster the development of these conventions and skills (see Chapter 17);
- not relying on the pictorial practice of the television industry.

These ideas are supported by the arguments presented in subsequent sections. The next section examines fundamental issues in matching signals to sensations, for person-to-person and conferencing applications of videotelephony.

7.4 SEEING, HEARING AND MEETING

7.4.1 Human abilities

The human visual system, through a combination of optics, muscle control and brain processing, achieves a seemingly contradictory combination of high resolution and wide-angle yet rectilinear imaging. The resolution of the eye on the optical axis is about 60 seconds of arc, but we are usually not aware that our peripheral resolution is very much lower. The angle of view of human sight is about 160° horizontally and 120° vertically.

We are very well adapted to the identification of faces and facial expressions from remarkably little visual information. The effectiveness of Baird's 30-line television depended on this fact. In face-to-face conversations, distance is usually related to intimacy. The subjective impression of distance in videotelephony is sometimes unlike face-to-face conversation, potentially creating inappropriate intimacy cues.

Human hearing has a very much lower accuracy of localization than the visual system, but its sensitivity extends behind, above and below the body. The visual and auditory systems can interact in a nonlinear way, so that sounds tend to be localized at visually likely positions and the perception of picture quality can be altered by suitable sounds.

We can easily engage in a conversation while reading a book, but we find it hard to deal with the interpretation of multiple sensory inputs in the same mode (the divided attention problem). However, divided attention is less serious for pictures of people than it is for sound or text, because movement is an attention cue. Our peripheral vision is especially sensitive to movement and plays a significant role in natural presence. Divided attention occurs naturally in face-to-face interactions with relatively little effect. Nonetheless, looking at simultaneous pictures of different locations inhibits a sense of natural presence if those pictures are very different in contrast, colour balance, viewpoint or perspective.

7.4.2 A sense of reality

In videotelephony, the full implications of the phrase 'natural presence' are very demanding. A communication system that could reproduce all the sensations of physical presence would be complex indeed. In common usage, natural presence means communication by sound and vision sufficiently detailed to allow a sense of subtle social involvement. When the reproduction of remote sound and vision covers a stage broad enough to replace most of the local environment in one direction, the qualifier 'immersive' can be added. In this definition, commercial cinemas are generally capable of natural presence and seats in an Imax cinema give an immersive experience. The term natural presence is also used to imply virtual reality, meaning an environment that does try to reproduce almost all sensory information.

Human face-to-face communication is multi-modal, with markedly nonlinear interactions between information in the different modes. Gesture and body language can reinforce, diminish or even contradict the semantic content of speech. Cultural and social conventions in speech, gesture and posture are a subtle yet potent means of moderating and directing group interactions, especially during argument and debate.

An exchange of notes, figures or scribbled drawings is a common and effective way of structuring debate and focusing decision-making in face-to-face meetings. In group interactions, we can signal attention, reflection, agreement, surprise, doubt, enthusiasm, distraction, boredom and many more states of mind — all without saying a word. Indeed, two-thirds of all communication is said to be non-verbal [32].

Natural presence in videotelephony means supporting this sort of multi-modal communication. In this sense, the word 'videotelephony' is misleading. Multi-modal telepresence is a better description of this expansion of telephony.

7.4.3 The eye of the beholder

If the information flow in face-to-face conversation is so complex, how is it that simple telephony works so well?

Communications engineers seldom acknowledge that the enormous success of telephony owes almost as much to the biology that links thoughts and words as to the technology of the telephone. When holding a handset, we have little difficulty in adapting our speech to suit the limitations of the telephone, in choice of phrase, speed of delivery, volume and intonation. The telephone makes it easy for us to consciously change and focus our behaviour in quite subtle and detailed ways. When we make a telephone call, we enter into a mutual pact with the distant person, to behave in a particular way. We unconsciously give a 'telephone performance', confident of the way in which the recipient will experience our behaviour. We can easily put ourselves into the position of the listener. Indeed, most communication failures in telephony are associated with the listener hearing something out of alignment with the expectations of the talker.

It is easy to forget that this psychological adaptation is an acquired skill. When Alexander Graham Bell asked the Chairman of a large American Bank to say a few words to his clerk in a nearby room, through Bell's new telephone, the Banker almost froze in horror. Unsure of how or whether he might be heard, but not wishing to give offence by saying nothing, this sober and solemn gentleman shouted an extremely uncharacteristic stream of meaningless gobble-dee-gook. The nervousness of some customers, when confronted with a videotelephone for the first time, is a similar sort of behaviour and may have the same cause.

In fluent, relaxed and natural communication, we all need to know how we are coming across. Achieving psychological confidence in a video connection — being able to enter the eye of the beholder — is much more difficult to do reliably than in simple telephony, because it depends on factors that are at present uncertain or insufficiently understood by most users.

7.4.4 Giving and taking

Traditionally, photography is something that we have done to us, or that we do to others. The common phrase 'taking someone's picture' implies that the subject lacks control of what happens or even that some kind of theft might be involved. The photographer controls the accuracy and relevance of the picture. In the world of entertainment, only the theatre allows actors some immediate control over how they are seen. In television, cinema and still photography, the visual representation of an actor is in someone else's hands.

This is the wrong metaphor for videotelephony. Simple telephony is so universally successful and is in such insatiable demand because its transactional and emotional metaphor is about reciprocal giving, not reciprocal taking. This metaphor exactly matches the social motivation of face-to-face conversation — we like mutual giving.

Few people in Western countries think about 'giving my picture', but that is what ought to happen in videotelephony, instead of thinking that someone is taking it. The pictorial motive in communication is to be seen by others as much as to see them. This difference of psychological attitude is why the pictorial culture of television is fundamentally inappropriate to videotelephony.

7.4.5 Active and passive

Listening is usually considered to be an active process, while looking is linked with the passive processes of reflective thought. This outlook is deeply rooted in the history of art and is reinforced by the way that pictures have come to dominate entertainment. We produce speech but we consume pictures. This attitude has led to the typical use of pictures in videotelephony, to add passive, emotional or social detail to the active speech channel [29].

However, a simple listing of the active and passive aspects of communication suggests that this divide is neither clear-cut nor logical. In Figure 7.6 the active aspects, that express attention, are grouped on the left side while the passive aspects, linked with inward reflection, are grouped on the right side. States of mind are shown in rectangles while affective actions are shown in ellipses. It is obvious that actions can be part of inward reflection and that states of mind can produce expressions of attention.

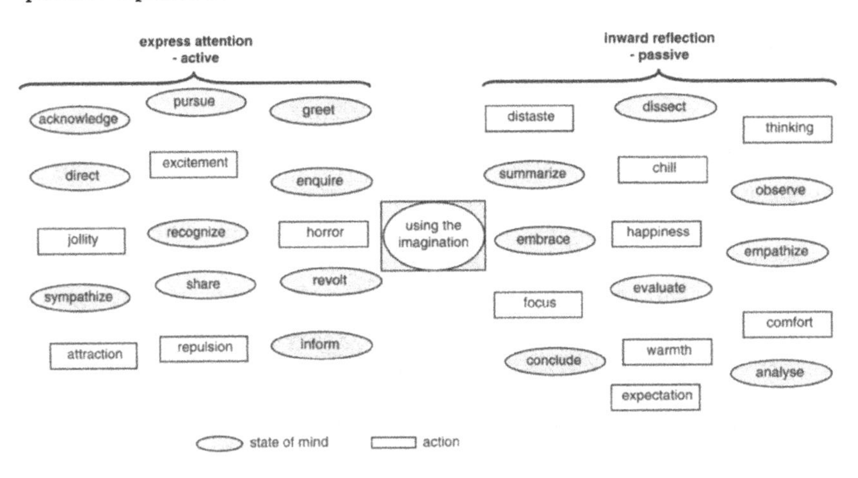

Fig. 7.6 Active and passive aspects of communication.

The use of pictures for communication only partially overlaps their use for entertainment. There is a different and more complex mapping of active and passive aspects on to thought and expression in communications. In particular, using the imagination can be either an inward reflection or an expression of attention, creating a state of mind or an action. The true purpose of pictures in videotelephony ought to be to foster imaginative thought, on an equal basis with the speech signal, not merely to add emotional colour. This places pictures in a new light.

7.4.6 Who's in charge?

Would users feel comfortable if simple telephony blocked or emphasized certain words? The idea that 'I control your surrogate ears (microphone) but I don't know what you hear' would quickly lead to confusion in simple telephony. Yet in videotelephony, the idea that 'I control your surrogate eyes (camera) but I don't know what you see' has been all too common. If the camera cannot be easily altered, in direction or field of view, it physically anchors and mentally constrains the users. In the worst case, nobody is in charge of the composition of the pictures, which greatly devalues their potential benefit.

In simple telephony, the user has essentially the same control of their acoustic representation — their speech — as during face-to-face conversation. For relaxed and comfortable videotelephony, this control should extend to the visual aspects. Users need to have as much control over how they are seen as over how they are heard. Only when users can put themselves into the eye of the beholder will they feel almost as comfortable with videotelephony as they do in face-to-face conversation. Successfully adding moving pictures to a telephone conversation means the telecommunications industry must define its own pictorial culture, in both engineering and social terms, so that users can 'give' their picture.

7.4.7 Points of view

The fundamental disparity of acoustic and visual representations arises because the perspective of a microphone and camera lens are very different. Most microphones are broadly omnidirectional, 'seeing all' in a perspective that is a soft-edged function of distance. In an audio teleconference, one omnidirectional microphone can be a surrogate for any number of distant delegates.

Lenses are highly directional and 'see partially' in a hard-edged function of angle. Using one camera in a videoconference means all distant delegates share the same view. They sit in the same remote chair. This fosters an 'us and them' seating layout, with a row of local delegates facing a row of distant delegates. For natural presence conferencing, different delegates might have different viewpoints [33]. This implies multiple cameras are needed. Should such cameras be controlled by the distant delegates or by the local delegates, so that each may give their picture?

The field of view of a lens may be narrow, wide or somewhere in between, but this is seldom obvious by looking at the lens. Visual perspective depends strongly on camera location. The quality of the lighting can have a dramatic effect on the visibility of fine detail and the emotional mood (low-key or high-key) of the picture. It is much harder to make a simple means of user control for a visual representation than for an acoustic one.

Remote control of the look direction and focal length of the camera lens can, with a self-view display, give users some control over their picture. However, few people are exactly laterally symmetric in appearance. For this reason, it can be mildly disturbing to see oneself without the left-right inversion of a mirror. As digital memory devices decrease in cost, it is increasingly likely that self-view pictures will use mirror-image format. However, a permanent self-view display may divide attention.

7.4.8 Richness and poverty

Between pictures and sounds, there are differences of perception as well as perspective. Pictures are unlike sounds, because they are more than one thing at a time and more than one time at a thing. We perceive sound as a single-stream process. In cognitive terms, our acoustic world is a cascade of transients, which are here and then gone. The cognitive world of our visual sense is inherently parallel, multi-stream and enduring, making us aware of the simultaneous existence of many objects. Even though one object is usually the focus of our attention, we can readily detect unexpected behaviour well away from that focus. The divided attention problem (see section 7.4.2) applies when interpretation is required — it does not imply limits to our simultaneous awareness of multiple objects.

This difference has enormous consequences for the creation of a sense of natural presence in pictures. When listening to speech, we prefer to avoid information overload and hear one talker at a time. In pictures, we prefer information abundance. We like to come back for more, and we like to choose when to do so. Pictures that contain too much to see at once, that are visually rich, seem closer to the real world and are more immersive.

The visual richness of the real world is why two people can go to the same theatre, sit in adjacent seats, yet see two different plays on the same stage. One metric for natural presence in pictures is whether this kind of simultaneous disparity can arise. Such pictures tend to be comfortable to look at, compelling to watch and suggestive of a sense of involvement. Visually impoverished pictures, that give viewers little choice of where to look or what to look at, tend to create a sense of distance or even alienation. In videotelephony, such pictures risk being ignored as unrepresentative or boring.

7.4.9 Gazing about in groups

When attending a lecture or watching entertainment, we are a passive consumer. Our visual attention is mostly directed at and around the talker. This is the model that has usually been used in videotelephony but it is the wrong model. When contributing to an interactive discussion, our visual attention is very mobile. Even when listening, we are almost as likely to look around the room or at other listeners as look at the talker. During face-to-face conversation, and in group discussions, we are not constrained to look fixedly at the head and shoulders of the person speaking.

In the multi-stranded simultaneity of our visual world, we direct our attention on to relevant or deliberately irrelevant objects as part of the evolving cognitive interplay between hearing, thinking and observing. We are, in a cinematic sense, our own cinematographer and director in one. We are not only free to direct our gaze but we need to choose where, when and how to look around, in person-to-person and group interactions.

The video signal is wholly unlike the speech signal, because it represents something that the viewer is accustomed to freely sift, select or ignore. The minimum standard of natural presence in videotelephony should mean duplicating just enough of this freedom to be intriguing. In telepresence, relevance is not what it seems and too much choice is almost enough.

7.4.10 Information and significance

"It is sufficient for images to resemble objects in but a few ways, and even that their perfection frequently depends on their not resembling them as much as they might."

René Descartes, Fourth Discourse on Optics 1637

Descartes was talking about the art of illusion and selective detailing in painting, exploiting our ability to see what is suggested rather than what is there. Despite great advances in videocoding algorithms, we cannot yet map information to meaning or significance as efficiently as an artist can. Many video-compression techniques have quantization errors that are distracting, although fractal coding creates visually plausible errors. Methods that segment the picture into areas of interest may approach the coding efficiency of the artist in future. Subjective richness in electronic pictures does not inherently require much more bandwidth or bit rate in the video signal.

Questions of perspective, viewpoint, presentation, perception, content and significance are all aspects that influence the impression of natural presence in a picture. The success of the entertainment industry may have conditioned our expectations, creating preconceptions about pictures that are hard to detect, let alone overcome. The next section examines some engineering conventions and artistic values that are often used in still photography, in the cinema and television to influence the impressions that viewers receive.

7.5 MAKING AN IMPRESSION

"Photography is a magic thing, with all sorts of mysterious smells, a bit strange and frightening but something you learn to love very quickly."

The seven year old Jacques Henri Lartique in 1901,
shortly after his father gave him a camera.

7.5.1 The power of a picture

During a visit to an American museum, one of the authors happened upon a display of black-and-white photographs taken by a group of disadvantaged school-children, most of whom had never used a camera before. Almost every image effortlessly captured a mood or suggested a state of mind. The pictures were simple yet startlingly good, as communication rather than decoration. The text of a display panel at the end of the gallery explained that many of the linguistic concepts and cultural ideas in native North American society were essentially mental pictures. These children found it easier to express themselves in pictures than with the letters and words of the white man. Perhaps our cultural emphasis on text might explain why so few of us, as children or adults, can take such startlingly good photographs.

7.5.2 Thinking in pictures

Every parent knows the importance of making pictures in a child's development. Drawings are the natural approach of children to organizing and understanding the world. Graphs, figures, sketches and mind-maps can be important tools for adult thinking and a vital part of working collaboratively. In legal, medical and scientific circles, pictures are widely and readily used. The success of facsimile is founded, in part, on our natural empathy with hand-drawn figures of every kind. Pictures, though undervalued, are a major aspect of cultural richness in our society [34].

But much of our intellectual activity is traditionally text-centric, especially in higher education. Linear means of expression have encouraged linear ways of thinking, in what has been termed the left-brain, linguistic or logical mode. This has been to the detriment of the more flexible and creative outlook that has been termed the right-brain, pictorial or parallel mode. This may partly explain why our typical astuteness of self-expression declines in changing from speech to text to pictures.

Cochrane [35] suggests a reorganization of schools and universities to exploit interactive multimedia technology for immersive, self-driven and self-questioning education. Kress [36] suggests that the growth of multimedia technology, that has been called the second 'printing' revolution, requires a reassertion of the importance of picture-making throughout education, to foster the creative imagination. Chomsky and Paiget have, in different ways, argued powerfully that language skills are innate. Yet young children make marks unprompted, on walls and paper, that they themselves describe as standing for objects not sounds. Is the making of language more innate than the making of pictures? If we are so good, as children, at making and interpreting pictures, what do we gain by neglecting or suppressing these seemingly intuitive visual skills?

7.5.3 Why are stills so moving?

It is strange how one, immobile, black and white photograph is sometimes capable of capturing the essence of an event more evocatively than several minutes of high-definition video footage in colour. A great photograph, in a newspaper or magazine, can have an enormous emotional impact, making us gaze long and ponder.

We have a deep cultural adaptation to painting and the expression of a whole world in a frozen moment. Because the image does not move, it allows detailed inspection and introspection. In a moving picture, something distracts us and both thought and picture are gone.

There is another reason for the potency of still pictures. Still photographers are keenly aware they have one chance to impress and that all their knowledge must go into an image, from technical issues through superficial layers of appearance to deeper levels of artistry and emotional content.

7.5.4 Commerce or convention?

> "An amateur is someone who thinks repeated mistakes are errors that make them more of an amateur. If you're a professional, you just call your repeated mistakes conventions and people think you're more of a professional."
>
> Lord Lichfield, professional photographer

Advances in electro-optical and electronic technology have made television cameras much lighter and smaller. Shoulder-mounted cameras, with the lens at eye level, are often used in television production today, even in a studio setting. Amateur camcorder owners seem to follow the same convention. Superficially, this position seems correct, matching the viewpoint of a surrogate observer. But we do not perceive optical perspective in images the same way as in life. When the camera is tilted down to show the whole body of a person, the perspective looks unflattering. If the camera is held level, either the top half of the frame is wasted or the lower part of the subject's body is excluded. To avoid this conflict, the cinema often uses dolly-mounted cameras, with the lens at chest or waist level. For the same reason, still photographers often crouch down a little when shooting pictures of people — the results look more natural.

When judged by results, an eye-level camera position seems to be a sacrifice of realism for convenience. Realism in making pictures is a subjective matter. The lesson for videotelephony is to always judge by results and not by an objective, geometrical analogy with a surrogate observer.

7.5.5 Hold me close, ever so close

A close-up view of a face can be a powerful way of expressing a person's character in neutral surroundings. But still photographers know that a close-up view is not automatically the best one. The same facial expression can attend several states of mind, in surroundings that are not neutral.

Figure 7.7 is a well-known photograph of Winston Churchill. Restrict this picture to the face and Churchill could be an impish old gentleman having a jolly time at a family celebration. Include the blurred background, the dark hat, coat and Churchill's shoulders and the impact of the picture is transformed. Churchill is now a tired and bowed but determined war leader, who could be thinking: 'If I can stand all this and smile, so can you'. The outlying picture area contains no detail of any individual significance and the emotional message of the picture is all about Churchill. Yet most of that message is a contradiction that is not portrayed in Churchill's face. What seems like more turns out to be less.

Fig. 7.7 Winston Churchill.

7.5.6 As if you're right there ...

The technical development of high-ratio zoom lenses and stabilized, body-worn cameras has let television become obsessed with the close-up view, as if people were only eyes and lips. This is most often seen in soap operas and sporting coverage.

For example, at a sporting event a cup is presented to the victor who, with an exultant yelp, lofts it above her head. The conventional practice in television, all around the world, is to zoom in and sacrifice her body language for the sake of every nuance of facial expression. Surely the greatest sense of 'being there' must be created by going right up close, by standing alongside? This kind of television thinks that 'in your face' is the best way of being there. The cinema knows better. A cinematographer would zoom out, sacrificing fine detail on the athlete's face but showing every sinew of her body, arms and legs. The pulse of exultation that was born in her knees when she touched the cup, that rippled along her spine, that leapt up her arms and threw itself out of her finger-tips was everything about being there, in that place, at that time.

Sacrificing the subtle emotion of body language for the more obvious drama of facial expression may be superficially attractive in entertainment; but this is the wrong approach for natural presence videotelephony. Feeling involved in a distant place means sharing an awareness of the seemingly inconsequential and irrelevant (see, for example, Fig. 11.7 on page 237). As section 7.4.2 remarked, the human visual system is very good at recognizing faces and their expressions from relatively few pixels.

Close-up views mean dicing with the devil. Used occasionally, such pictures are powerful and impressive. Used permanently, they deny our natural tendency to want to look around, creating feelings of remoteness or alienation. Small pictures are probably less of a barrier than small minds in creating a sense of natural presence.

7.5.7 All about bodies

Most films from the Golden age of the cinema make very sparing and selective use of close-up views, limited to points of high drama or significance. Most scenes use medium or wide angle views, to show body language and the surroundings. John Ford's *The man who shot Liberty Valance* [37] is an excellent example of this tradition. The diffident turmoil of James Stewart's lawyer or the rough diamond of John Wayne's drunk are expressed more revealingly by their body language than by their dialogue or through their relatively stereotyped facial expressions. Ford's formative years as a director were in the days of silent pictures and his cinematic style remained more visual than wordy throughout his career.

Even at quiet moments, body language is an important aspect of human communication and a very important part of creating a sense of surrogate presence or involvement in a remote location. It allows us to empathize with other people and to use our imagination to put ourselves in their place. Body language helps to make face-to-face contact scalable, so that communication is almost as easy in a group of thirty people as a group of three, when meeting face-to-face.

Visibility of body language can be more important than the visibility of lip synchronization, especially in a 'pending interjection' situation of group interaction. If all users in a group are shown in permanent close-up view, it is difficult to signal a pending interjection without seeming rude, unless the users know each other well. The alternatives are silence or interruption, either of which degrades the naturalness and effectiveness of the social interaction.

7.5.8 Sounding realistic

Cinema and television agree on the importance of high-quality sound in creating a sense of being there. The addition of sound to the cinema in the 1920s was a major challenge, in both technical and artistic terms. Multi-channel stereo sound is now a successful part of the cinema experience, that places particular emphasis on the value of subjectively accurate localization of sound and on the reproduction of low-frequency sounds. The techniques of NICAM coding and Dolby™ surround sound have recently extended these techniques to television.

The telecommunications industry has traditionally placed more emphasis on clarity and intelligibility than on naturalness in sound. Many designs of videotelephone have followed this bias, to the significant detriment of a sense of natural presence. As Fluckiger [38] says: '... pictures make telephone users less tolerant of technical and semantic distortions in the sound channel'. The effective use of high-quality sound requires care, since low-frequency background noise can become more annoying. Videotelephony often uses hands-free speech, which demands good echo control [39].

7.5.9 The craft of the cinema

> "I used to think that drama was when the actors cried. But drama is when the audience cries."
>
> Frank Capra, film director

The authors thought that young people would find black-and-white films like 'The Thirty Nine Steps' or 'Casablanca' to be tired and passé. Yet some of today's teenagers do not think so. Children whose grandparents were born after these films were first shown can be captivated and entranced by them, despite loud competition from modern 'blood-and-gore' fantasies. What do these old films have that makes them timeless?

The answer is probably connected with natural presence. Such cinema classics often pay great attention to picture composition, choice of angles of view, depiction of body language and environmental context, careful planning of camera movement, subtle control of lighting effects and a general sense of quality and meticulousness. That is why today's teenagers can be moved by a plain and slow tale like 'Casablanca'. For a short while, they can imagine that they too are in Morocco, in wartime and in love.

If fictional images can tell the truth, factual images can lie. Historians now agree that none of the events in Eisenstein's *October*, a film about the Russian revolution, took place in the way shown. Music and the mass choreography of 100 000 extras depicted myth not reality. Nevertheless, we apply a critical attitude to the written and spoken word but tend to accept pictures as true. Is this another reason why we feel uncomfortable about being seen through an inappropriate picture in videotelephony? This view of pictorial truth reflects an unsophisticated attitude to pictures, in a working and learning world dominated by text.

It is significant that the cinema industry calls the creation of pictures 'cinematography' not 'camera work' and that composition in photography is still a matter of debate. Attention to composition in the creation of pictures can be found in television productions, such as travel programmes or high-budget dramas, when creating a sense of being there is important. These programmes show that even a small picture with mono sound can be so captivating that we lose track of time, if skilfully created.

7.5.10 Fit for purpose

The motives of television tend to make it, in general, a poor exemplar for videotelephony. Video technology can achieve a strong sense of natural presence, yet this is uncommon in broadcast television. A unique combination of commercial competition and technology may well encourage pictorial staleness and uniformity in television entertainment. The medium faces the pressure of an audience that can be instantly fickle, abandoning one programme for another, in a way that cinema or theatre audiences cannot do. Film and theatre directors know that they have the indulgence of a largely captive audience.

The pictorial traditions of the cinema and theatre are therefore more relevant to videotelephony, because their commercial and artistic conditions set them free to exploit subtlety, slowness and the seemingly dull, in building a feeling of involvement. For example, the theatre usually uses scenery, lighting and sound effects to support the performance of actors, rather than for pure show. In videotelephony, the performance of the 'actors' is everything and paramount.

The different purpose of the picture in entertainment and communications has powerful repercussions on the choice of pictorial culture. This means looking at the art as well as the science. Few videotelephone calls need aspire to Oscar nomination for their visual artistry, yet even a mundane picture need not lack all art. We choose our words without reluctance in a telephone call. So why not do the same with pictures? The technology ought to make those pictorial choices easy.

Fortunately, videotelephony operates in well-defined circumstances and does not need to create convincing impressions over a wide range of environments.

Few calls are likely to take place in deep space or under a waterfall. Unfortunately, videotelephony faces new issues that arise from its bi-directional interactivity, that no electronic picture medium has yet had to address. These new issues are complex and varied, so the next section outlines only some of the developments ahead.

7.6 TOWARDS NATURAL PRESENCE

7.6.1 Task, tenor and testing

To practice what we preach, this section starts by standing back. There is unlikely to be a single telepresence machine, in the way that there has been a single telephony machine. Telepresence means different things to different people and will have different embodiments, varying substantially in design and operation to fit their purpose. It means modest improvements to tomorrow's desktop videotelephones as well as multi-sensual, fully-immersive, ultra-realistic, virtual-reality, shared-space, communicating societies of the future.

Yet these wide-ranging applications share a common tenor or direction — that of improved communication through enhanced transmission of sensations, closer to the sensory interplay that lies behind the subtle power of face-to-face interaction. Following such a direction and putting task before technology means focusing on fully understanding why it is important to do something before exploring how it might be done.

Two tests for future telepresence systems, with or without pictures, were suggested in sections 7.4.6 and 7.4.7. Namely, that different observers should sometimes experience simultaneous disparity of perception and that attention should be able to be directed away from the relevant. When exploring issues in the signal-to-sensation interface, a further test is suggested, using a form familiar to economists.

- Who benefits, why and when?
- Who loses, why and when?
- Who provides, what, when and how?
- Who pays, in what way and how much?

Nikias [31] sees four fundamental issues in the engineering and technology of the 'creator/computer/consumer' interface. He speaks about unidirectional multimedia systems, but the same issues are relevant in real-time and bi-directional telepresence systems too:

- symbiosis — a balanced combination of meaning, presentation and design;

- heterogeneity — the interoperability of diverse technologies;
- immediacy — low latency in robust information delivery;
- representation — the expression of information at various levels of abstraction.

7.6.2 Understanding surrogacy

Finding solutions to the design issues of better videotelephony amounts to acquiring a deeper and more profound understanding of surrogacy, meaning a sense of mutual presence by proxy at remote locations. Figure 7.8 shows some of the dimensions of surrogacy in communications. Neither signal nor sensation matters in isolation, it is the appropriateness of the linkage that creates surrogacy. As described in section 7.4, human senses in face-to-face conversation and group interaction are multi-dimensional and nonlinear. Modelling and matching these sensations with signals in realizable systems is a challenging task.

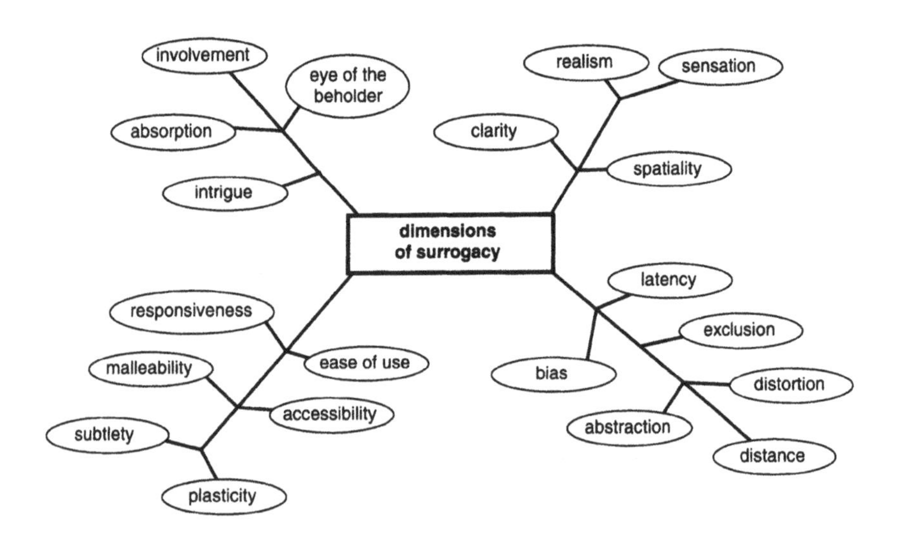

Fig. 7.8 Some positive and negative dimensions of surrogacy.

Conventional television provides passive surrogacy, as an observer. Videoconference systems provide active surrogacy when the room layout, user expectations and motivations, equipment configuration, and performance combine effectively. Speech alone provides a surprisingly powerful sense of involvement, yet the 'people on the box' can be ignored if there are deficiencies or imbalances in the technical issues that Nikias [31] describes as symbiosis and immediacy.

The technical dimensions in Fig. 7.8, such as distortion, clarity and latency, are well understood. But the more subjective or affective dimensions, such as intrigue or accessibility, are much less well-defined or explored in communications engineering. Yet the entertainment industry regularly deals with subjective or vaguely defined qualities. Film or theatre critics would probably not hesitate to rate a production in terms of most of the dimensions in Fig. 7.8. Convergence may lead to a sharing of language and concept as well as technology.

7.6.3 Fundamental issues

A fresh approach to the interface between signals and sensations in videotelephony is likely to be market-specific and should start with a fundamental re-examination of the purpose of the picture. What should it be a picture of, and why?

Figure 7.9 shows pictorial purpose in communication, classified by content, situation, task and transaction. Any of the aspects shown can be represented in a simple telephone call or an audioconference. Videoconference equipment is usually designed for all the types of pictorial content but videotelephony has focused on facial and social content, with little emphasis on the other kinds. The transaction types are all expressible in speech, but some are much more intuitively expressed by body language.

7.6.4 Layout and impact

In the past, the large size of video cameras and displays left little flexibility for arrangement. In today's videotelephones, convenience usually dictates that camera and display are collocated (see Fig. 7.10), but does this arrangement give the best performance? Video cameras are becoming steadily smaller and flat displays are increasingly affordable, so that the optical geometry of videotelephony should be re-examined.

The system magnification is a function of both camera and display parameters. What is the relationship between camera distance, display distance and overall magnification for optimum naturalness in different applications? Answering this question will require a mixture of geometrical optics, affective understanding and subjective knowledge. This will be essential to achieving the vision shown in Fig. 7.11, where the participants are not constrained by the bezel of a monitor or the view of a camera.

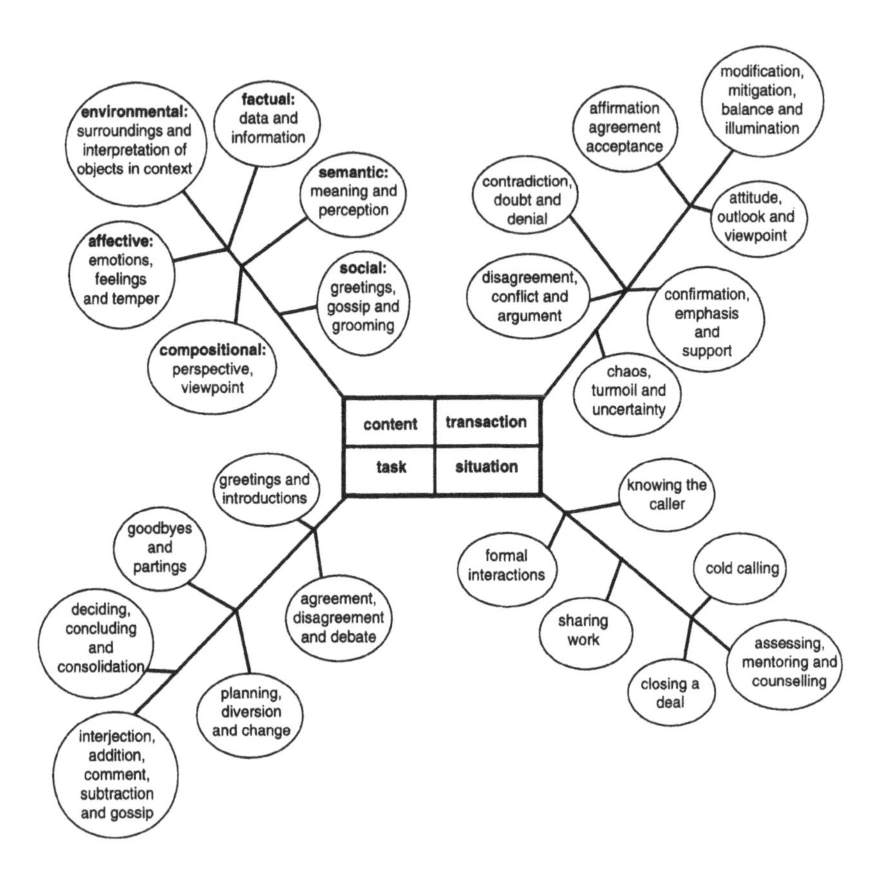

Fig. 7.9 Pictorial purpose in terms of content, situation, task and transaction.

7.6.5 Function and flexibility

If video signals are to share the flexibility and significance of speech signals, then users need to have easy and intuitive control over the picture-capture process, instead of accepting the output of a single, fixed camera.

Fig. 7.10 Desktop videotelephony in 1996.

Fig. 7.11 Our vision of immersive telepresence.

Table 7.2 shows the likely use of six compositions, differing in viewpoint and angle of view.

Table 7.2 Camera views and their likely uses in videotelephony.

Name	*Composition*	*Likely purpose*
Room	Wide angle or plan view of location	Overview of delegates and events
Person	Wide angle, showing whole person	Listening and talking
Torso	Medium angle, showing head and hands	Talking and listening
Portrait	Close view of head and shoulders	Emphasis, especially when formal
Poster	Medium view, from directly above	Explanation with large objects
Hand	Close view of hands, from above and side	Explanation with small objects
Face	Close up of face	Emphasis or strong emotion

It is important to see, at some time and in some way, almost the whole of the distant location or room. All participants need to feel a sense of sharing in what is happening, including the social impact of distractions and interruptions that are out of sight in a close-up view. A whole-room view also influences the scalability of group interaction and is also a factor in the perceived physicality of virtual environments.

The 'portrait' composition is widely used in television to show the head and shoulders. In videotelephony, this view shows less detail than a 'face' composition yet conceals the body language of the arms and hands. Shoulders can be shrugged, but few people make extensive shoulder manipulations without radically changing their whole posture, when the 'person' composition would be more appropriate.

The use of user-settable or automatically adaptive framing, selected via cordless user buttons or menu selection, is one approach to such picture control. Additionally, pre-prepared still pictures or moving sequences of delegates might be useful for transmission to participants on low-bandwidth connections or if non-visibility is preferred at some point.

7.6.6 Choices, choices

As discussed in section 7.4.7, different camera locations are needed for optimum eye contact and naturalness in group interactions. Current research at BT Laboratories is studying how the view of a camera, virtually located in a display screen, might be synthesized [33]. The use of multiple cameras with different perspectives means that several screens or split-screen presentations are needed for display. Alternatively, single displays could be switched in viewpoint, a fixed medium-view could be selected or a tokenized map of the room might be displayed. Finding a suitable means of multiple display is harder than capturing such views and exacerbates the problem of displaying multipoint links, especially with immersive and continuous-presence pictures. Larger, more detailed and wrap-around displays have been suggested to help reduce these problems (see, for example, Fig. 7.5 in section 7.3.8).

7.6.7 Putting ourselves across

Cochrane [40] remarks that the optimum orientation of a display depends on the task, for physiological and psychological reasons. For example, a small whiteboard or 'electronic paper' display is much easier to use when placed horizontally. Orientation also matters in videotelephony, though in a different way.

There are two kinds of body language. The absolute kind, such as gesture, which means the same from any angle and the relative kind, whose meaning depends on the angle of view. Conversations held face-to-face are not always body-to-body, with people sitting or standing directly opposite each other. When people do meet in this way, it often signifies some kind of privacy, because of argument, intimacy or a location that would otherwise imply open participation.

Photographs of people taken 'square on' to the camera signify formality or a position of authority. We usually face computer screens at nearly zero degrees, giving our whole attention to the display as to a document. If videotelephone users adopt this orientation, the resulting picture may imply a formal or authoritative social position, or a private social interaction.

Informality and relaxation in most photographs is signified by an angled disposition of body and limbs. In most peer interactions, faces are turned towards the talker, but bodies are seldom set exactly 'square on'. When peer interaction or informality is intended, a slightly angled body position, with the head facing the camera, gives a more appropriate social impression in videotelephony. The requirements of informal desktop videotelephony do not precisely match those of desktop computing on the same display.

7.6.8 Automating the director

If multiple cameras and changes of look direction and lens focal length become the norm in videotelephony, then perhaps this 'performance' needs to be directed in response to the interplay of human conversation. In the limited range of videotelephony environments, perhaps this direction could be made automatic?

Word spotting and a hybrid mix of voice-activity detection and speech-recognition techniques might allow the reliable classification of emotions such as anger, excitement, boredom and puzzlement. Voice-operated cameras have already been used in multipoint conferencing and automatic location of eyes and faces has proved feasible [41]. Simple kinds of gesture and body language detection also seem feasible [42]. Information from other sensors, such as directional microphones, thermal movement detectors and participant remote buttons, could aid this orchestration of picture and presentation.

Related techniques have been proposed (see Chapter 10) for virtual conferencing over very low bandwidth connections, controlling the appearance of realistic avatars with a few hundred bits per second (see Fig. 7.12). This kind of object-based video coding will form an extension to the MPEG 4 standard.

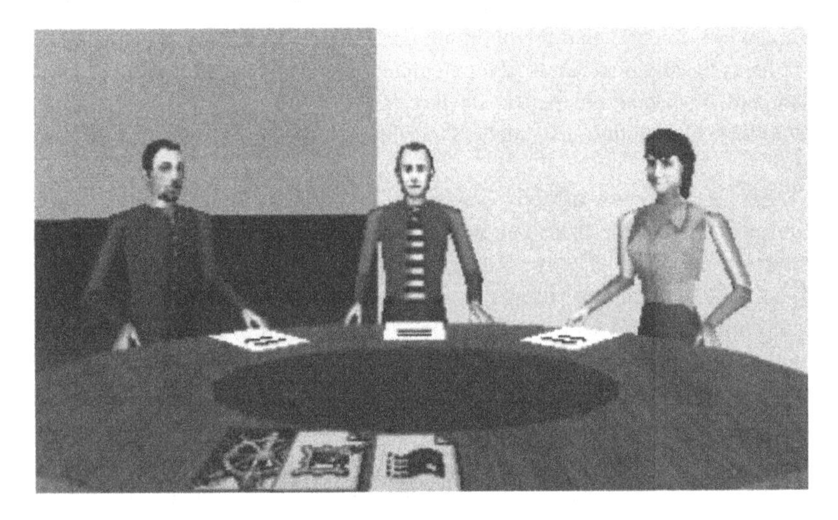

Fig. 7.12 A virtual conference using avatars.

7.6.9 A leap too far

There is a serious risk in the temptation to extend such ideas from virtual to live conferences, for automatic direction. Even less is known about the optimum use of automatic direction than about the appropriate use of pictures in videotelephony. Making inappropriate decisions about picture selection and control could make videotelephony less acceptable to users.

The idea of automating the director for high-quality videotelephony seemed like a serious proposal but was a leap too far. The idea focused on technology not the task. As section 7.6.1 described, exploring how to do something before understanding why it should be done is the opposite of what should happen. The first step in the orchestration of better videotelephony should be to give users more control of the picture-making process; that would be immune to a lack of precise knowledge about why or when such control might be most important.

7.7 QUESTIONS

The development of videotelephony with a strong sense of surrogate presence, using novel camera and display arrangements to convey a wide range of content, tasks and transactions, raises the following questions.

- What is the relationship between camera distance, overall magnification and display distance for optimum subjective effect in different applications?
- How does the purpose of the picture change with time and the interplay of conversation?
- How does the relative value of close-up and wide-angle views change, depending on the importance of body language in the interaction?
- Are there typical patterns in these changes of pictorial purpose and value of view?
- How might user controls allow the socially comfortable 'giving' of pictures?
- Should the choice of view or camera be controlled locally, by the distant participants, or by a combination of both approaches?
- What is the subjective value of a given amount of bandwidth, used to carry telephone-band speech, whiteboard data, 3-D environments, spatial sound, computer data, pictures of whole rooms, of whole people or of whole faces?

The nature of multi-modal interplay in face-to-face contact and group discussion, described in section 7.4.2, means integrating pictures with other information, such as spatially localized sound, whiteboard, cordless pointer and computer workspace data. This integration will be influenced by the cross-modal interactions of human perception and by the answers to the questions listed above. The answers are likely to be complex and dependent on the content, task and transaction, but they will be vital to the future of both videotelephony and of human interaction in 'virtual-world' multimedia environments.

7.8 SUGGESTIONS

The main suggestions of this chapter for the development and use of electronic pictures in communications are that:

- the purpose of the picture should be re-examined, in a task-specific way, to improve the interface between technical signals and human sensations — this includes issues of pictorial culture, the use of still, moving and adaptive-view pictures, the integration with sound and other signals, and the study of cross-modal interactions;

- the importance of the audio channel should be given greater emphasis compared with the video channel, both for information transfer and in creating a convincing sense of surrogate presence;
- the metaphor in videotelephony should be the mutual giving, not taking, of pictures;
- videotelephones should offer users more control over how they are seen, with standardized and intuitively obvious manipulations of camera selection, look direction, focal length and lighting — the option of non-visibility should be provided by the use of prepared still or moving picture sequences;
- the visibility of body language is an important part of conveying a feeling of natural presence and can be more important than the visibility of lip-synchronization in group interactions;
- the permanent use of head-and-shoulder or close-up views damages the feeling of natural presence, by eliminating body language and environmental context;
- the appropriate use of multiple cameras, with multiple channels, can aid a feeling of natural presence with simultaneous detail in wide-angle and close-up form.

7.9 CONCLUSIONS

In answering the question posed in section 7.1.4 — 'Do minds meet more easily when voices are disembodied ?' — this chapter has explained why the story and status of broadcast television is a deeply misleading guide. Television is widely successful as entertainment, yet its pictorial values are a bad model for communicating in pictures. It is important not to think of videotelephony as telephones with television pictures. In America, native Indian culture is undervalued today, yet their disadvantaged children can probably teach more about the art of communicating subtle messages in simple pictures than can television.

If television is a distorting mirror it is surely through choice. The lack of natural presence artistry in much broadcast television is not connected with inherent limitations in the technology of electronic pictures. However these artistic and emotional limitations arise, they probably do decrease the public expectation of the benefits of natural presence videotelephony. Indeed, the application of an inappropriate pictorial culture is partly responsible for the historical unpopularity of videotelephones.

Video pictures are capable of delivering a sense of atmosphere and of place, in an inclusive and involving way, if the artistic and technical aspects of picture composition are adequately managed. There is good reason to say that adding video pictures to telephony can enhance the effectiveness of human telecommunications, if done appropriately.

The technical, ergonomic, cultural, social and psychological requirements of more natural human telecommunications are not the same as those of entertainment. A marked divergence of design is required, with models and metaphors about real-time, bi-directional communication. Far too little is currently known about what constitutes appropriateness in videotelephony to allow clear foresight in system design. This chapter has only said what is not appropriate and suggested some approaches, tests and questions for future exploration.

It is unlikely that videotelephones will ever completely replace simple telephones or that videoconferencing will become a wholescale replacement for face-to-face contact. Videotelephones can be more powerful and effective than simple telephones, but they are nowhere near as simple to get right.

Unlike simple telephony, the social norms and psychological adaptations appropriate to videotelephony are poorly developed or not well understood at present. If videotelephones are to be used as fluently and casually as simple telephones are today, their design and deployment must evolve so that users feel confident of how their electronic image will be interpreted by others. Furthermore, that image must have sufficient pictorial quality to portray their state of mind or emotional disposition at least as accurately as the simple telephone does today.

The future of telephony is likely to be a plural mix of different but interoperable systems. The devil will be in the details and developing these details will need a strongly interdisciplinary mix of skills in interface design. More profound and insightful development will be needed of operating conventions, ergonomics, hardware and software that is specific to the requirements of greater naturalness and ease of use.

This implies that the telecommunications industry must define its own pictorial culture, looking at the best practice of the theatre, still photography and the cinema, to design that culture into workable equipment. Only then will videotelephones begin to emulate the success and wide social acceptability of the simple telephone.

The barriers to greater popularity of videotelephony also impede the future of multimedia and computer-generated environments as a whole. If the use of videotelephony is to grow, a greater level of visual awareness and familiarity with expressive picture-making will be required in society. Such an evolution might already be under way, with the growing use of multimedia in education and through the symbiosis of the World Wide Web, merging traditional text-centric information with picture-centric illumination and fun.

7.10 CODA

Long ago, an amateur was someone rich enough to do a thing for love not money. Calling someone an amateur was a high compliment, since professionals had to earn a living and had lower standards. Today, a resurgence of the enthusiastic and inspired thinking that gave birth to television is directing new applications of electronic pictures. Much of this resurgence is amateur, in that it comes from outside the established professions. Such enthusiasm is valuable and should be embraced for the new tricks it might teach.

The power of the written word eclipsed the power of pictures relatively recently in human history. Less than a thousand years ago, on innumerable walls and parchments, pictures were the mechanism that transported the human imagination in time and space. Not all societies have forgotten their picture skills. The rest of us need to re-learn those abilities, to handle the intellectual and social demands of the second 'printing' revolution that is known today as multimedia information technology.

It would be wrong to advocate vague dreams instead of academic proficiency, yet these are just the qualities that gave birth to the telephone and television. The authors do call for less single-minded analysis and more multi-minded synthesis. That little boy, from the start of this chapter, may have been jaded by the years but he still believes in the power of teleporting the imagination.

For the many challenges that lie ahead, open and inventive minds will be needed, as free of established convention as were the minds of Baird and Farnsworth in the early years of television.

We have much to learn and many opportunities to take.

REFERENCES

1. 'The future of television — in your face', The Economist (7 June 1997).

2. Shankar B: 'Breaking the ice in Greenland', Telecommunications, pp 57-59 (July 1997).

3. Gibbs W W: 'Taking computers to task', Scientific American, pp 64-71 (July 1997).

4. Bell Laboratories Record (complete issue devoted to the Picturephone), 47, No 5 (May/June 1969).

5. Noll A M: 'Anatomy of a failure: Picturephone revisited', Telecommunications Policy, pp 307—316 (May/June 1992).

6. McClelland S: 'An interactive world', supplement to Telecommunications, pp S1-S3 (September 1997).

7. Walker G R: 'The Mirror — reflections on inhabited TV', British Telecommunications Eng J, 16, No 1, pp 29-38 (April 1997).

8. Bridgewater T H: 'A A Campbell Swinton', Royal Television Society monograph (1982).

9. Moseley S A: 'John Baird: the romance and tragedy of the pioneer of television', Odhams, London (1952).

10. Herbert R: 'Radio amateurs and early television', Radio Communication, pp 17-19 (October 1996).

11. Fisher D E and Fisher M J: 'Tube — the invention of television', Counterpoint, Washington, DC (1996).

12. Farnsworth E G: 'Distant vision: romance and discovery on the invisible frontier', Pemberly Kent, Salt Lake City (1990).

13. http://songs.com/noma/philo/index.html (The Farnsworth chronicles)

14. McLean D F: 'Computer-based analysis and restoration of Baird 30-line television recordings', Journal of the Royal Television Society, 22, No 2, pp 87-94 (April 1985).

15. http://members.aol.com/mcleandon/tv_index.htm

16. Rechtin E: 'The synthesis of complex systems', IEEE Spectrum, 34, No 7, pp 50-55 (July 1997).

17. Whybray M W et al: 'Videophony', BT Technol J, 8, No 3, pp 43-45 (July 1990).

18. Catchpole A and Davies S: 'VideoPBX computer telephony integrated applications', British Telecommunications Eng J, 16, Part 2, pp 93-99 (July 1997).

19. Waddington S: 'Television futures', Electronics and beyond, 16, No 117, pp 8-15(September 1997).

20. http://www.dtv.org/ (Digital Television Team, Microsoft, Intel and Compaq)

21. http://www.teecom.com/dvb.html (Digital Video Broadcasting, European)

22. http://www.atsc.org/ (Advanced Television Systems Committee, American)

23. Seachrist D: 'See and be seen over IP', BYTE, 22, No 9, pp 104-108 (September 1997).

24. Kao S: 'Digital images', BYTE, 22, No 9, pp 21-24 (September 1997).

25. Dettmer R: 'Conjugal lights', IEE Review, 42, No 4, pp 135-137 (July 1996).

26. Younse J M: 'Mirrors on a chip', IEEE Spectrum, 30, No 1, pp 27-31 (November 1993).

27. Bonfield P: 'From Marconi to multimedia', Executive Voice (November 1996).

28. Anderson R: 'Value extracting services — a better deal for customers', BT internal publication (July/August 1997).

29. Dix A, Finlay J, Abowd G and Beale R: 'Human computer interaction', Prentice-Hall, London (1993).

30. Chellappa R, Chen T and Katsaggelos A: 'Audio-visual interaction in multimodal communication', part of 'The past, present and future of multimedia signal processing', IEEE Signal Processing, 14, No 4, pp 37-38 (July 1997).

31. Nikias C L: 'Riding the new integrated media systems wave', part of 'The past, present and future of multimedia signal processing', IEEE Signal Processing, 14, No 4, pp 32-33 (July 1997).

32. Birdwhistle R: 'Kinesics and communication', Univ of Pennsylvania Press (1970).

33. http://webserver.bt-sys.bt.co.uk/visionvideo

34. http://www.photo98.com/

35. Cochrane P: 'The desert and the oasis', British Telecommunications Eng J, 16, No 2, pp 168-169 (July 1997).

36. Kress G: 'Before writing — rethinking the paths to literacy', Routledge, London (1997).

37. Ford J (Dir): 'The man who shot Liberty Valance', Ford Productions (1962).

38. Fluckiger F: 'Understanding networked multimedia: applications and technology', Ch 10, Prentice Hall, London (1995).

39. Milner B, Lewis A V and Saeed V: 'Signal enhancement', in Westall F A, Johnston R D and Lewis A V (Eds) 'Speech Technology for Telecommunications' Chapman & Hall, pp 376-405 (1997).

40. Cochrane P: 'Why it's better to work flat out', The Telegraph (24 December 1996).

41. Machin D and Sheppard P J: 'A computer vision system for natural communication', British Telecommunications Eng J, 16, No 1, pp 45-48 (April 1997).

42. Whybray M and Shackleton M: 'Image processing for telecommunications in the 21st century', British Telecommunications Eng J, 14, Part 4, pp 315-316 (January 1996).

8

VIDEO CODING — TECHNIQUES, STANDARDS AND APPLICATIONS

M W Whybray, D G Morrison and P J Mulroy

8.1 INTRODUCTION

Teleconferencing has developed through the stages of audio, and video, and now data-conferencing (including fax, file exchange, shared-document editing, etc), and is poised to move on to virtual reality (VR) methods. VR may involve the mixing of real and synthetic video (for example, placing the real video images of peoples' head and shoulders round a synthetic conference table), or even just purely synthetic video (for example, using completely synthetic representations of people — 'avatars').

In the near future at least, a vital component for teleconferencing will continue to be the ability to compress the real video component, which commonly utilizes by far the majority of the bandwidth. This chapter reviews the fundamentals of video compression, the various standards that have emerged to cater for different application areas, and what the future might hold.

8.2 VIDEO COMPRESSION TECHNIQUES

Analogue video connections for broadcast television consume in the region of 5 MHz bandwidth, but for teleconferencing purposes which require switching and long-distance transmission, digital video is the only feasible approach. For digital operation, straightforward pulse code modulation of studio-quality video according to the sampling rates and quantization law specified in ITU-R Recommendation BT.601 [1] requires a bit rate of 216 Mbit/s.

Clearly, when the above rates are compared to the 64 kbit/s of telephony circuits, the likely tariffs would stifle all but the most enthusiastic and wealthy customer's interest in videoconferencing and videotelephony services! Fortunately, once in a digital form video signals are amenable to compression.

Virtually all the compression algorithms in use today are waveform coders. These attempt to re-create, at the decoder, a replica of the original image waveform by using only the properties of a low-level representation. For example, an image is sampled to produce a set of picture elements, termed pixels, the brightness and colour of each being represented by numbers. Compression is achieved by exploiting properties both of these numbers and of the human visual system. No higher level understanding is undertaken.

An analogy is the difference between text transmitted by a facsimile machine and text transmitted as ASCII characters. The latter is at a much higher level and hence is a more efficient representation. Today, image compression is still the equivalent of the facsimile representation.

If the numbers are reproduced exactly, the coding scheme is said to be lossless and in this case is operating purely on statistical redundancy. Techniques which remove visual redundancy reconstruct images which differ objectively from the originals and are classified as lossy. The errors are either subjectively invisible or at an acceptably low level. There is no rigid definition of how much distortion is permissible because the acceptability is subjective, depending on many factors, such as the picture material, the viewing conditions, the viewer's eyesight, the use to which the pictures are being put and, not least, the cost.

8.2.1 Removal of statistical redundancy

Statistical redundancy is not unique to image data processing. Many computer users, on finding that the capacity of their hard disk has become insufficient, have installed software which allows more data to be stored, or used file compression utilities to send files more efficiently by e-mail. Similarly most PSTN modems now incorporate data compression [2] which increases the apparent transmission speed.

Many of these techniques are based on the Ziv-Lempel algorithm [3] which recognizes repetitions of parts of the data and sends a pointer to earlier occurrences of that data, rather than repeating the data in full. Depending on the characteristics of an image (natural or synthetic, background noise level, etc), the Ziv-Lempel algorithm may provide compression in the range just above unity up to three times.

However, the above algorithm does not make effective use of the fact that image pixels tend be highly correlated with their neighbours in space or time,

8.2.1.1 Predictive coding

The basis of predictive coding is to use pixels or data already received to form an estimate or 'prediction' of the next pixels or data to be transmitted. A basic form of predictive coding is illustrated in Fig. 8.1. Pixels are fed into the system in raster scanned order (i.e. pixel by pixel scanned left to right across the image, line by line down the image, and picture by picture in a moving sequence). Differences are formed between each pixel and a nearby one temporarily stored in the delay element. The delay element is fixed and corresponds to selecting the pixel immediately to the left, the pixel immediately above (both intraframe predictions) or, in motion video, the pixel in the same place in the previous picture (interframe prediction).

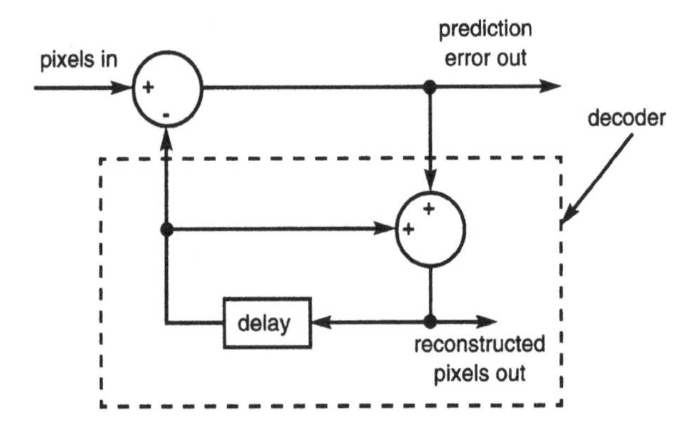

Fig. 8.1 Basic predictive coder.

The difference or 'prediction error' is seldom exactly zero because even in regions with no detail or in stationary areas the camera system usually adds some noise. However, the variance of the prediction error is generally substantially less than that of the original, and further techniques such as entropy coding (see section 8.2.1.3) can be used to represent the data with fewer bits than the original pixels. The original pixels can be reconstructed from the prediction errors by a decoder which consists of just the components in the outlined box.

As well as using just one previous pixel as the predictor, more complex schemes employing several previous pixels together may be used, e.g. a local average of a few spatially adjacent pixels may provide a better estimate of the new one by averaging out noise, or estimating and extrapolating the local slope of the image function.

When predicting from one picture to the next in moving video, the single pixel predictor is of course very good in regions of the image where there is no change (or 'motion'), but often very bad where there is actually motion.

Unfortunately the latter is precisely the area where most data needs to be sent and compression is most needed! This observation led to the introduction of conditional replenishment, whereby in areas of the picture where there is no temporal change at all, it is sufficient for a decoder to simply display the pixels from the previous frame again. In changed areas, new pixels are transmitted using prediction from spatially adjacent pixels. Omitting the redundant transmission of unchanged pixels gives a big gain in compression.

To achieve higher compression in the changed areas, motion compensated prediction was developed.

8.2.1.2 Motion compensation

This technique is an extension of the above prediction method, in which the delay for each pixel is varied dynamically throughout each still image to find the place in the previous still which is the best prediction. The corresponding horizontal and vertical offsets to refer to that best prediction are included in the coded bit stream. Ideally these offsets would be sent for every pixel but the overhead of doing this is much more than the savings obtained from the smaller prediction error. Instead, a single pair of offsets is applied to many pixels, a 16×16 block of them being a common compromise.

Motion-compensated prediction works fairly well on videoconference type scenes where the gross movements of people's heads and bodies are represented to a first approximation by simple translations of their positions from the previous image. However, real-world objects also undergo rotation, deformation, occlusion and so on, which are not well modelled by the above block-based motion compensation. More complex schemes have been devised to model these higher order changes in image structure, including representing the motion by complex warping functions, and delineating the boundaries between areas with different motions.

8.2.1.3 Entropy coding

Entropy coding is a general term for lossless data compression methods which rely on the statistics of a set of 'events' to be compressed. In most practical cases this means the overall frequency of occurrence of the various events in a set. The length of the codeword used to convey a particular event is matched to the likelihood of it occurring. Shorter codes are used for the frequent events and longer codes for those appearing less often, hence the term 'variable length coding'. For example, in Morse code a single dot represents the frequently occurring letter 'e' whereas the rarer letter 'q' is encoded as dash, dash, dot, dash.

It can be shown [4] that the optimum length of a codeword for an event of probability p is $\log_2(1/p)$. If all possible events in the set are assigned codewords according to this formula, the overall bit rate required to send events from that set with the given probabilities will be minimized. Thus for an event with a probability of occurrence of 1/8 the optimum codeword length is 3 bits. Unfortunately in practical cases the formula will yield non-integer values for codeword lengths, and a means is needed to optimally assign integer length codewords to events. Huffman [5] devised an algorithm to do this, hence the term Huffman coding (which is often wrongly used as a term for variable length coding, which need not, in general, be optimal).

Huffman's method assigns integer length codewords to events, but this results in a loss of efficiency since the theoretically optimal codeword lengths are non-integer. Arithmetic coding [6] is a technique which overcomes this by not having a one-to-one mapping of events to codewords, and thus comes closer to the optimum compression. However, it is also more difficult to resynchronize the decoder in the presence of errors.

Entropy coding gives very little compression if applied directly to image signals because the distribution of the brightness levels is fairly uniform. However, prediction errors have a very peaked distribution centred about zero, and variable length coding is very worthwhile.

8.2.2 Removal of perceptual redundancy

The degree of compression obtainable from lossless techniques is modest, typically rather less than 5:1. To achieve lower bit rates, it is necessary to employ lossy compression methods. Most complete algorithms incorporate both types.

8.2.2.1 Subsampling

A straightforward approach is to reduce the number of samples in the originals. After decoding, the missing samples can be replaced by repeating, or, better still, interpolating from, those that were coded. This lowers the spatial or temporal resolution but may not materially affect the usability of the resulting pictures. Unfortunately, the coded bit rate does not decrease linearly with the sample rate because the fewer samples have less correlation and are less amenable to compression.

Colour images are usually represented by three colour components. Red, green and blue are used for image capture and display purposes, but for transmission and compression an alternative format known as YUV is preferred. RGB can be converted to YUV and back by a linear matrix operation. The Y or luminance component represents the brightness, whereas U and V are 'colour difference' signals, which both have zero value in black, grey and white parts of the image, and are non-zero where colour is present[1]. Because the human eye has a higher visual acuity for the luminance component of images compared to the colour component it is possible to reduce the spatial resolution of the U and V signals significantly compared to the Y component with no perceptible loss of quality. In videoconference applications this has led to the use of image formats such as common intermediate format (CIF) where the Y resolution is 352 by 288 pixels, and U and V are reduced to 176 by 144 pixels each.

8.2.2.2 Quantization

Quantization reduces the number of discrete values a variable can take, reducing the number of bits needed to encode it. Accuracy is lost — a number of different input values will be the same at the output. Recommendation BT.601 allocates 8 bits each to the Y, U and V components, this being near the minimum necessary to avoid a 'painting by numbers' effect called contouring (although greater numbers of bits are often used during editing and manipulation stages to ensure adequate resolution is retained in the final result). There is not much scope for further quantizing the original pictures directly without introducing visible degradation.

This barrier does not apply to prediction errors, which are generally largest at edges or fine detail in the originals. The human visual system is much less sensitive to errors in high-detail regions — a phenomenon called spatial masking. This can be exploited by applying a nonlinear quantizer to the prediction error so that large errors are encoded less accurately than small ones.

8.2.2.3 Transform coding

Transform coding is a mathematical operation which transforms a set of numbers to another set with more favourable properties, while retaining the same information. The discrete cosine transform (DCT) is one of the more popular transforms for image compression.

The DCT is usually applied to a block, typically 8×8, of original image pixels or prediction errors, and produces an equal number of coefficients which are in many ways similar to Fourier transform coefficients. They contain, in order, increasingly higher frequency components of the original data.

For natural images, these DCT coefficients have the useful property that, over a large ensemble of blocks, the probability distributions of the coefficient values are highly peaked around the value zero, and can be efficiently compressed using

[1] To be more precise, Y, U and V are usually represented as 8-bit numbers, which are interpreted as the range 0 to 255. U and V take the mid-point value of 128 in black/grey/white regions, and fluctuate above and below 128 in colour regions, so 128 is their notional 'zero' value.

entropy coding. Additionally, the variance of the probability distributions decreases with increasing spatial frequency of the coefficients such that after quantization, many of the coefficients are zero and can be discarded. This is illustrated in Fig. 8.2, where a complex dark-to-light shading across an 8 × 8 pixel block results in only a few DCT coefficients having significant non-zero values. Furthermore, the eye is less sensitive to the higher spatial frequencies allowing them to be more coarsely quantized, resulting in additional compression.

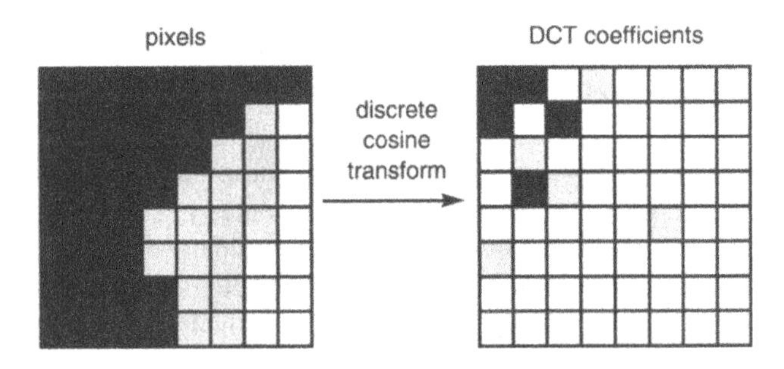

Fig. 8.2 Discrete cosine transform of an 8 × 8 block of pixels.

8.2.2.4 Wavelets

Wavelets [7] are another way of transforming an image into an alternative representation based on spatial frequencies, in a similar manner to the DCT. However, whereas the DCT is performed on a block of pixels at a time to produce a block of DCT coefficients representing the spatial frequencies within the block, wavelets are normally used on the complete image to split it into two spatial frequency bands in each direction (usually known as the L or low, and H or high bands), yielding a total of four sub-bands.

These four bands are designated LL, LH, HL, and HH. The LL band is essentially the original image reduced in scale by a factor of two in both directions, whereas the other bands represent higher frequency details associated with horizontal, vertical and diagonal edges or other detail. The LL band is usually subject to one or more further rounds of wavelet decomposition, resulting in a hierarchy of bands as illustrated in Fig. 8.3.

As with the DCT, the wavelet transform itself gives no compression directly, but it maps the image into a domain where the significant data is structured in a way which is amenable to compression. Thus, apart from the LL band, the resultant wavelet coefficients have highly peaked distributions around zero (as with the DCT coefficients), and entropy coding can be used. Secondly, the

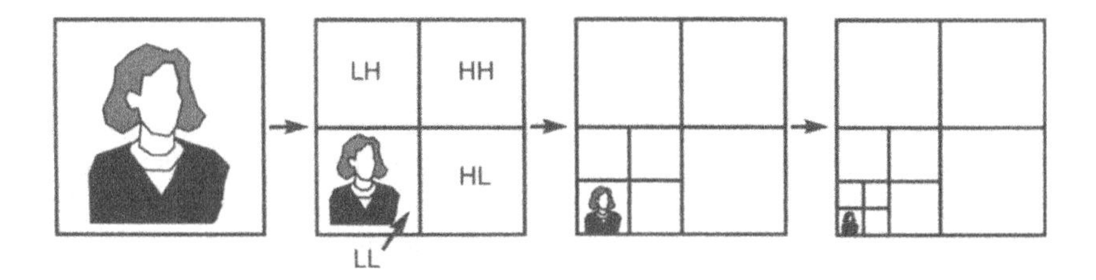

Fig. 8.3 Progressive stages of wavelet transform of an image.

locations of non-zero coefficients in the various bands are usually spatially aligned and quadtree type addressing schemes can efficiently indicate those coefficients to be transmitted with low overhead [8]. Thirdly, the hierarchy of spatial frequencies provided by the wavelet transform are a good match on to the spatial frequency sensitivity of the human eye, allowing coarser quantization to be used on the higher frequencies.

Because the wavelet transform operates equally on all pixels in the image rather than grouping them arbitrarily into blocks, the coding distortions produced by excessive quantization tend to be less visible and objectionable than those produced by block transform coders, where the block structure itself often becomes visible at high compression.

8.2.2.5 Vector quantization

Wavelets and the DCT exploit the correlation of adjacent pixels by generating coefficients that have small or zero value if adjacent pixels are similar. Another way to exploit this correlation is vector quantization (VQ) [9]. A block of pixels or prediction errors is quantized as one unit (a vector), rather than a pixel at a time. Typically, 4×4 sized blocks might be used, and a codebook of hundreds or thousands of possible example block patterns would be searched to find the pattern giving the best match. Only the index number of that entry in the codebook need then be transmitted.

VQ has a simple decode operation, being merely a table look-up, but the encoding process can be very demanding because at worst each block will require a full search of the codebook. In practice, various ways of pre-normalizing the data, and structuring the codebooks can reduce this problem. VQ can also be used to quantize the data resulting from other processes such as DCT or wavelet transformation, and both theoretically and practically should perform better than an equivalent number of independent scalar quantizers. However, there are some problems concerning construction of codebooks to adequately cover the range of vectors to be quantized, and of the processing power required to perform the codebook searching at the encoder.

8.2.2.6 Fractal image coding

Fractal coding is another way of exploiting redundancy in images. It is similar to VQ except that no explicit codebook is required. It relies on the fact that in any given image it is usually possible to find one part of the image which, when suitably scaled down in size, rotated and grey-scale adjusted, provides a very close approximation to another part [10, 11]. By dividing an image into small regions (usually square blocks) and finding the appropriate 'mapping function' for each region, the whole image can be represented purely as a set of mapping functions. The surprising thing is that as long as these mapping functions are contractive (meaning essentially that the size and grey level must always be scaled down in the mapping) no actual image data at all is required to initialize the process. By starting with any image and applying the mapping functions repeatedly, the resulting image will finally converge to one close to the original image used to select the mappings. Depending on the scalings this would typically take less than ten iterations of each region mapping to converge to the end-point.

Fractal coding can achieve very high compression in specially selected cases, but for typical images it is comparable to other methods. However, it has some other useful properties:

- it retains edge sharpness well;
- on natural images it can generate textures that although not faithful to the original image are quite often subjectively acceptable;
- the mappings found during compression can be used to scale images up or down in pixel resolution — although scaling up cannot accurately regenerate information previously discarded, the results can be subjectively pleasing;
- the decoding process is simple.

A major drawback is that the encoding process is very computationally intensive and to achieve real-time encoding it is currently necessary to compromise the compression efficiency by simplifying the process.

8.3 STANDARDS AND APPLICATIONS

The preceding section has highlighted most of the commonly used image compression techniques. This section reviews the historical development of image compression standards.

After some proprietary videoconferencing codecs operating at 6 Mbit/s from US and Japan, the European COST 211 project had developed by 1984 a 2-Mbit/s codec [12] which was subsequently adopted by the CCITT as Recommendation H.120. This was based on conditional replenishment, changed areas being updated using an intraframe pixel prediction method.

A few years later more sophisticated techniques were applied in the development of ITU-T Recommendation H.261 [13]. This provides the motion video component of ITU-T Recommendation H.320 [14] for videophone and videoconferencing services at total bit rates between 64 and 1920 kbit/s, and embodies the basic principles from which subsequent algorithms including MPEG 1 and 2, and H.263 were developed through a process of progressive refinement and addition of new features.

8.3.1 H.261

The picture format used for H.261 is either CIF (see section 8.2.2), or quarter CIF (QCIF) where the numbers of pixels in each dimension are half those of CIF. Remembering that the colour components U and V are at half the spatial resolution of the Y pixels, Y, U and V pixels are grouped separately into 8 × 8 blocks, and then four Y blocks and the spatially corresponding U and V blocks are associated into a 'macroblock' as shown in Fig. 8.4.

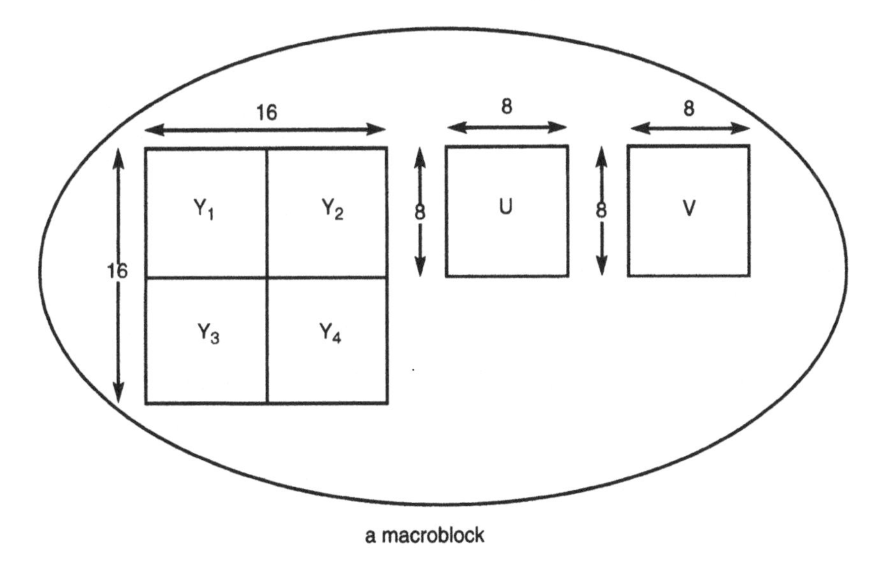

Fig. 8.4 A macroblock composed of four Y blocks and a U and a V block.

The structure of an H.261 encoder is shown in Fig. 8.5 and can be seen to combine several of the techniques discussed in section 8.2.

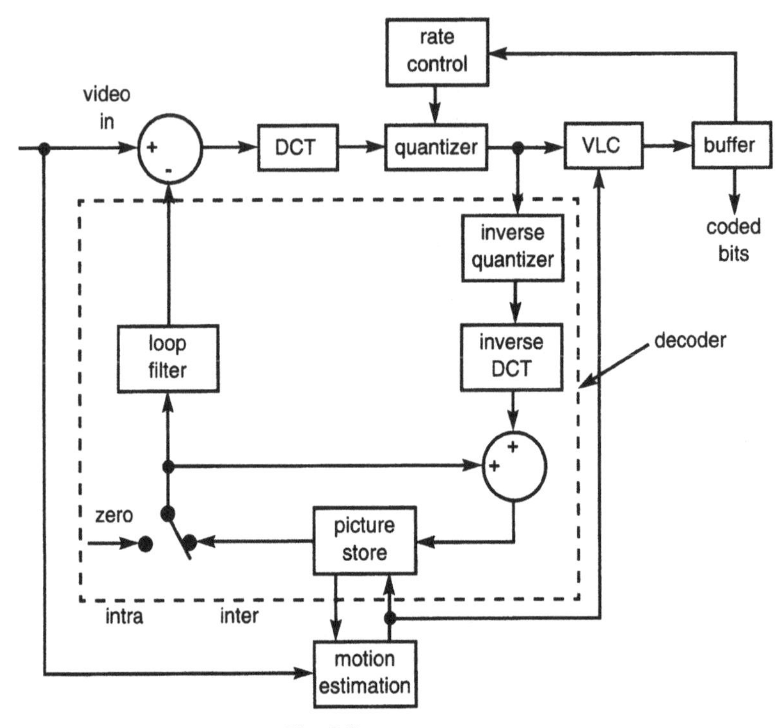

Fig. 8.5 H.261 encoder.

The encoder works as a predictive coding loop by analogy with Fig. 8.1. It uses past data to form a prediction of the current input data and quantizes the prediction error for transmission. It then performs the 'decoder' functions of inverse quantization and summation with the prediction to reconstruct an approximation to the input data.

To understand the coding loop in more detail, begin with the picture store, which holds the previously decoded picture. The motion estimation unit compares the incoming picture, a macroblock at a time, with this stored picture to determine the best motion vector to use for each macroblock. Having done so, it directs the picture store to output pixels (an 8×8 block at a time) to the rest of the coding loop, using the previously calculated motion vectors to select the appropriate parts of the stored picture. When the switch is in the 'inter' position, a block passes through a low-pass spatial filter which has been found to improve coding performance, and is then used as the prediction for the corresponding input block. After subtraction to yield the prediction error, the block undergoes a discrete cosine transform, the DCT coefficients are quantized, and the non-zero coefficients addressed and passed to the variable length coder. An inverse quantizer generates the appropriate reconstruction levels for the quantized coefficients and passes them through the inverse DCT to give an approximation to the prediction error, with some added noise due to the quantization process. On adding this to the prediction, the reconstructed picture is produced, which is close to the input picture, but with the added quantization noise. As this noise is generated in the transform domain, it appears in the image as a fairly evenly distributed random component, though, at high compression, blocking and ringing artefacts are also visible.

The encoder is completed by a buffer, which receives data generated by the coding loop at a variable rate and outputs it at a usually constant channel rate and a rate control mechanism, which adjusts the quantizer and also usually the coded picture rate, to keep the long-term output bit rate the same as the channel rate and the short-term variation within the range of variation that the buffer can absorb.

The H.261 decoder consists of the components within the box in Fig. 8.5 depicted by a broken line, preceded by an input buffer and variable length decoder. As the H.261 decoder does not have to perform the motion estimation function or the forward DCT, it requires much less processing power than the encoder.

The prediction generated within an encoder must match that generated by a decoder, or else the contents of the two respective frame stores will progressively diverge as more frames are coded. Ideally the 'decoder' part of the encoder loop would be defined to be mathematically exactly equivalent to all decoders. However, in practice there are several ways of implementing the DCT which, because of the effects of rounding errors, can result in small differences in the numbers produced in different systems. Although the magnitude of this error is specified and constrained by the H.261 specification of the DCT accuracy, it can still accumulate and cause encoder/decoder mismatch. As a means of constraining this divergence, in addition to the predictive- or inter-coding mode, the encoder can also use the switch in Fig. 8.5 to set the prediction to zero for a macroblock or a complete picture, in which case the input signal is coded directly in 'intra'-mode. Since this mode does not rely on previously transmitted data it is used to initialize the encoder and decoder loops to the same state at the start of a session, to clear any transmission errors that may have resulted in the encoder and decoder loops getting out of step, and to control long-term build-up of rounding errors in the DCT. Intra-mode is not used more often than necessary as it does not provide as much compression as inter-mode.

The H.261 coder is only formally specified to work on CIF and QCIF pictures, and at bit rates in the range 64 kbit/s up to 1920 kbit/s. Although it can be used outside this range, other codecs such as MPEG 1 and 2 are generally more suitable at higher rates, and H.263 at lower rates. The main application of H.261 is still videoconferencing using the H.320 Recommendation. On a 2B channel ISDN connection the speech is usually coded with G.728 at 16 kbit/s leaving 108.8 kbit/s for the video (after deducting multiplexing overheads), at which rate H.261 provides adequate quality to cope with two or three people at each terminal with moderate amounts of motion. For higher numbers of people per terminal, or for longer conferences, a higher bit rate provides higher quality and less user strain. H.261 is now also being used in other areas including streaming of video from Web sites, and videoconferencing using local area networks (H.323).

8.3.2 MPEG-1

The H.261 coding methods were further refined in ISO MPEG-1 [15] and MPEG-2 [16], which are aimed at a wider range of applications including those where storage is an important aspect.

MPEG-1 added the technique of half-pixel motion estimation to H.261, which used only integer pixel estimation. This allowed motions to be more finely tracked and improved the prediction and hence compression performance. The second main change was to add the concept of 'B-Pictures'. These are pictures in which predictions can be formed from a picture occurring later in time than the current picture, instead of just from the previous picture as was the case with H.261. Using such bi-directional predictions (hence B-Pictures) improves the overall performance. The ability to look ahead in time is accomplished by storing up a short set of pictures in frame buffers, coding the latest stored picture in the normal forward prediction (P-Picture) mode, then going back and coding the remaining stored ones in B-picture mode (Fig. 8.6). Although B-Pictures offer significantly higher compression than P-Pictures (by a factor of typically 2 to 3), which in turn have higher compression than I-Pictures (by a similar factor), there is a penalty of increased picture delay, and more complex processing requirements. The delay of several pictures (equivalent to typically 100 ms) is usually only a problem for real-time conversational services such as teleconferencing, for which the delay can be minimized at the expense of compression by using few or no B-Pictures. The compression achieved by MPEG-1 is around 30% greater than that of H.261, though this is somewhat dependent on picture material and bit rate.

Figure 8.6 also illustrates the regular insertion of intra-coded pictures (I-Pictures) in a typical MPEG sequence. Since these can be decoded independently of any preceding or following frames they are used to ensure rapid recovery from any transmission errors, and allow random access (i.e. jumping into the compressed video stream at any point). They also provide a means of playing the video sequence backwards as is necessary to duplicate video cassette recorder functionality such as fast reverse play — essential for video-on-demand type applications and for general browsing of compressed video. MPEG-1 was originally developed for use at single-speed CD ROM bit rates of around 1.2 Mbit/s and CIF resolution. It is, however, suitable for other applications not requiring interlaced pictures (see section 8.3.3), such as the interactive multimedia services trials run by BT [17] where a video bit rate of 1.6 Mbit/s was used to deliver roughly VHS quality video to homes for a mix of entertainment, education and information.

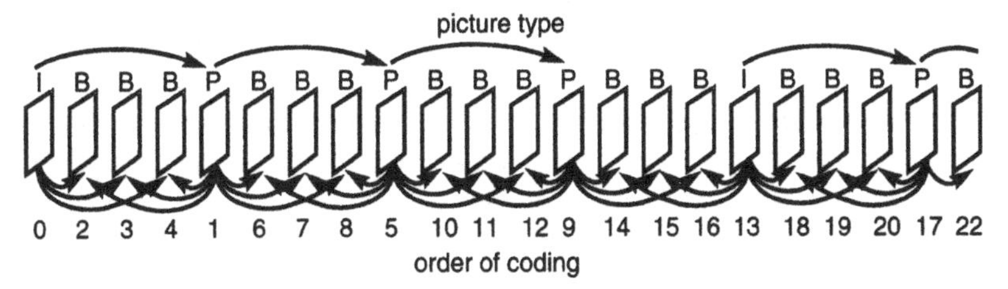

Fig. 8.6 Order of coding and prediction paths of I, P and B-pictures.

8.3.3 MPEG-2

MPEG-2 was developed in collaboration with ITU-T and is also known as ITU-T Recommendation H.262. It extended MPEG-1 to deal with the interlaced scanning of conventional television systems. Interlace is a hangover from analogue transmission and display systems which addressed the problem that for a good rendition of motion only 25 to 30 pictures per second are required, whereas at least 50 screen refreshes per second are required to avoid annoying flickering. In cinema this is done by displaying each picture several times. In analogue television there was no means of storing a picture at the receiver for repeated display, so each picture or 'frame' was scanned as two 'fields', each using half the number of lines in the frame, with the lines spatially interlaced to give double the number of lines for the frame, effectively increasing the vertical resolution. So 50 (or, in the USA-developed NTSC system, 60) fields per second are delivered to reduce screen flicker, at 25 (or 30) frames per second. When digitally coding such pictures it is important to realize that the two fields were captured at different times, and thus any moving object appears at different positions in them. If the fields were coded as a single frame there would be some very complex high-spatial frequencies to deal with as the alternate lines of the fields contain uncorrelated information. MPEG-2 deals with this by having extra prediction modes which can predict from one field to another in appropriate ways to extract the useful correlations remaining. By coding interlaced TV signals, MPEG-2 can be used to deliver broadcast (BT.601) resolution, at around 4 to 8 Mbit/s. MPEG-2 can also be used to encode higher resolutions such as HDTV, which was to be the subject of a separate 'MPEG-3' phase, but was subsumed into MPEG-2.

As with all video coding schemes, the quality achieved depends on the complexity and motion in the pictures, and on the available bit rate. Since historically the networks available have operated at fixed bit rates it has been usual to operate video compression schemes with a buffer and control system to meet that fixed bit rate. However, the result is that the video quality actually varies through the sequence — dropping in the more complex parts and perhaps using more bits than is necessary for acceptable quality in others.

With the arrival of networks that can support variable bit rate transmission such as ATM, it should be possible to operate with constant picture quality by allowing the bit rate to vary as necessary. Similar ideas can be used on satellite or cable delivery systems where a total bit rate budget can be shared adaptively between several video channels according to their relative need, achieving a higher overall quality than a fixed bit rate budget for each. Thus in reality it is difficult to put a single figure on the bit rate required for 'broadcast quality video', and the 4 Mbit/s mentioned above would be adequate for most purposes, but will not prevent some visible impairments on complex scenes such as sports with rapid camera and player motions.

Another new feature introduced by MPEG-2 is 'scalability', whereby one video bit stream can be used to bootstrap the quality of pictures provided by a second video bit stream, and so on. An alternative view is that one compressed stream contains within itself sub-streams which can also be decoded to give lower quality pictures than the full stream. This scalability involves using the decoded pictures of the lower bit rate stream as sources of prediction for the next layer, which can be configured to provide a higher spatial resolution, a higher temporal resolution (more pictures/second), higher quality (less distortions), or a combination of these. A drawback is that the quality achieved by using all the layers is usually lower than that achieved in a single layered encoding at the same bit rate.

There are several applications for scalability. One is where a network may offer several levels of service such that a lower bit rate stream is guaranteed timely arrival, whereas one or more higher rate streams may suffer some data loss. The lowest layer guarantees that a picture is always available for display, while the other layers enhance the quality in a way that is resilient to occasional data loss. This scenario can apply in ATM and in Internet protocol (IP) networks.

A second application is multicasting on IP networks, where a single server sends out the multiple layers, and clients receive as many layers as network conditions allow them to, adding and dropping layers dynamically as network congestion changes, to receive the best quality video possible at any one time.

8.3.4 H.263

With the advent of faster modems, especially V.34 yielding 28.8 kbit/s [18], V.34 extensions to 33.6 kbit/s (full duplex) and 56 kbit/s (though not bi-directional at this rate), the possibility of providing useful motion video over the PSTN has emerged. The experience gained from H.261 and MPEG has been further augmented by development of the H.263 video compression algorithm [19]. The main aim for this algorithm was higher compression, given the limited bit rate available on the PSTN, yet low delay as the main application was conversational services. This aim was largely achieved as it delivers around twice the compression of H.261, though, as with the gain provided by MPEG-1, this is dependent on picture material and bit rate.

A significant departure for H.263 was to make some of the coding enhancements optional, to be negotiated between an encoder and a decoder. These optional modes appear as annexes to H.263. Those appearing in Version 1 of H.263 (1996) are listed below.

- Annex D — unrestricted motion vector mode

  This allows motion vectors to point outside the normal picture boundaries by extrapolating from the edge pixels, and is particularly useful if the camera or whole scene is moving. This mode also doubles the allowed range of motion vectors from 16 to 32 pixels.

- Annex E — syntax-based arithmetic coding mode

  This substitutes arithmetic coding for the usual variable length coding procedure, and achieves slightly higher compression efficiency.

- Annex F — advanced prediction mode

  This allows motion compensation on 8×8 blocks as well as the usual 16×16 ones, and also uses a technique called overlapped block motion compensation, whereby a degree of overlap of adjacent compensated blocks is used to smooth out some of the block edges otherwise generated by block-based motion compensation.

- Annex G — PB-frames mode

  This allows the use of bi-directional prediction (see section 8.3.2) for increased compression. In contrast to the MPEG B-Pictures, here a P and a B picture are combined together into one data structure.

Figures for the amount of extra compression achieved by and the subjective effect of each Annex on different types of sequence are given in Whybray and Ellis [20]. Figure 8.7 shows a comparison between H.261, the base level of H.263, and H.263 with all four optional annexes (as described above) turned on.

The intended application for H.263 was the H.324 PSTN Videophone Recommendation, which was ratified by the ITU in 1996, with products appearing in early 1997. Many of these H.324-compliant videophones were implemented as PC applications, requiring only a suitable modem and camera interface card to convert a multimedia PC into a videophone, with the video-

Fig. 8.7 Stills from sequences coded at 28.8 kbit/s, 12.5 pictures/second. The four pictures are: (a) original frame (b) H.261 (c) H.263 base level and (d) H.263 with options D, E, F and G on [20].

coding algorithm implemented in software on the PC. H.263 has also found favour with organizations providing streamed video on the Internet, where again the high compression efficiency is a benefit.

Further enhancements to H.263 were developed and 'Decided' by the ITU in February 1998, under the title of H.263, Version 2, also known informally as H.263++. These enhancements appear largely as optional additional annexes.

- New picture types:

  — scalability — use of enhancement layers of H.263 coded video to build higher resolutions/qualities at high bit rates in an embedded manner;

  — custom source format — enables coding of pictures with arbitrary frame sizes, and arbitrary pixel aspect ratios;

  — improved PB frames — improved version of an existing picture type for higher compression.

- New coding modes:
  - — advanced intra-coding — improved compression of intra-frames;
  - — deblocking filter — reduces blocking artefacts to improve subjective quality;
  - — slice structure — groups macroblocks for improved error resilience;
  - — reference picture selection — speeds up recovery from channel errors by use of a back channel;
  - — reduced resolution update — helps maintain a high frame rate where there is high motion;
  - — independent segment decoding — improves error resilience;
  - — alternate inter-VLC — slightly improves inter-compression;
  - — modified quantization — allows more rapid control of the quantizer, and increases quantizer range.

The large number of these optional modes extends the application areas for H.263, but has caused concerns about compatibility. The existing scenario in conversational services is that the end-points should negotiate (normally using ITU Recommendation H.245) which modes are common and can be used, though the complexity of this negotiation may be getting out of hand. For other applications where there is limited scope for negotiation, e.g. streaming of pre-compressed video from a server, the decoder must be able to handle all modes used by the encoder.

8.3.5 JPEG

Regarding still pictures, in the 1970s, several telcos had introduced or were considering videotext services. Using similar formats to teletext, these were character based, providing text or block mosaic graphics. There was a desire to add 'photographic quality' pictures, but a full screen image at Recommendation BT.601 quality is more than 800 Kbyte. At the then typical modem rate of 1200 baud, such a still would have taken about 100 minutes to receive. Even with smaller images and some relaxation of their quality, the waiting times would have been unacceptable. With photographic videotext as a catalyst, the usefulness of a generic compression standard for still images was widely appreciated. As a result the ISO and ITU-T jointly developed the JPEG algorithms [21, 22] which are equally suited to both storage and transmission applications. The basic JPEG algorithm is very similar to the intra-frame coding modes of the moving picture standards, being based on an 8 × 8 DCT. In teleconferencing, JPEG will often be used as the compression method for sending still images between conference participants as part of some higher level teleconferencing application.

8.4 VIDEO TRANSCODING

Compression algorithms are normally used to compress video to a known target bit rate, either for immediate transmission at that bit rate, or for storage and later play-back. Once the compressed bit stream has been produced, it is usually not possible to directly access and decode it at a different bit rate. However, in the future it will increasingly be necessary to do this to support services such as:

- video databases which can be accessed at multiple bit rates;
- interworking adapters between networks carrying video at different rates, for example PSTN/ISDN;
- continuous-presence multipoint videoconference units, where the input and output bit rates may differ.

The latter two are particularly important for teleconferencing applications.

In general, changing to a lower rate can only be accomplished by decoding the bit stream back to images and then re-encoding. It was previously assumed that the delays of the two tandemed pairs of coding and decoding operations must be additive and therefore may become unacceptable in some interactive applications. However, investigations by BT have shown that transcoders can be constructed such that the total delay from the input to the first encoder, through the transcoder, to the output of the final decoder is substantially less than the sum of the delays through two equivalent but independent encode-plus-decode pairs [23]. In fact, the level of delay can decrease to that of the original encoder plus decoder.

The key lies in recognizing that the vast majority of the unavoidable delay, that which cannot be lessened by more powerful hardware, in a coder plus decoder combination, arises from the rate-smoothing buffers. These are needed to match the varying rates of bit generation by the coding kernel and bit consumption by the decoding kernel to constant rate transmission channels. These kernels and buffers are illustrated in Fig. 8.8. In Fig. 8.9 the buffers in the transcoder are rearranged to make them contiguous, by operating on uncompressed instead of compressed video. By precisely controlling the coding kernel in the transcoder so that the fill levels of the two buffers are complementary, the total delay through them becomes constant and they can be removed.

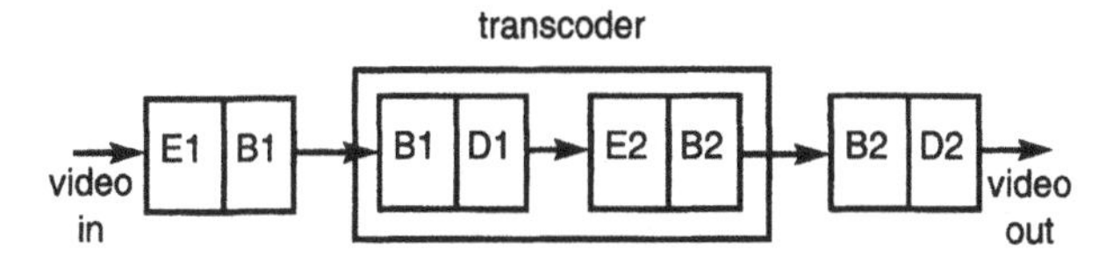

Fig. 8.8 Encoder (E) and decoder (D) kernels and their buffers (B).

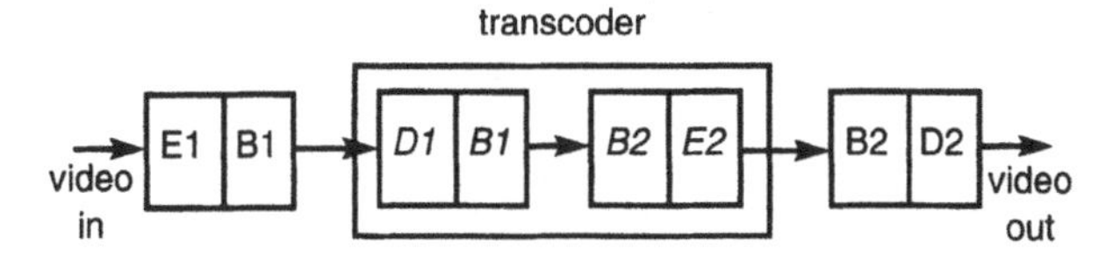

Fig. 8.9 Encoder (E) and decoder (D) with adjacent buffers operating on uncompressed video data.

In an extension of this technique the coding kernel in the transcoder is greatly simplified by using the motion vectors from the originating encoder. The bit rate reduction is achieved by the requantizing of DCT coefficients. However, this causes problems for predictive schemes because the prediction loops at the original encoder and at the final decoder are operating with different data, allowing the loops to diverge and distortion to build up at the final decoder. BT has devised the technique, shown in Fig. 8.10, of introducing a correction signal at the transcoder. This does not eliminate the error but continuously attempts to reduce its size, thus counteracting its growth. The result is a bit stream which has been further compressed, but with very little extra delay, with low complexity compared to a full decode and re-encode, and with a quality only slightly lower than if the full compression had been performed in one pass. As the new services mentioned above become more widely used, transcoding will become increasingly commonplace.

8.5 FUTURE CODING ALGORITHMS

There is general agreement in the image processing community that existing algorithms are all approaching a limit of compression dictated by purely statistical analysis of images, and exploitation of the human visual system's characteristics. Any large improvements in compression ratios will now require a deeper consideration of the nature of images, in effect using prior knowledge to enable a higher level of abstraction of the image data to be achieved. For example, most images are formed as the projection by a lens of objects in a 3-dimensional world into a 2-dimensional image. There is a direct correspondence between real-world objects and regions in the image. Since the

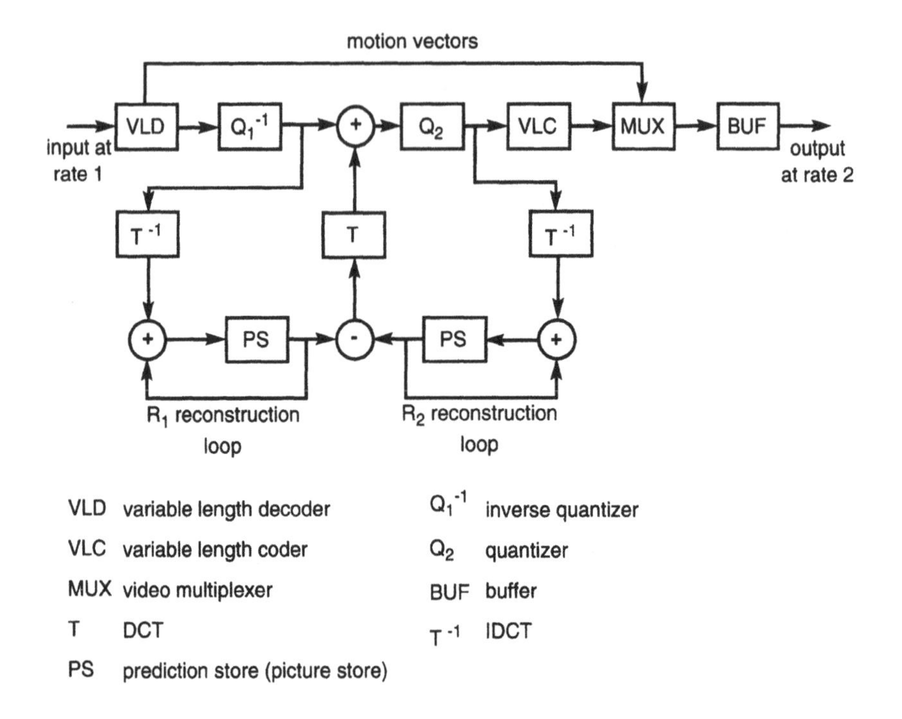

Fig. 8.10 Transcoder for bit rate reduction with drift correction.

original objects often have some continuity of shade, motion, texture, shape and so on, this continuity is mapped into the resultant image. Conventional coding algorithms such as predictive coding and motion compensation take some advantage of these properties, but do not make full use of the object-like nature of the real world. Admittedly, not everything is 'objects' — water, grass, skin for example are better described as textures, but can usually be segmented into discrete regions in a 2-dimensional image.

Algorithms are being developed which both segment each image in a video sequence into regions relating as much as possible to real-world objects, and, by analysing the motions of these regions, can describe the video sequence in a very compact form [24]. At present this relies typically on warping the objects in one frame into their new shapes and positions in the next frame, then sending whatever small additional information is necessary to correct any areas not well handled by this method — such as new surfaces revealed by object rotations. It is anticipated that this technique will develop to include full 3-dimensional modelling of the original objects, allowing more accurate reconstruction and higher compression.

Some extreme examples of this form of coding have already been developed in specialized areas. For example, 3-dimensional computer models of human heads have been used to enable synthesis of realistically animated images of human heads talking, moving around naturally, and with facial expressions [25].

Particularly life-like results are achieved if the image of a real person is mapped on to the model. The bit rate required to animate the head is of the order of only a few hundred bits per second. Although synthesis of a human head is now fairly straightforward, analysis of an image of a person's head to extract the motion, mouth shape and facial expression accurately enough to drive the model remains only partly solved.

Current standards work on advanced coding algorithms is centred around MPEG-4.

8.5.1 MPEG-4

MPEG-4 is an emerging ISO/IEC multimedia coding standard that, in addition to higher compression, aims to support new, content-based tools for communication, access and manipulation of digital audiovisual data. There has been considerable technical convergence between the main users of digital media — broadcast and telecommunications industries on the one hand and the computer and Internet industries on the other, with elements that had previously belonged exclusively to one being introduced into the others. Also there are major trends towards wireless communications, interactive computer applications and integration of audiovisual data into more and more applications. MPEG-4 is seeking to address the new expectations and requirements emerging from these developments — for example, to exploit the opportunities offered by low-cost, high-performance computing technology and rapid proliferation of multimedia databases.

Early work on requirements for MPEG-4 video identified eight key functionalities which were not thought to be well supported by existing or other emerging standards. These were grouped into the three fundamental categories of 'Content-Based Interactivity', 'Compression Functionality' and 'Universal Access Functionality' (see Table 8.1).

Table 8.1 MPEG-4 video categories.

Categories	*Functionalities*
Content-based interactivity	Content-based multimedia data access tools Content-based manipulation and bit stream editing Hybrid natural and synthetic data coding Improved temporal random access
Compression	Improved coding efficiency Coding of multiple concurrent data streams
Universal access	Robustness in error-prone environments Content-based scalability

The full MPEG-4 standard will comprise six parts, of which systems, audio and visual are the most advanced at the time of writing. Other parts include a conformance-testing specification, a technical report on reference software implementations and a multimedia integration framework definition (DMIF). The systems, audio, visual and DMIF specifications are currently at committee draft (CD) status. These will form the core of MPEG-4 Version 1 which is expected to progress to full international standard (IS) status in December 1998. There will also be a (backward compatible) Version 2 to follow one year later.

Audio (CD 14496-3) and visual (CD 14496-2) will define a standardized coded representation of audio and visual content, both natural and synthetic, called 'audio-visual objects' or AVOs. Systems (CD 14496-1) will standardize the composition of these objects to form compound AVOs (e.g. an audiovisual scene), and multiplex and synchronize the data associated with individual objects, so that they can be transported over networks at appropriate quality of service levels.

The visual standard will include all the coding tools relating to visual data, natural and synthetic. Currently there are two verification model (VM) specifications — the video VM for natural video coding (frame based or arbitrarily shaped), and the synthetic/natural hybrid coding (SNHC) VM for synthetic 2-D/3-D graphics tools. As the tools are tested and proven within these VMs, they will be incorporated into the visual CD, and ultimately, if widely agreed on, in the IS.

For the highest efficiency in video compression, the video VM is building on the work of ITU-T Recommendation H.263. Extensions have already been made to the core coder, most significantly the ability to code shape and transparency information of so-called video object planes (VOPs) (see Fig. 8.11). The final coder will be capable of coding arbitrarily shaped video at rates from 10 kbit/s to more than 4 Mbit/s.

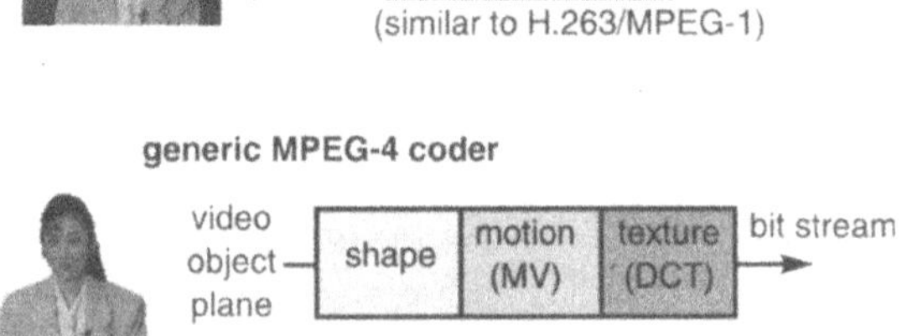

Fig. 8.11 Video verification model stages.

Transcoding between MPEG-4 and MPEG-1/H.263 elementary bit streams should be straightforward in the frame-based (or rectangular VOP) operating mode, e.g. when using the proposed real-time communications profile. The addition of arbitrary shape coding in the generic coder enables a range of new content-based functionalities, such as content manipulation in the compressed domain, and content scalability.

Shape coding is performed using a bitmap-based technique known as context-based arithmetic coding, but the visual standard will not specify how shape and alpha (transparency information) planes are to be generated as this is a producer or encoder issue. MPEG philosophy is that only decoder issues should be specified to guarantee interworking — enabling competition between companies as to which can provide the best encoder engine, and also, in this instance, provide the most useful shape and alpha data. Blue-screen techniques, already in widespread use in the television industry, are one source for this information. The video group is considering automatic video segmentation algorithms which may well form an informative annex to the standard to help content producers. Other automatic and semi-automatic segmentation techniques are also being investigated within the framework of the European COST211 project [26].

The synthetic video components of the SNHC VM currently include media integration of text and graphics (MITG), face and body animation, texture coding (generic and view-dependent textures), and static and dynamic mesh coding with texture mapping. MITG provides the capability to overlay and scroll text, images and graphics on coded video backgrounds. Work on this is currently proceeding and is likely to be included soon in the visual CD.

Face animation will allow definition and animation of synthetic 'talking heads' as shown in Fig. 8.12. Only the face definition parameters (FDPs) and the

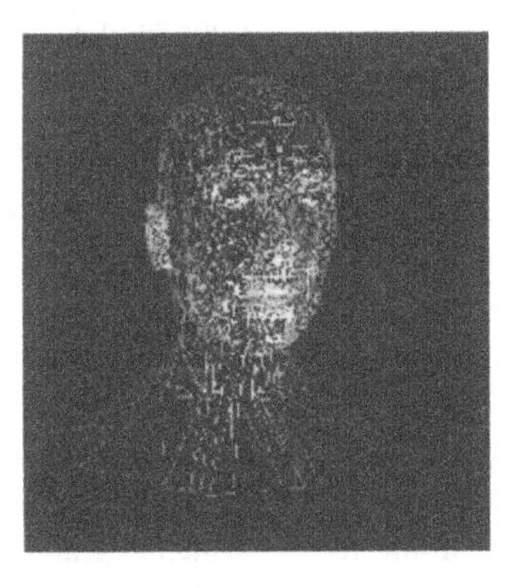

Fig. 8.12 Wireframe head model.

face animation parameters (FAPs) need standardizing here. Work at BT Laboratories, combining this with texture mapping (Fig. 8.13), has shown how realistic these synthetic personae can actually be (see Chapter 10). General 3-D model representations and body animation issues have been worked on to a lesser extent, but will form part of the Version 2 IS. One proposal in this area known as JACK [27] has already been input to the group.

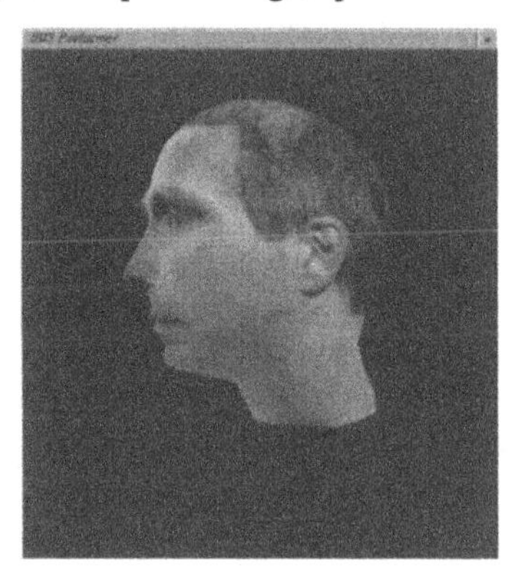

Fig. 8.13 Texture-mapped model.

8.5.2 MPEG-7

The title of the MPEG-7 activity is 'Multimedia Content Description Interface' [28]. The focus of this new work item is no longer that of efficient compression of audio/visual content, but rather its representation for searching and browsing purposes. This is a very real need with the rapid deployment of digital multimedia databases including sound and images on the World Wide Web (WWW), video-on-demand systems, and corporate image databases — hence the ambitious time-scale of this activity to just two years after ISO/MPEG-4 is finalized.

What MPEG-7 does inherit from previous MPEG activities is the requirement that only the minimum needed for interoperability is to be standardized. For example, it may specify particular sets of audio and video features that can be used as descriptors, such as shape outlines, colour histograms, frequency components, and how they are associated with the original content. However, it will not specify how to extract these video or audio features required for content description, in the same way that the analysis and encoding stages are not specified for the earlier compression-based standards. Nor will it specify a search engine or user interface — which might in practice take the form of a web browser or an intelligent agent. MPEG-7 will concern itself solely with a standardized description of audio/visual content features and will thereby be able to benefit from future progress in automatic feature extraction research.

Both information extracted from the scene (e.g. colour, shape, texture) and external information about the scene (e.g. textual annotations, script) are expected to be represented, although the work is at a very early stage at the time of writing.

The standard will have many application domains but a major one is likely to be WWW search engine functionality on audio, video and still pictures. Nearly all WWW searching is currently done on a text basis, although image-based searching is starting to appear in a basic form [29]. MPEG-7 is currently in a requirements capture phase of development, but is scheduled to reach international standard status in the year 2001.

8.6 CONCLUSIONS

Video coding has been one of the success stories of the last decade, enabling a host of applications including PC-based multimedia, videotelephony, highly flexible video-on-demand, and digital broadcasting, that were not possible with the previous analogue formats or uncompressed digital video. The need for compression comes from the limited bit rates of transmission channels, and the cost of digital storage. Although both of these will continue to increase in capacity and reduce in price, we still seem to be some way from the time when the cost will be so low that compression is not needed. Even then, some means of delivery such as radio and satellite channels will continue to have restricted bandwidths.

The successful market development in many areas has been dependent upon agreed standards, principally from the ITU and ISO/MPEG, although, particularly for PC-based applications, proprietary algorithms have also found a place.

Conventional compression algorithms seem to be reaching an asymptote, and although MPEG-4 will provide more functionality in terms of being able to define and code objects within a scene separately, in terms of absolute compression it will not provide very much gain over MPEG-2 or H.263. New techniques in MPEG-4 for synthetic/natural hybrid coding will push compression further, but only in limited domains such as human head/body modelling, and synthetically generated video. Nevertheless, video coding remains an active area of research with, for example, the full potential of wavelet-based schemes yet to be realized, and it is expected that compression ratios will continue to inch upwards.

Using real video in a teleconference has the distinct advantage that within the bounds of coding distortion there is no direct possibility of misrepresentation of the participant's appearance including facial expressions and body language, and brings a feeling of trust to the teleconference. A purely synthetic avatar in a virtual teleconference could be made to do anything, but the complete independence from one's surroundings and the physical world that an avatar brings makes new teleconferencing paradigms possible.

REFERENCES

1. ITU-R Recommendation BT.601: 'Encoding parameters of digital television for studios'.

2. ITU-T Recommendation V.42 bis: 'Data compression procedures for data circuit terminating equipment (DCE) using error correction procedures'.

3. Ziv J and Lempel A: 'A universal algorithm for sequential data compression', IEEE Transactions on Information Theory, 23, pp 337-343 (May 1977).

4. Jayant N S and Noll P: 'Digital coding of waveforms', Prentice-Hall Signal Processing Series, pp 146-148 (1984).

5. Huffman D: 'A method for the construction of minimum redundancy codes', Proc IRE , pp 1098-1101 (1962).

6. Witten I H, Radford M N and Cleary J G: 'Arithmetic coding for data compression', Communications of the ACM, 30, No 6, pp 520-540 (June 1987).

7. Proceedings of the IEEE, Special issue on Wavelets, 84, No 5 (April 1996).

8. Shapiro J M: 'Embedded image coding using zerotrees of wavelet coefficients', IEEE Trans on Signal Processing, 41, No 12, pp 3445-3462 (December 1993).

9. Gersho A and Gray R M: 'Vector quantization and signal compression', Kluwer Academic Publications (1992).

10. Barnsley M F and Hurd L P: 'Fractal image compression', A K Peters Ltd, Wellesley, Massachusetts (1991).

11. Beaumont J M: 'Image data compression using fractal techniques', BT Technol J, 9, No 4, pp 93-108 (October 1991).

12. Duffy T S and Nicol R C: 'A codec system for worldwide videoconferencing', Globecom '82, 3, pp 992-997 (1982).

13. ITU-T Recommendation H.261: 'Video codec for autiovisual services at p × 64 kbit/s'.

14. ITU-T Recommendation H.320: 'Narrow-band visual telephone systems and terminal equipment'.

15. International Standard IS 11172-2: 'Coding of moving video and associated audio at rates up to about 1.5 Mbit/s, Part 2: Video'.

16. International Standard IS 13818-2 and ITU-T Recommendation H.262: 'Generic coding of moving pictures and associated audio, Part 2: Video'.

17. Kerr G W: 'A review of fully interactive video on demand', Signal Procecessing: Image Communication, 8, pp 173-190 (1996).

18. ITU-T Recommendation V.34: 'A modem operating at data signalling rates of up to 28 880 bit/s for use on the general switched telephone network and on leased point-to-point 2-wire telephone-type circuits'.

19. ITU-T Recommendation H.263: 'Video coding for narrow telecommunications channels'.

20. Whybray M W and Ellis W: 'H.263 — video coding recommendation for PSTN videophone and multimedia', IEE Colloquium 95/154 (June 1995).

21. International Standard IS 10918-1 and ITU-T Recommendation T.81: 'Information technology — digital compression and coding of continuous tone still images, Part 1: Requirements and guidelines'.

22. Penbaker W B and Mitchell J L: 'JPEG still image data compression standard', Van Nostrand Reinhold (1992).

23. Morrison G: 'Video transcoders with low delay', IEICE Trans (June 1997).

24. Musmann H G, Hotter M and Ostermann J: 'Object-oriented analysis-synthesis coding of moving images', Signal Processing: Image Communication, 1, No 2, pp 117-138 (October 1989).

25. Welsh W J, Searby S and Waite J B: 'Model-based image coding', BT Technol J, 8, No 3, pp 94-106 (July 1990).

26. Mulroy P: 'Video content extraction: review of current automatic segmentation algorithms', to be presented at COST211ter Workshop on Image Analysis for Multimedia Interactive Services (WIAMIS '97), UCL, Leuven-la-Neuve, Belgium (June 1997).

27. 'JACK Human Body Modelling Software', Centre for human modelling and simulation, University of Pensylvania, http://www.cis.upenn.edu/~hms/jack.html

28. Koenen R: 'Overview of MPEG-7 goals and objectives', COST 211ter Workshop in Image Analysis for Multimedia Interactive Services (WIAMIS '97), UCL, Louvain-la-Neuve, Belgium (June 1997).

29. Yahoo Image Surfer, http://isurf.yahoo.com/

9

EVALUATING THE NETWORK AND USABILITY CHARACTERISTICS OF VIRTUAL REALITY CONFERENCING

C M Greenhalgh, A Bullock, J Tromp and S D Benford

9.1 INTRODUCTION

Recent years have seen a growing interest in the technology of collaborative virtual environments (CVEs) — distributed multi-user virtual reality systems that support mutual awareness and communication among their inhabitants [1]. The essence of CVEs is that multiple users share a computer-generated virtual world. Each user is able to freely navigate their own viewpoint within this world, is directly represented to other users through a process of embodiment, can communicate with other users employing a range of media (such as audio, text, graphics and video), and can directly manipulate objects within the world. There might be many potential applications of CVEs including virtual meeting rooms, shared 3-D information visualizations, battle and emergency simulations, and various kinds of entertainments.

BT's own research into CVEs has included the Virtuosi project, a three-year DTI/EPSRC-funded project which was led by BT and which developed CVE demonstrators for the manufacturing and fashion industries [2], and The Mirror, a project between BT, Illuminations Television, the BBC and Sony, to stage a public trial of a 'dial-up' CVE alongside the television series *The Net*. There are several potential benefits of this CVE technology [3].

- Support for natural, spatial communications skills

  Humans have developed a range of subtle but important communications skills based on the use of space as a 'shared resource'. Key examples include the importance of gaze direction in conversational turn-taking and the deployment of a wide range of spatial gestures such as pointing, facing and many other more subtle accompaniments to speech. CVEs may allow users to employ these skills to an extent that is not possible with other communications technologies.

- Dynamic group formation

  Many social situations, especially those involving more than a handful of participants, involve a degree of dynamic group formation — conversational groups come and go, groups split into sub-groups, sub-groups merge into larger groups and participants move between groups.

  In contrast to previous teleconferencing systems which often assume a relatively static and predefined group membership, CVEs actively encourage 'mingling' behaviours and highly dynamic group formation.

- Scaling to larger numbers of participants

  Following on from the previous observation, a common goal of CVEs is to support large numbers of simultaneous participants. CVEs support scalability in several ways — the perspective view offered by 3-D interfaces allows many participants to be seen at once, the use of attenuation with distance or even fully 3-D sound might help participants cope with many simultaneous audio streams, and the application of scoping techniques (such as 'aura', described in sections 9.2.1 and 9.5) allows users to limit their personal scope of communication in a densely populated environment. Current systems claim anything between ten and a hundred simultaneous participants, depending on the level of communication supported (e.g. audio or text), and other factors such as the level of user activity (indeed, factors affecting scalability are a key topic of this chapter).

- Peripheral awareness

  Computer-supported co-operative work (CSCW) research has indicated the importance of users establishing 'peripheral awareness' of the presence and activities of others [4]. By situating users within a 3-D space, CVEs implicitly support users in conducting locally focused communication, while at the same time 'keeping an eye' on more distant (i.e. peripheral) events. Some CVEs, such as the MASSIVE-1 system discussed in this chapter, go further by introducing explicit mechanisms for establishing different levels of awareness within a shared virtual space.

- Partially overlapping media

  An interesting property of human perception in the physical world is that we do not perceive the same objects at the same levels of detail in all of the available media. Typically, we can see some things that we cannot hear (e.g. distant objects) and hear some things that we cannot see (e.g. objects that are behind us). These 'partial overlappings' of media give rise to the potential for peripheral awareness. For example, we can listen to one person while keeping an eye on someone else who is approaching from the distance. Unlike other teleconferencing technologies which tend to depict the same objects across all media, CVEs naturally support the idea of trading off directed and peripheral awareness across different media.

- Representation of users and information within a single display space

  Many teleconferencing technologies fail to adequately integrate communications channels with shared information. For example, shared drawing tools and editors are usually displayed in windows that are separate from the representations of the participants. In contrast, CVEs locate participants and information within a single shared display space, thereby enabling participants to see who is attending to, or working with, what aspects of the information, from spatial cues such as body position, gaze direction and representations of virtual limbs.

However, to date there has been little formal experimentation aimed at verifying or refuting these various claims. There is also little understanding of the network traffic characteristics of CVEs, i.e. what kind of traffic will they generate or, conversely, what limits on scalability and complexity will be imposed by the use of different underlying networks (e.g. what could realistically be achieved using ISDN, ATM or the Internet). Answering these last questions will be a key step in provisioning any future CVE-oriented network services.

This chapter presents the results of a programme of experimentation aimed at answering some of these questions. This programme has been carried out in the 'Inhabiting the Web' (ITW) project, a BT-funded project within the BT/JISC (Joint Information Systems Committee of the Higher Education Funding Council for England) Research Collaboration Programme. ITW ran from October 1995 until March 1997 and involved a collaboration between BT Laboratories and five UK universities (Nottingham, Lancaster, Manchester, Leeds and UCL). The project ran a series of network experiments of Nottingham's MASSIVE-1 VR teleconferencing system [3] over the SuperJANET network in order to:

- construct a predictive model of network traffic valid for CVEs which are based on a peer-to-peer unicast network architecture — this involved the statistical analysis of system logs and network traffic measurements captured from virtual meetings;

- identify key social issues inherent in the use of CVEs, especially potential limitations of the technology or new emergent social phenomena — this involved a programme of 'usability' studies.

A secondary effect of the project has been to refine the methodologies required for dealing with these kinds of issues.

Section 9.2 of this chapter provides a brief description of the MASSIVE-1 system and of the underlying spatial model of interaction upon which it is based. Section 9.3 describes the structure and organization of the experiments. Sections 9.4 and 9.5 then focus on the network traffic analysis and model, while section 9.6 discusses the social issues that have been raised. Finally, section 9.7 describes the design of the new MASSIVE-2 system and introduces two new projects, both involving BT and the University of Nottingham, which build on this work within the context of so-called 'Inhabited Television' applications.

9.2 OVERVIEW OF MASSIVE-1

Firstly, a brief overview of the MASSIVE-1 VR-teleconferencing system is provided, describing both its functionality and its underlying network architecture. MASSIVE-1 was implemented as part of EPSRC-sponsored PhD research. The system runs on Silicon Graphics workstations over standard Internet protocols. A detailed description can be found in Greenhalgh and Benford [3].

9.2.1 Functionality of MASSIVE-1

MASSIVE-1 supports multiple virtual worlds connected via portals. Each world may be inhabited by many concurrent users, where each user can independently navigate their own viewpoint within the world. These users can communicate using combinations of graphics, audio and text media. The graphics medium renders objects visible in a 3-D space and allows users to navigate this space with six degrees of freedom. The audio medium supports both real-time conversation and the play-back of preprogrammed sounds. The text medium provides a plan view of the world via a window (or map) which looks down on to a 2-D plane across which users move (similar to some multi-user dungeons). Text users may interact by typing messages to one another or by emoting (i.e. smile, grimace, etc). These different media may be arbitrarily combined according to the capabilities of a user's terminal equipment. Thus, at one extreme, the user of a sophisticated graphics workstation may simultaneously run the graphics, audio and text clients (the latter providing a map facility and allowing interaction with non-audio users). At the other extreme, the user of a dumb terminal (e.g. a VT-100) may run the text client alone.

MASSIVE-1 is a specific VR-conferencing application and many of the features that one might expect of a general VR platform are not supported (e.g. full immersion and the ability to directly manipulate objects in a virtual world). The core functionality that is provided includes:

- moving with six degrees of freedom;
- communicating — speaking, triggering simple graphical gestures, and exchanging text messages;
- adopting a range of different camera views (e.g. out of body, over the shoulder, bird's-eye and mirror views) with the added ability to zoom the virtual camera in and out;
- posting text messages on simple message-board objects located within the virtual world.

Each user may specify their own graphics embodiment via a configuration file. In addition, some default embodiments are provided which are intended to convey the communications capabilities of the users they represent (i.e. ears suggest audio capability, eyes suggest graphics capability, and so forth). A simple mouth object is also provided which appears whenever a user speaks above a specified volume threshold.

The most sophisticated aspect of MASSIVE-1 is the way in which all communications media are driven by the so-called 'spatial model of interaction' [5]. The spatial model defines a set of high-level mechanisms called 'aura', 'awareness', 'focus', 'nimbus' and 'adapters' which enable users to negotiate social interaction in terms of different levels of mutual awareness. These mechanisms also drive underlying network connectivity and level of detail for the different media.

The spatial model assumes the existence of some kind of shared space (i.e. a shared co-ordinate system), populated by objects which are exchanging information over a set of media types (e.g. graphics, audio, text, video).

As implemented in MASSIVE-1, aura is a bounding volume on a user's personal presence. Auras are medium specific, so that a user might have a different-sized aura for each medium. Consequently, different media have different scopes of interaction (e.g. a user might be seen at a greater distance than they are heard). A collision between two users' auras (as notified by the MASSIVE-1 aura collision detector process) results in a network connection being established. Figure 9.1 shows a space containing five objects which, as a result of their auras, are currently connected into two separate conversing groups for this particular medium (there might be other patterns of connectivity in other media).

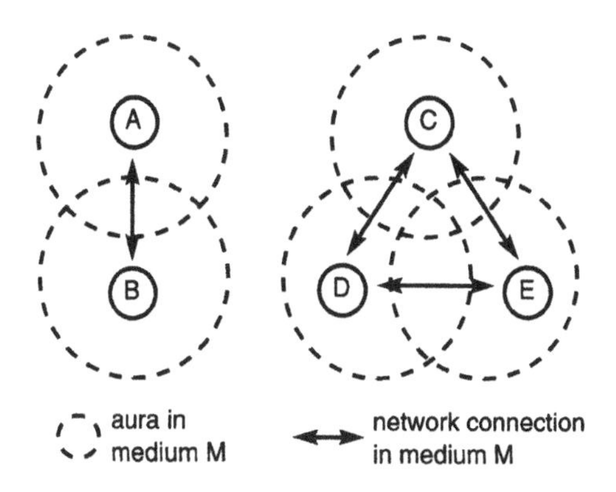

Fig. 9.1 Auras used to establish network connections.

Once connections have been established in this way, the exchange of information across these connections is determined by mutual levels of awareness. Awareness in MASSIVE-1 is a quantifiable measure of the relevance that one object has to another at a given moment in time. Awareness can loosely be interpreted as an expression of desired quality of service for a given connection. Awareness is continually renegotiated as objects move around space through the further mechanisms of focus and nimbus. Focus models an observer's allocation of attention across space and nimbus corresponding to the projection of their presence into space. The more that an object is within your focus the more you become aware of it. The more an object is within your nimbus the more it becomes aware of you. As with aura, focus and nimbus can be defined on a per-medium basis.

The awareness that an observing object, *A*, has of an observed, *B*, is therefore a function of *A*'s focus on *B* and *B*'s nimbus on *A*. Figure 9.2 shows how different arrangements of focus and nimbus can result in different levels of awareness. The figure is only concerned with *A*'s awareness of *B* and hence only involves *A*'s focus and *B*'s nimbus. In Fig. 9.2(a) *A* is focusing on *B* and *B* is nimbusing on *A*, resulting in *A* having full awareness of *B*. In Fig. 9.2(c) the converse is true, so that *A* has no awareness of *B*. Figure 9.2(b) shows the two cases where peripheral awareness occurs — in one of these *A* is focusing on *B* but *B* is not nimbusing on *A* (perhaps *A* is monitoring *B*?), in the other *B* is nimbusing on *A* but *A* is not focusing on *B* (perhaps *B* is trying to interrupt *A*?).

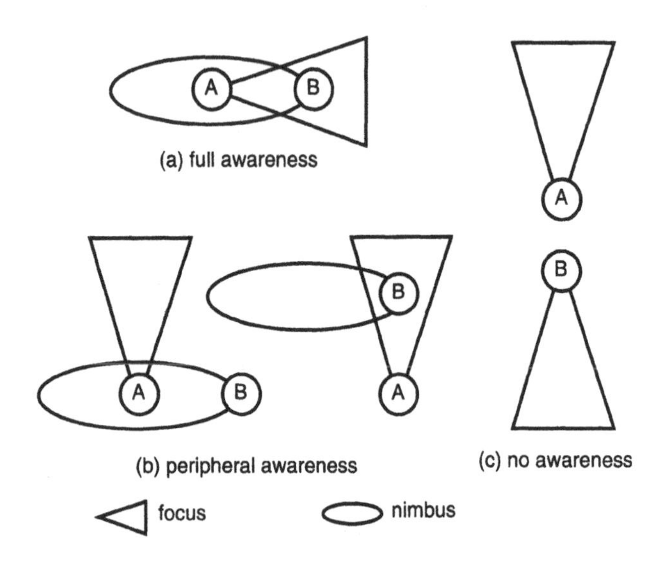

Fig. 9.2 Focus, nimbus and awareness.

Users can manipulate aura, focus and nimbus, and hence awareness, implicitly by moving and re-orienting themselves within space or explicitly by choosing between different settings or even directly changing their shape and extent, e.g. retracting their aura in order to drop some network connections. Various adapter objects can also be used to dynamically alter aura, focus and nimbus. Examples of adapters in MASSIVE-1 include a podium which increases the range of aura and nimbus and a conference table which gives its users common auras, foci and nimbi, centred on the table.

MASSIVE-1 applies these concepts to each of its three media so that audio volume, graphical level of detail and the display of text messages are all awareness driven. For example, audio awareness levels are mapped on to volume with the net effect that audio interaction is sensitive to both the relative distances and orientations of the objects involved. Unlike Fig. 9.2 where focus and nimbus are simple containment volumes of space, MASSIVE-1 defines them as scalar functions across space. The shapes of these functions are such that focus and nimbus are at a peak directly in front of the user and then gradually drop off with distance from the user and with angle from their direction of face. Consequently, a range of different levels of awareness are possible.

In the current implementation, users can choose between three predefined settings for focus and nimbus — normal (intended for general conversation), narrow (intended for private conversation) and wide (intended for browsing). The wide setting is the default and is used most of the time.

Figure 9.3 shows how a virtual world appears in the graphics interface. Three embodiments can be seen, of which the one in the foreground ('Chris') is our own seen from an out-of-body camera view. A message-board can also be seen in the background. Figure 9.4 shows an example of the text interface. Top left of Fig. 9.4 is a plan view of a virtual world; top-right is the key to the symbols used in this plan as well as an indication of current awareness levels between the user and objects in the world; at the bottom is the on-going text dialogue.

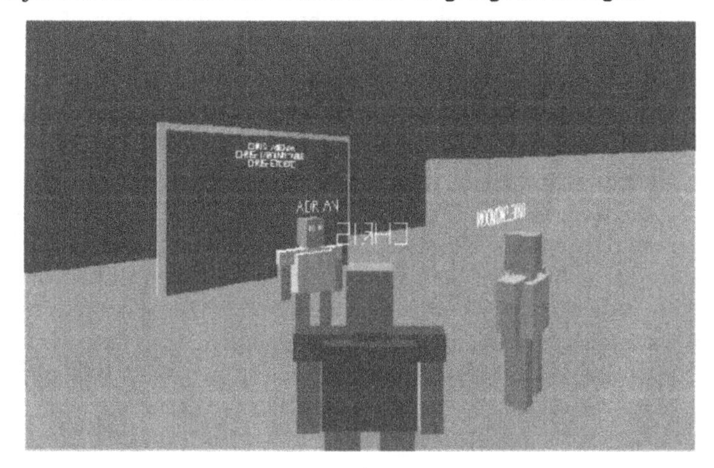

Fig. 9.3 Three MASSIVE-1 embodiments.

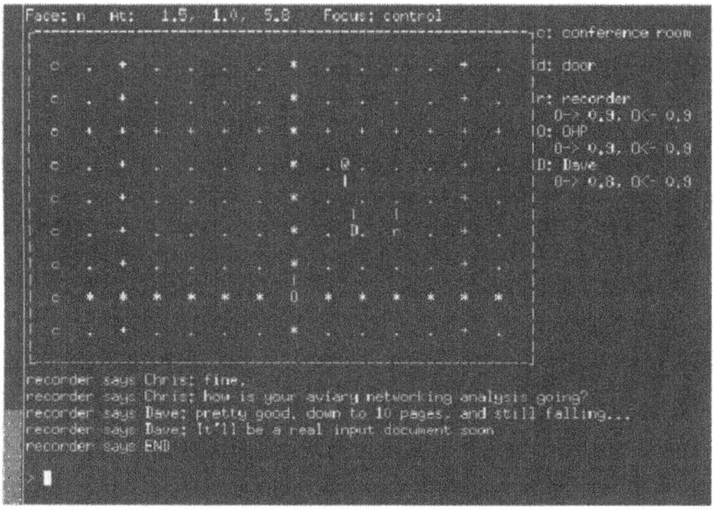

```
Face: n  At:  1.5, 1.0, 5.8   Focus: control
c: conference room
d: door
r: recorder
  0-> 0.9, 0<- 0.9
O: OHP
  0-> 0.9, 0<- 0.9
D: Dave
  0-> 0.8, 0<- 0.9
recorder says Chris: fine.
recorder says Chris: how is your aviary networking analysis going?
recorder says Dave: pretty good, down to 10 pages, and still falling...
recorder says Dave: It'll be a real input document soon
recorder says END
>
```

Fig. 9.4 The text interface.

9.2.2 Architecture of MASSIVE-1

This section focuses on the underlying network architecture of MASSIVE-1 as some knowledge of this is required in order to understand the analysis of network traffic later on (see Greenhalgh and Benford [3] for a detailed description). The processing model used in MASSIVE-1 is of independent computational processes communicating over typed peer-to-peer connections. Each computational process may support any number of interfaces, each of which is characterized by a combination of remote procedure calls (RPCs), attributes and streams in order to support a mixture of distributed VR functions (e.g. collision detection), distributed system functions (e.g. trading), and continuous media (e.g. audio).

In the interests of scalability, responsibility for detecting aura collisions and establishing connections is distributed. On entering a world, an object contacts the 'master trader' for this world and declares its interfaces (media). The master trader returns the addresses of the aura collision detectors responsible for these interfaces. At present there is a separate aura collision detector for each type of interface. These detectors monitor for aura collisions and, when they occur, pass peer interface references (combinations of IP address, port and a local interface ID) to the objects involved, allowing them to establish a connection.

Once connected, the calculation of mutual awareness levels using focus and nimbus is the responsibility of the peer objects. In the current implementation, objects are described solely by a sphere located in space, and focus and nimbus are described by scalar fields over space, which are sampled at the location of a peer object to generate focus and nimbus values. The current awareness function is multiplicative (i.e. focus and nimbus are multiplied to give awareness). This gives equal control to the observer and the observed, and is subtractive in nature, i.e. either party can force zero (no) awareness, but neither party can force awareness against the other's wishes. Current focus and nimbus functions describe conical volumes projecting forwards from an object with some additional transition volume where focus or nimbus drops off to a specified background level. The operation of focus and nimbus may therefore be controlled by a few key parameters such as conical angle, maximum radius, background level and transition rate, allowing variation from generally spherical to tightly focused regions.

Adapters exist in their own medium, complete with aura, focus and nimbus. As an object moves about, it may connect to an adapter as a result of an aura collision in the adapter medium. When the object's awareness of the adapter subsequently crosses some threshold the adapter is triggered. The implementation of adapters exploits the ability to parameterize focus and nimbus functions; when triggered, an adapter passes some new parameters back to the object across the adapter medium. These new parameters replace the object's current parameters. Thus, an adapter either may extend the range of focus and nimbus, may change their shape (i.e. conical angle), or may alter the way in which they fade to a background level. The object restores its normal parameters, when it moves away from the adapter.

9.3 OVERVIEW OF THE EXPERIMENTS

The 'Inhabiting the Web' project staged twenty virtual meetings using the MASSIVE-1 system. These meetings involved up to ten participants from BT Laboratories and the five universities listed above (section 9.1). On average, the meetings were held once a fortnight. A suite of meeting worlds and associated tasks were specifically designed for the project. The default workstation used by participants was a Silicon Graphics Indy, although a few participants had access to higher specification machines such as Indigo 2s. The meeting worlds were most often hosted on a twin processor Onyx 2 at Nottingham which was used for its computational power, not for its graphical rendering performance. The SuperJANET network provided connectivity between the partners, with appropriate tunnelling through the firewall at BT. Four kinds of data were captured from the meeting:

- direct video output from participants' machines at Nottingham and BT Laboratories;
- MASSIVE-1 log files created at each site, where each file recorded and time stamped all MASSIVE-1 events for a given local client (e.g. all movements, speech, gestures and so forth for an individual user);
- output from the 'tcpdump' command which recorded details of packet type (e.g. UDP, TCP), size and IP source, and destination addresses for packets which were observed on the local network (i.e. shared Ethernet) at Nottingham;
- completed questionnaires which were designed to capture information about users' backgrounds and experiences and which related to specific experimental tasks.

The twenty virtual meetings can be divided into four distinct phases — familiarization and training, stabilization, experimental tasks, and post-project social events. Familiarization and training involved three initial meetings with standard MASSIVE-1 worlds, where participants were given some tutorial instruction on how to navigate and control their embodiments. They also involved some exploration of possible experimental tasks. Stabilization focused on creating a stable experimental infrastructure in terms of the MASSIVE-1 software, virtual worlds, and also user skills and procedures (e.g. data capture techniques). During this time, the Inhabiting the Web virtual worlds were created and refined and a number of problems with software and networking were identified. Technical problems included configuration of the firewall at BT, problems with variable network delays and setting appropriate time-out values in MASSIVE-1's protocols, problems with overly complex worlds and/or under-powered server machines, as well as more general bugs which had to be fixed.

The experimental phase of the project involved six meetings in which different tasks were carried out. The tasks were chosen to exercise a range of different MASSIVE-1 capabilities (e.g. movement, speech, interaction with the board and the resizing of focus) as well as different modes of co-operation (e.g. team games which required co-operation within teams and also competition between them). The content of these six meetings included:

- team and individual word games (e.g. teams pairing off categories of words which were written on two boards and writing the answers on another board) — this exercised navigation, conversation, dynamic group formation (e.g. to discuss answers in teams), board usage and peripheral awareness (e.g. overhearing the other team's discussions);

- searching for objects (e.g. fragments of text) or solving a sequence of clues located in different virtual worlds — this exercised the use of portals;

- dynamically altering audio focus so as to identify individual sound sources from among a large collection of sources — this exercised focus manipulation;

- the 'balloon debate' (a hypothetical discussion of which character should be thrown out of a rapidly descending balloon) — this exercised conversation among a relatively static and focused group;

- playing hide and seek in a virtual maze and completing a virtual obstacle course — these exercised navigation and awareness;

- disco dancing — this exercised rapid navigation and movement.

Finally, at the request of the participants (several of whom had never met face-to-face), several post-project meetings have been held. To date, the focus of these meetings has been the staging of a dramatic sketch (Monty Python's Fruit Sketch).

Figures 9.5-9.8 show different scenes from the ITW meetings and virtual worlds. Figure 9.5 shows several users gathering around the table in the central meeting space where Jolanda is giving a talk. Figure 9.6 shows several participants interacting with three message-boards as part of a team game. Figure 9.7 shows the obstacle course located in the end-of-project party world. Figure 9.8 shows the final scene from the 'Fruit Sketch'.

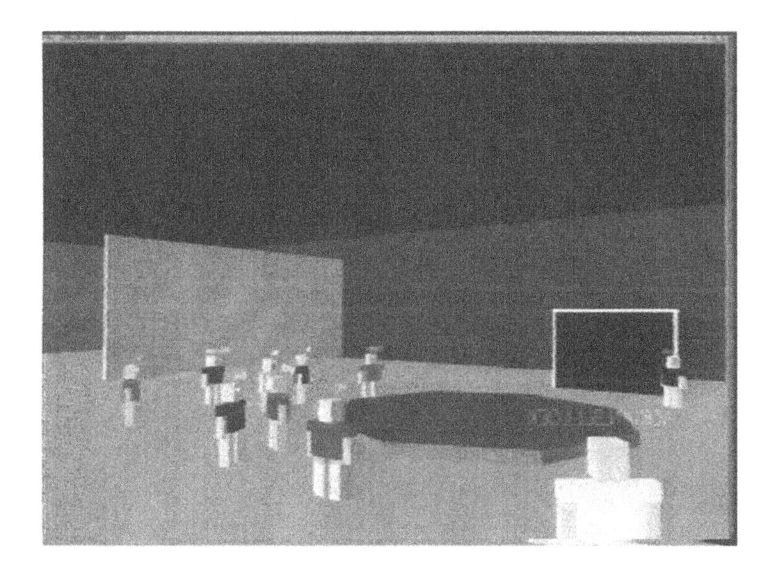

Fig. 9.5 Gathering around the meeting table.

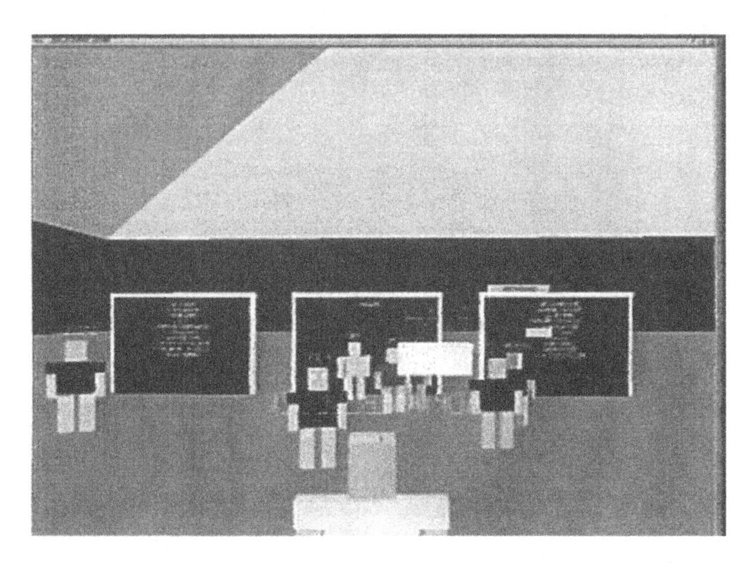

Fig. 9.6 A word game on the message boards.

Fig. 9.7 The obstacle course and disco world.

Fig. 9.8 Enacting the 'Fruit Sketch'.

9.4 NETWORK TRAFFIC ANALYSIS — USER PROFILING

A primary goal of the ITW project was to explore the characteristics of the network traffic generated by MASSIVE-1 and, as a generalization, by peer-to-peer unicast-based CVEs. More specifically, the aim was to be able to predict what volumes and patterns of traffic might be expected under different circumstances (e.g. for different numbers of users) and, conversely, what constraints on scaling would be imposed by the use of different network technologies.

The analysis proceeds in two stages. Firstly, a profile of typical user behaviour is developed which allows predictions of how an individual participant might be expected to behave (e.g. how much time they might spend moving, how often they might talk, and so forth) and also to what extent are different users' actions correlated with one another (e.g. whether participants tend to move together in groups). Secondly, this profile is combined with knowledge of the MASSIVE-1 protocols in order to predict network traffic for different numbers of users. The first part of this analysis might be expected (within limits) to apply to a wide range of CVEs. The second part is specific to MASSIVE-1, although the general pattern that emerges might apply to the class of peer-to-peer unicast CVEs and a similar process could be repeated for other systems given sufficient knowledge of their protocols. The data presented in the following is based upon six experimental ITW meetings.

Quite early in MASSIVE-1's development, facilities were added to all programs, though especially the user client programs (the visual, audio and text clients), to generate time-stamped logs of key events. The events which can be recorded include:

- starting the application;
- moving to a new virtual world (through a portal);
- moving and changing orientation within a virtual world;
- updating the graphical view (i.e. rendering a single frame);
- changing focus and nimbus settings;
- making graphical gestures;
- speaking (or, more specifically, whether the user's graphical 'mouth' is displayed);
- sending and receiving network data.

Each program generates its own independent log file, and one of the problems of analysing this information is relating the information from different hosts, especially since each machine's system clock is typically different.

9.4.1 Analysis of movement data

The area of user behaviour which has proved most amenable to analysis based on the available MASSIVE-1 data is that of user movement. In particular, from a user client application record the MASSIVE-1 event logs that user's moment-by-moment position and orientation within the virtual world, and also any change of world (due to passing through a portal). Firstly, an overview is provided of the movement data available, which is then used to address three issues for a subset of the meetings held using MASSIVE-1. The things that need to be established are:

- the fraction of time which people spend moving rather than stationary, and whether participants move simultaneously or independently;
- whether participants move through portals singly or in groups, and, if moving as groups, how long the transition takes;
- whether participants return to worlds which they have already visited, and, if so, after what period of time.

The implication of each of these is then considered:

- network and computational requirements for handling movement within virtual worlds;
- the scope for using multicast to handle inter-world transitions by a group's participants;
- whether 'world caching' (i.e. the user's local machine retaining a copy of the world for some time after they have left it in case they return again) would be effective and, if so, on what time-scale.

Figure 9.9 is a visualization of movement-related information from the ITW meeting held on 25 September 1996.

Time is shown along the horizontal axis, while a combination of participant and world is presented on the vertical axis. More specifically, the plot is divided into six horizontal bands which correspond to the six virtual worlds which were visited by participants during this meeting. The top-most band corresponds to the 'presentation world', the next to 'balloon world', and so on. Each participant is allocated a consistently placed slice within each of these bands; when a participant is present in a particular world this is represented by a horizontal bar in that participant's slice of the world band. When participants change worlds, this is represented by vertical dashed lines between the worlds in question. Finally, the width of the horizontal line which represents a participant's presence within a world is varied according to whether that participant is moving or not; a thick line or box indicates that the user is moving, whereas a thin line indicates

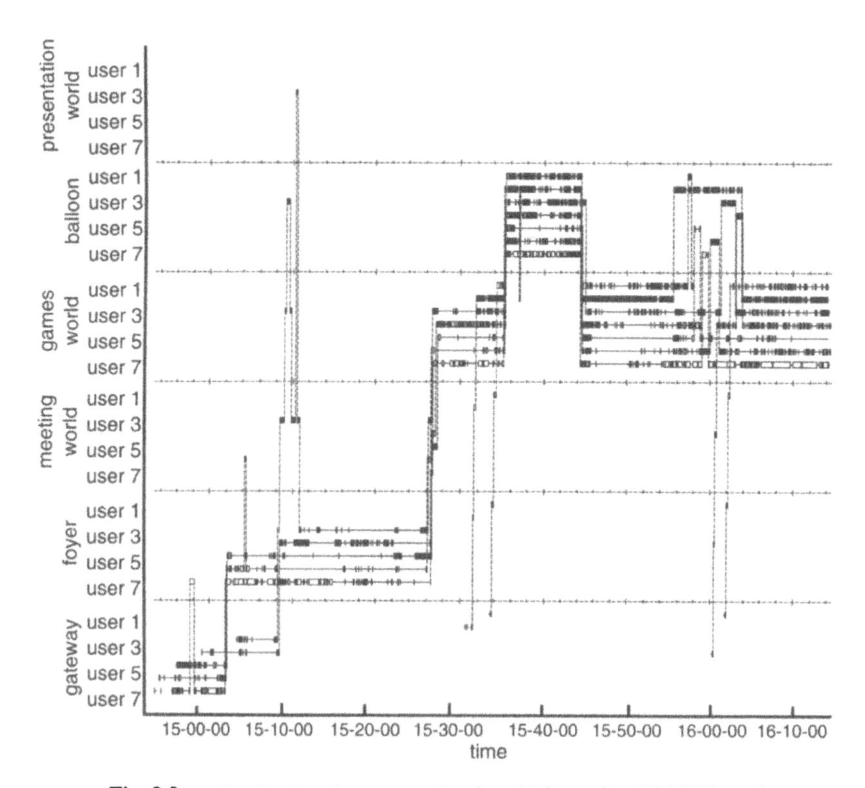

Fig. 9.9 Visualization of movement data from 25 September 1996 ITW meeting.

that they are stationary. So, considering the top-most band of Fig. 9.9, it can be seen that participant number 3 jumps from 'meeting world' to 'presentation world' at approximately 15:10, moves about for a short time (probably checking if anyone else is there) and soon returns to 'meeting world'.

9.4.1.1 Time spent moving

Figure 9.10 shows the percentage of time which participants spent actively moving within a world. Each participant was considered independently, and for each visit to a world the time for which they were moving and the total time for which they were present was established. In Fig. 9.10 the horizontal axis corresponds to the percentage of time spent moving, while the vertical axis indicates the number of person-seconds for which that level of movement was observed. Both cumulative and point distributions are shown with 1% bucket sizes.

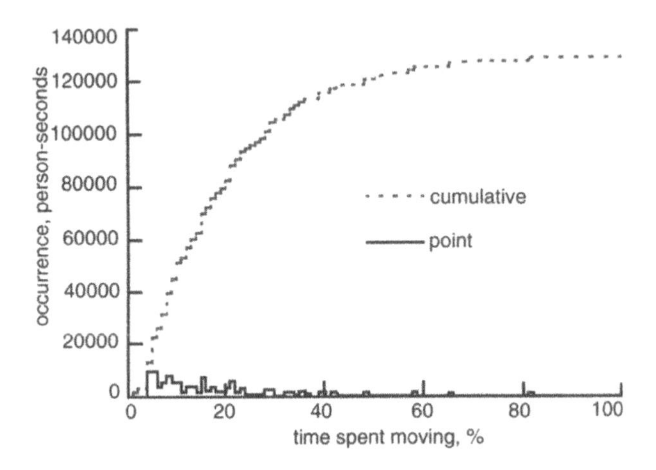

Fig. 9.10 Percentage of time present in a world which was spent in motion.

The average percentage of time spent moving for all participants and worlds is 19.6%, though it is clear from Fig. 9.10 that this measure is highly variable. Independent of world visited, the figure for each participant varies from 7.2% to 28%. On the other hand, for each world, averaging over all its visitors, the percentage of time spent moving varies from 7.5% (in the classical music world of the ITW end-of-project virtual party) to 54.6% (in the disco world, home of the disco dancing competition); in the main meeting world the average value was 16.7%.

The figures above are based on analysing each participant independently. In addition each world was analysed over the duration of the meetings to determine whether participants tended to move at the same time, or independently. Figure 9.11 shows the distribution of the number of participants in the same world at the same time who were moving. The solid line shows the observed distribution, while the dotted line shows the distribution that would be expected if all participants ignored each other and just moved when they felt like it (based on the same overall time spent moving, as previously determined). It is apparent from the graph that participants do co-ordinate their movements to some extent — the deviation from random activity is significant at the 99.9% level. However, the overall shape of the graph is very similar, and large numbers of participants moving simultaneously is possible — but less likely — in either case.

The time which participants spend moving and the amount of correlation between participants will have implications in terms of the network and computational requirements of communicating, processing and presenting each participant to the rest of the virtual world. Specifically, when a participant is moving, position and orientation updates need to be sent over the network and

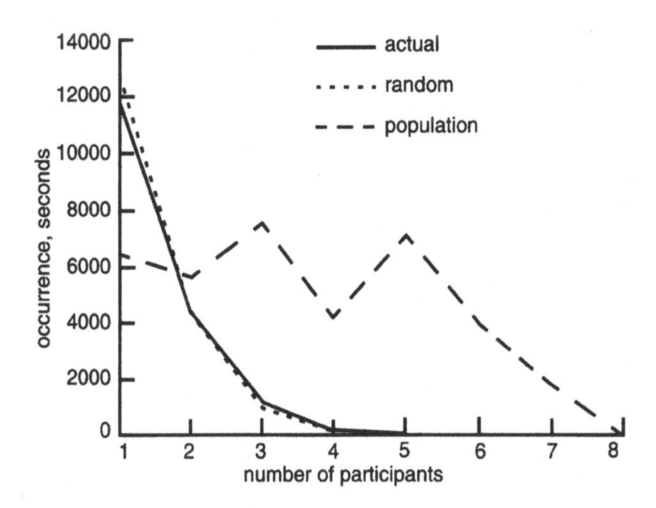

Fig. 9.11 Distribution of numbers of virtually collocated participants moving simultaneously.

received and processed by each observing process. When dealing with non-deterministic agents (e.g. people), techniques such as dead reckoning may — or may not, depending on the application — reduce the number of such updates; however, they can never entirely eliminate the need for some of them. Once a participant is stationary no further positional updates are necessary until they start to move again. Consequently the amount of time which participants spend moving can be a significant factor in assessing network and computer requirements.

The overall average percentage of time spent moving observed in the six meetings being analysed here was 19.6%, as noted above. This may be taken as a base-line value. However, there is also a large variation between individuals and also a significant task dependence, as indicated by the differing averages in different worlds. Consequently, the figure arrived at here should be treated as something of a rule of thumb, rather than a definitive answer. Additionally, when considering the combined instantaneous load due to several participants, it has been noted in this analysis that it is not valid to assume that their activities will be independent and uncorrelated. Rather, at least when participants are involved in a common task, there is a small but highly significant element of correlation, statistically, between their activities.

This argues for additional caution when considering, for example, the scope for exploiting statistical multiplexing of movement-related traffic for larger numbers of users (i.e. being able to require or reserve less bandwidth on the basis that while some users are moving — and generating network updates — many other users will not be).

9.4.1.2 Group world transitions

Having discussed one key aspect of movement within a world, two aspects of moving between worlds are now considered. Firstly, the likelihood and form of co-ordinated inter-world transitions by groups of participants is considered, and, secondly, whether and in what circumstances participants return to previously visited worlds.

It would be expected from some of the activities organized in the meetings being analysed that group world transitions would occur. For example, participants would typically gather initially in the gateway world and wait for the others to arrive. Then at some point the meeting organizer would invite everyone to go through to the meeting world for the formal start of the meeting, and all of the participants would move — in a vaguely co-ordinated fashion — through to the meeting world. The purpose of this aspect of the analysis is therefore not simply to discover whether such transitions occur — it is already known that they do; rather, it is the significance and character of such transitions that needs to be assessed.

For the purpose of the automated analysis, a group world transition is defined as an event in which two or more participants who are in a world at the same time move via a single portal-jump to another world so that they are together again. Figure 9.12 shows the incidence of singleton and group world transitions in the meetings analysed. The solid line shows the number of incidents, while the dashed line shows the total number of participants involved in those incidents.

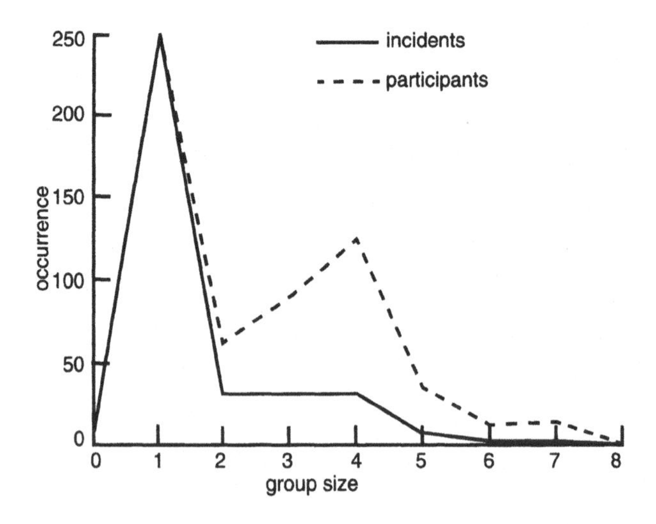

Fig. 9.12 Distribution of group size for group world transitions.

The underlying data can be summarized as follows:

- participants jumped to new worlds on a total of 584 occasions, of which 337 (58%) were in groups of two or more;
- individuals or groups made world transitions on 350 occasions, of which 103 (29%) were group transitions and the average size of those groups was 3.27 participants.

Figure 9.13 shows the distribution of world entry delay for participants involved in group transitions, i.e. for each member of a group (excepting the leader) it shows how much time elapsed between the group leader and the group member reaching the destination world. The range of delay shown on the graph, up to 30 seconds, accounts for 203 (87%) of the 234 non-leading participants to make group world transitions; 104 (44%) of these occur within 5 seconds, while 159 (67%) occur within 10 seconds.

Group world transitions are important when considering the design and requirements of this kind of system, because moving to a new world is a significant event which will almost always involve an exchange of data between the participant's processes and the rest of the system. In MASSIVE-1, for example, on entering a new world the user's client applications terminate connections to objects in the old world, are informed of the existence and identity of nearby objects in the new world, establish network associations with each other, and exchange general and medium-specific information such as location, awareness, name, graphical appearance, etc. This can result in a significant but transient burst of network traffic and implied load on other processes.

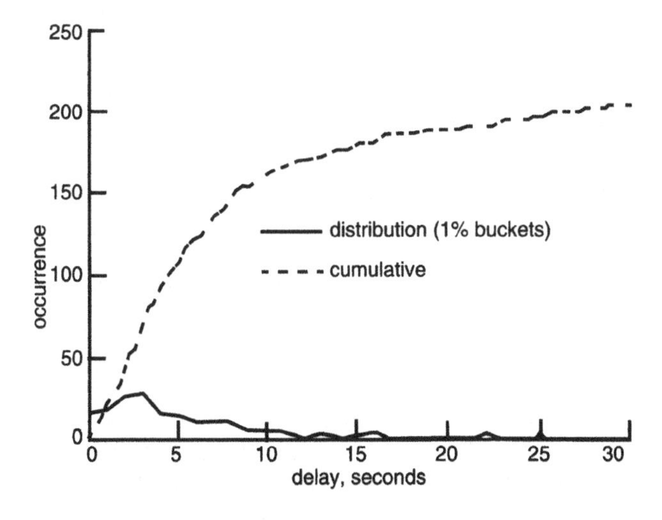

Fig. 9.13 Distribution of group member arrival delay for group world transitions.

Different systems will organize this information in different ways and obtain it from different sources, but there will still be some requirement for the participant's applications to learn about the new world. This makes the occurrence of group world transitions important in two respects:

- a co-ordinated movement by a large group could generate a much greater transient load than might be expected if inter-world movement were assumed to be independent and uncorrelated;

- specialized system support for group world transitions (for example, based on the use of network-supported multicasting of new world information to all group members) could not only alleviate this problem, but also reduce the total load relating to world transitions when compared to a model of independent and uncorrelated movement.

For example, for the meetings analysed, imposing a world transition delay of 10 seconds would both ensure that state transfers for the same world did not need to be performed more than once in any 10-second period, and would require approximately 325 unicast and 100 multicast state transfers rather than 584 unicast state transfers. Imposing longer delays on world transitions would increase the effectiveness of group transfers, by allowing more transitions to be effectively grouped, whereas shorter delays would include fewer transitions.

Before moving on to consider participants returning to worlds, a little must be said about the general applicability (or otherwise) of this result. As was noted at the beginning of this section, group world transitions were an organized aspect of the activities being analysed — will they occur in other applications and situations?

Such an question cannot be answered definitively without gathering a great deal more data about a wide range of different applications and scenarios. However, some more subjective and tentative observations can be made:

- MASSIVE-1 is not alone in adopting a multiple world model with portals between worlds (see, for example, DIVE [6]);

- the world and portal model was widely accepted and effectively employed by participants.

It may be argued from these observations that a multi-world structure is a more generally useful and appropriate virtual design style. So it may be anticipated that group world transitions will occur in many applications involving formal or informal co-operation or interaction, e.g. as common interest groups form and dissolve, or as time-linked activities such as performances and meetings begin and end.

9.4.1.3 Returning to worlds

The final aspect of participant movement to be considered in this analysis concerns the incidence and character of return visits to virtual worlds, i.e. when a participant visits the same virtual world on more than one occasion in the same meeting.

Figure 9.14 shows the cumulative distribution of time elapsed between consecutive visits by individual participants to any world. There are a total of 353 return visits in the data analysed compared with 231 first visits. The average time lapse between leaving and re-entering the same world is just over 7 minutes, while half of these return visits occur within 3 minutes.

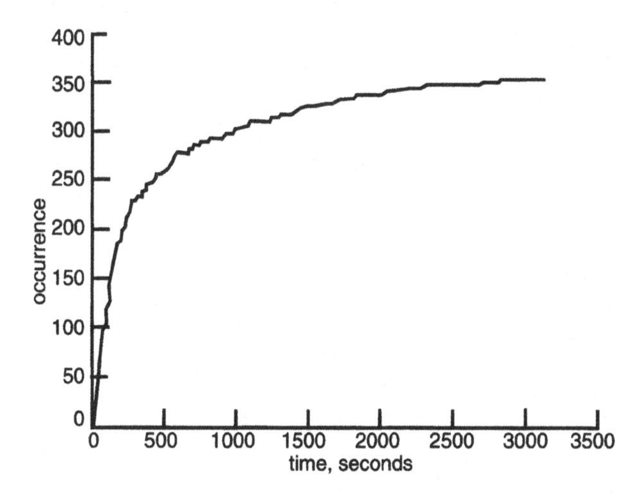

Fig. 9.14 Distribution of time elapsed between consecutive visits to the same world.

The occurrence of revisiting worlds is important for the same fundamental reason as group world transitions were considered — because changing worlds is a significant event which implies an exchange of information with corresponding requirements for network and computing resources. In particular, world return visits are important because of the implications which they have for the possibility and utility of world state caching in (or near) participant's applications. For example, for the data analysed, if each user client maintained a cache of world state for at least 6 minutes after visiting a world and if the worlds did not change on this time-scale then approximately 40% of new world state transfers could be satisfied from the cache.

The demonstration in the above example of the potential utility of world state caching must be qualified in two areas — the extent to which return rates can be generalized from the specific meetings analysed, and the extent to which worlds can be expected to remain static over the time periods in question and how this can be established by the applications.

Dealing with these issues in turn, it is clear that, if return visits were more common or sooner, caching would become increasingly effective, and conversely, if return visits were rarer or more delayed, caching would become less effective. As with group world transitions, it may be argued that most world return visits in the meetings considered were peculiar to the style and organization of these meetings and not generalizable. However, there is again an admittedly limited counter-argument that these worlds and meetings were not established to demonstrate these effects, but that these effects emerge from natural choices of world and meeting structure which may also occur in other systems and applications. For example, the worlds for the ITW meetings were structured (for the most part) in a hierarchy, which is a very common method for organizing related objects. A natural consequence of this is that the worlds closer to the root of the hierarchy were visited more frequently and were returned to more often, e.g. by participants in the process of moving from one task-oriented world to another; so it might be expected that revisiting of worlds will be a general effect.

Secondly, the potential validity of cached data — and establishing its validity or otherwise — is more clearly a general issue. In MASSIVE-1 worlds, the background content is static while the participants are highly dynamic. In other systems and applications the differences may be less clear cut. In any case, for caching to be effective there must be (significant) elements of world state which do not change between visits, and there must be some well-defined method of establishing what has not changed, and of efficiently combining cached and new information. This is left as an exercise to the reader.

9.4.2 Analysis of audio data

In addition to the analysis of movement and world transitions presented above, there has also been an analysis of audio-related activity in the same six MASSIVE-1 meetings. The figures that were available relating to participants' audio activity were records of audio data packets captured in the network traffic log, and events in the MASSIVE-1 user client log files indicating when the visual 'mouth' was visible.

Each of these sources of data corresponds to a simple threshold test of audio volume. The visual mouth appears when the sound captured by the participant's microphone exceeds an experimentally chosen level, while the audio service sends audio packets (which are recorded in the network traffic log) when the sound captured by the microphone exceeds a lower experimentally chosen level. The aim here is to be generous in transmission (i.e. to transmit anything that might possibly be speech), but to only provide visual feedback via the 'mouth' when there is a high volume level so as to encourage people to speak up.

These sources of data will each be considered in turn before joint conclusions are drawn.

The data for logged network audio traffic from 'tcpdump' should include all audio in the same world as the principal meeting organizer. It extends to 33 198 person-seconds of audio from 125 571 person-seconds of apparent presence, i.e. the average speaking proportion is 26.44%. Apparent presence of audio participants is deduced from periodic audio timing packets which are sent by the audio service even when full audio data is not being transmitted. Figure 9.15 shows the number of simultaneous speakers for the data set (solid line) and the expectation if speaking was an independent event, with no correlation between virtually collocated participants (dotted line). The dashed line shows world population; if all the participants spoke continuously then this curve would result. The observed distribution differs from the uncorrelated case at the 99% significance level, but the difference at each data point is small — the difference is only significant because of the very large sample size.

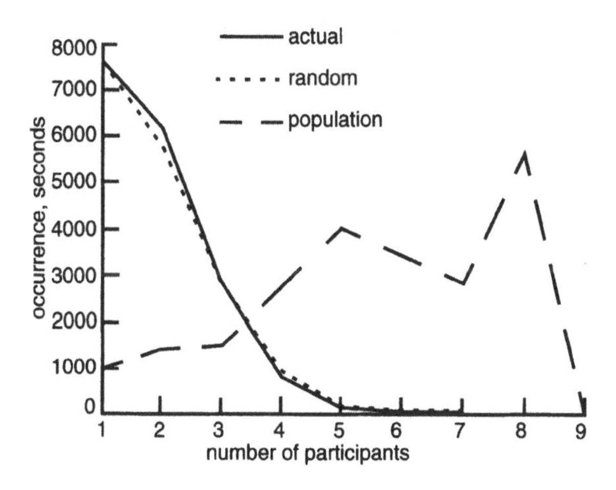

Fig. 9.15 Correlation of network audio packets.

The data concerning the visual representation of a 'mouth' (actually more of an abstract speech balloon) is available where participants have correctly enabled MASSIVE-1's own logging facilities and have returned the event log files. The total participant time covered, 129 550 person-seconds, is similar to the network audio data. However, the amount of time for which the mouth is shown is much less — 9355 person seconds, or 7.22%. Figure 9.16 shows the number of simultaneous 'speakers' (actual and uncorrelated expectation) and world population for this data set.

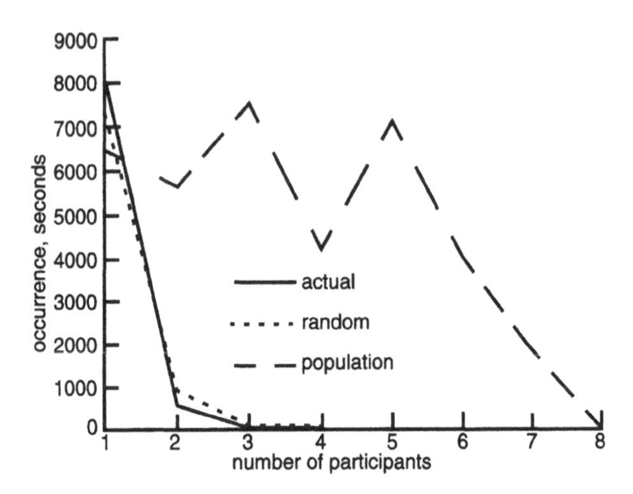

Fig. 9.16 Correlation of visual 'mouth' data.

As for the network audio data the deviation of the observed number of simultaneous speakers from an uncorrelated distribution is significant at the 99% level, but small in percentage terms. Complete event logs for all participants are not generally available, and this is reflected in the shift of the world population distribution towards the left (fewer simultaneous participants) when compared to the network audio data. The distributions for speaking drop off much more steeply than for the network audio data partly because of this, but primarily because of the much smaller amount of speaking recorded.

Before the analysis, it was anticipated that the data for speaking might exhibit significant anti-correlation, reflecting the assumption that one participant would be less likely to speak when another participant was already speaking. In fact this was seen to a limited extent in the visual 'mouth' data, but not at all in the network audio data. At first sight this might suggest that the participants are ignoring one another, or that the network delays might cause a break-down in normal conversational turn-taking. However, a subsequent comparison between segments of the logged data and the video recordings revealed a more mundane reason — the majority of the apparent 'speaking' was actually due to background noise (e.g. typing, other activities in the office, breathing noise) and feedback from open speakers. This was especially true for the more sensitive network audio data. A number of observations and reflections can be made:

- use of open speakers (at least without sophisticated echo-cancellation techniques) can increase the apparent number of speakers and the resulting network load by replaying sounds back into the system;

- using more aggressive silence detection algorithms could reduce network traffic (moving the 26% speaking rate for network data towards the 7% speaking rate for the 'mouth'), but at the risk of missing significant quiet utterances and non-speech noises (e.g. hmms, grunts and ahs);
- whatever the participant's 'real' behaviour in terms of speaking (as determined by a human expert, for example), it is the behaviour that may be deduced automatically (as here) that is significant for assessing network and computational requirements.

Elements of the above analysis will be used in the next section which develops a network traffic model for MASSIVE-1.

9.5 A PREDICTIVE MODEL OF NETWORK TRAFFIC

This section describes the development of a model of the network traffic required to support a MASSIVE-1 teleconference. A key element of this analysis is consideration for the way in which requirements change with the number of users. MASSIVE-1 has not been used for very large numbers of simultaneous users — the largest meeting to date involved ten participants. However, this traffic model allows us to consider the effects of much larger scale usage. Firstly, there is a description of the key elements of the user model employed and the additional assumptions required, and then briefly the main factors contributing to expected network traffic; these are combined with the user model to give a simple approximation of expected total network traffic. The section concludes by reflecting on the resulting traffic characteristics.

The key elements of the model and additional assumptions required are listed below:

- all users have a full complement of text, graphical and audio clients and use standard embodiments;
- the most significant user events are moving and speaking (text messages and gestures are ignored) — for these, users move 20% of the time (section 9.4.1) and speak (or rather send network audio data) 25% of the time (section 9.4.2);
- the average frame rate is 6 Hz (observed in use on a Sun 10ZX, and adequate for normal use);
- users move between worlds or groups of users such that they change the peers with which they interact once per minute (this value is somewhat higher than might be the case in the experiment meetings, but it will be seen that the related traffic component remains a relatively insignificant part of the total);

- all users interact with (on average) M other users, e.g. users might organize themselves into variable and changing groups of average size $M+1$ participants;
- the contribution of background objects is much less than that of users, and can be ignored (in MASSIVE-1 a background object generates approximately one tenth the traffic of a participant).

In MASSIVE-1 the main causes of network traffic are co-ordinating multiple user clients, updating the aura collision manager (due to movement, world transition or aura adaptation), establishing new associations with other users and objects upon aura collision, and interacting with other users and objects while in aura range. Of these four items, the first two are independent of the number of other participants using the system; the third — establishing new associations — depends on the rate at which groups or associates change; the fourth — interaction — depends on the number of other users and objects within aura range at any time (denoted by M).

The three key events considered are user movement, speech, and the arrival of a new interaction partner (i.e. an aura collision and subsequent data exchange). Table 9.1 shows the basic traffic generated by each of these events independent of any particular assumptions about user behaviour; each has a component which is per user only (the upper row), and a component which is also dependent on the number of peers which each participant has (the lower row). This is combined with the user model (average movement rate of 1.2 Hz, speaking rate of 25%, peer change rate of once in 60 seconds) to give the average resulting network traffic bandwidths in Table 9.2.

Table 9.1 Network traffic resulting from key events.

	Movement (kbyte/step)	*Audio (kbyte/sec)*	*New peer (kbyte/peer)*
Per user	1.2	0	2.1
Per peer per user	2.1	8.3	13.2

Table 9.2 Average network bandwidths.

	Movement (kbyte/sec)	*Audio (kbyte/sec)*	*New peer (kbyte/sec)*	*Total (kbyte/sec)*
Per user	1.4	0	< 0.1	1.4
Per user per peer	2.5	2.1	0.2	4.8

9.5.1 Traffic model

From the final column of Table 9.2 an overall expression is obtained for the total expected network bandwidth (in kbyte/sec) as a function of the total number of participants, N, and the average number of other participants in aura range, M:

$$B = N\,(4.8\,M + 1.4) \qquad \text{... (9.1)}$$

For constant group size (i.e. constant M) this will be proportional to N, the total number of participants. On the other hand, if all users are in aura range (i.e. effectively a single group) then $M = N - 1$ and the maximum possible bandwidth is obtained:

$$B_{max} = 4.8\,N^2 - 3.4\,N \qquad \text{... (9.2)}$$

This is equivalent to not making use of aura or multiple worlds to structure and restrict interaction. Figure 9.17 shows the resulting total bandwidth against the number of simultaneous participants for different group sizes (different values of M).

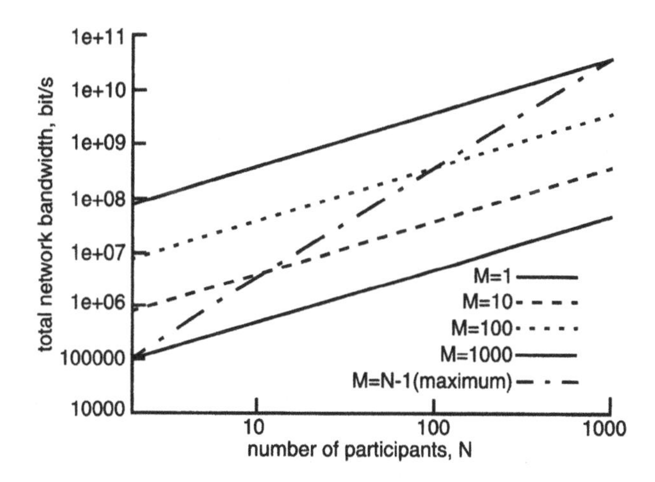

Fig. 9.17 Total network bandwidth against number of participants for a range of group sizes.

Figure 9.18 shows the trade-offs between total number of participants and average group size for a range of total network bandwidths.

The assumptions that underlie this simple model of total network traffic have been made explicit, and, of course, they limit the accuracy or appropriateness of the model in other situations or scenarios and should be borne in mind when

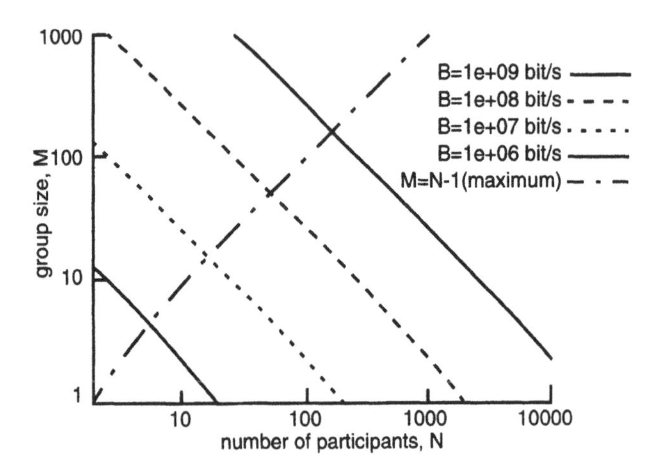

Fig. 9.18 Trade-off between number of participants and group size for a range of total network bandwidths.

applying these results. The main observations concerning the scope of applicability of the model are listed below:

- different levels of user activity will result in different numbers; however, the overall form of the results will be the same;

- different systems will use different protocols, giving rise again to different exact numbers — in particular, MASSIVE-1 may be seen to be rather inefficient (e.g. duplicating movement messages for each medium); however, it is relatively uniform in its inefficiency, so that again the same kind of behaviour would be observed, albeit for slightly greater numbers of participants or group sizes;

- as noted, MASSIVE-1 uses unicast-based peer-to-peer networking — the model and inferences are specific to this approach;

- the impact of background objects has been ignored, which may be appropriate in a teleconferencing application, but would not be appropriate in a database visualization with thousands or millions of non-user objects — to apply the model in this kind of situation these objects would need to be taken into account;

- a certain rate of change-over of peers and of peer complexity (e.g. geometry size) has been assumed — with these assumptions the contribution to total traffic of establishing new peer associations is very small (less than 5%); however, a combination of rapid change-over and much greater peer complexity could dramatically increase this traffic component and the total requirement.

Nonetheless, the model (and MASSIVE-1 itself) remains representative in form, if not fine detail, of any CVE system architecture based on unicast peer-to-peer communications in conjunction with spatial trading.

9.5.2 Use of aura

The model and expected network traffic demonstrate the potential of auras to reduce network requirements and/or increase the potential number of participants when compared to a system without auras. Most significantly, the use of auras reduces the total network bandwidth from $O(N^2)$ (equation (9.2)) to $O(NM)$ (equation (9.1)). The management traffic overhead for interacting with the trader is not large (it is one component of the per-user traffic in Tables 9.1 and 9.2) and is only $O(N)$. Using auras to scope (and limit) interaction in this way has the same effect as enforcing a reasonably even distribution of participants between different worlds. However, auras and spatial trading have significant advantages over the simplistic approach of splitting participants according to world in that they naturally include division by world as a special case, they allow gradual, natural and visible transitions between groups within the same world, as opposed to sudden and discrete jumps between worlds, they avoid the need for invasive system intervention such as barring access to busy worlds, and they support flexible control and graceful degradation through interactive (and potentially automatic) modification of aura, e.g. pulling in one's aura to reduce system and network load in a busy area.

9.5.3 Scalability of unicast-based CVEs

Even with spatial trading to manage the number of active peers, it is clear from this traffic model that peer-to-peer interaction is the dominant component of network bandwidth, even with relatively small numbers of active peers (this is the per-peer, per-user component in the model, which is $O(NM)$). In a CVE there are elements of interaction which are specific to a single pair of peers; this is especially the case in a CVE based on mutually negotiated awareness, as MASSIVE-1 is. However, the normal concept underlying a CVE is of a shared and primarily objective virtual world — participants can agree about the locations and properties of objects and each other. This objectivity and agreement are significant for enabling co-operation. Consequently, much if not all of the information being communicated between participants' processes is objective in character, i.e. observer independent.

The unicast-based networking employed in MASSIVE-1 restricts each network packet (each message or block of information) to be delivered to a single destination. Consequently, to send the same information to a number of destinations (such as a number of peers) requires one packet to be sent to each destination. These packets will normally travel over the same network links and segments for some or all of their journeys. It is this duplication of packets which gives rise to the $O(NM)$ limiting term in the network traffic model.

However, network multicasting allows the same packet to be delivered to potentially many recipients. With shared physical medium networking technologies, such as Ethernet and FDDI, the same packet can be directly received by any number of machines on that network. The addition of multicast routers allows a single packet to be distributed to many recipients over wide-area and inter-networks. Because the multicast routers defer duplication of packets until the last possible moment (i.e. when the network divides), duplicate packets can be avoided on all network links and segments [7].

Ideal multicasting could reduce the total network bandwidth (in kbyte/sec) to:

$$B = N\,(0.2\,M + 6.0) \qquad \text{... (9.3)}$$

This is shown in Fig 9.19, which is directly comparable to Fig. 9.17 for unicast-based networking.

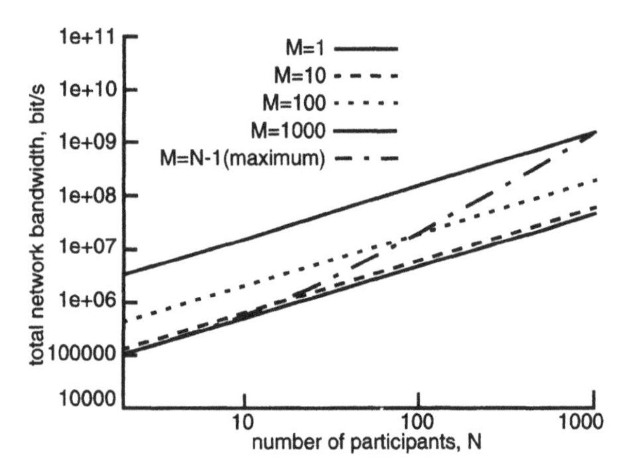

Fig. 9.19 Total network bandwidth against number of participants for a range of group sizes assuming ideal multicasting.

All updates (movement and audio) are assumed to be ideally multicast. The remaining much-reduced $O(NM)$ term corresponds to forming the initial association with a new peer and exchanging information, such as appearance. There are a number of optimistic assumptions in this new model which would tend, in reality, to increase the $O(NM)$ term:

- this ignores the possibility of specifically peer-to-peer elements of information exchange which is inherently $O(NM)$ (this might include aspects of awareness negotiation, for example);

- it also ignores any multicast management overhead associated with joining or leaving groups;
- it assumes reliable multicasting — for at least some updates, though probably not audio — with no overhead or per-receiver growth in traffic;
- it assumes (as does the model already presented) that total network traffic is the most important factor — this assumption would seem to be more realistic for the case of unicast network architectures;
- it depends strongly on the rate at which new peer associations are formed, which will be highly application and context dependent.

Nevertheless, the potential improvement is dramatic — for an average group size of ten it gives a six-fold reduction in required bandwidth (or equivalently six times as many participants in the same bandwidth); for larger group sizes the benefits are greater still, up to a factor of more than 20 times. So for an objective CVE the potential scalability of a unicast-based system will typically be much less than for a multicast-based system, as far as total network bandwidth is concerned. This observation has been a key factor in motivating the development of MASSIVE-2 as discussed at the end of this chapter (section 9.7).

9.6 USABILITY

A brief summary of the usability issues raised by the ITW experiments is now presented. A replicated time-series design was followed for the usability analysis, in which the repetitive collection of data involved constructs such as satisfaction, experience, ease of use, group involvement and awareness. In addition, each session was used to explore new constructs such as dealing with multiple virtual embodiments, distributed awareness and switching between the virtual environment and the real environment. Human behaviour and experiences were captured using questionnaires, and video-recordings were made of each experiment. Over one hundred questions were asked in total during the experiments. The participants were all researchers in the area of virtual environments and so had a high level of familiarity with the underlying concepts. Several of them had considerable prior experience with MASSIVE-1 but a few had never seen the system prior to the project.

The questions were collected through the World Wide Web, and consisted of attitude statements with Likert-scales, and open-ended questions, e.g. after an experiment aimed at learning a feature of the interface called 'focus', which allows a user to change their degree of awareness of each other and the environment: 'How easy or difficult do you find it to use the 'focus' commands? Please score on a scale of 1 to 7, if 1 = very difficult, and 7 = very easy,' and

'Please explain how you would change the commands for the 'focus' feature to improve its ease of use.' Interviews were used as a follow-up for some of the more interesting observations. The remainder of this section focuses on three of the key issues that emerged from this work — user embodiment, navigation, and the use of the spatial model — drawing out some implications for future CVE system design.

9.6.1 User embodiment

Many of the findings come down to the suggestions for improvements of the degree of control of users' embodiments, of missing non-verbal interaction mechanisms, and of the lack of natural mappings to embodiment controls. The need for improved embodiments expressed by the subjects in the ITW study suggest that automatic embodiment behaviours ought to be introduced, in order to allow higher level interaction abilities. For the purpose of this chapter, the suggested improvements can be summarized as follows:

- supporting automated actions, e.g. gestures for waving, nodding, running, walking and facial expressions, and also support for higher level automated actions, such as the actions a chairperson would want to use to open a meeting;
- objects should react to the user, instead of the user to the object in order to interact with it, e.g. having the object's functions triggered and made available by approaching the object;
- buttons for often-used behaviours, e.g. having an 'applaud' button which produces the sound of hand clapping — the more users press the button at the same time, the louder the applause;
- transparency of control, e.g. it should be obvious who is the current speaker, or who is currently manipulating an object, and a user should be able to go against the automatic behaviour of an object when appropriate;
- passive capture of often-used behaviours, e.g. in a virtual meeting users want to take their usual place around the table, and adjust their view to be able to see all other participants around the table.

The changes to MASSIVE-1 which are needed to satisfy these demands are best addressed at the level of the user embodiment. By providing the user with a programmable interface to the embodiment, the users should be able to create their own automatic behaviours, control their embodiment's reactions to objects, and create sets of automatic actions which they feel are appropriate for certain recurrent tasks. Users should also be able to share such sets of behaviours with each other. This creative editing and inheriting of automatic actions should also contribute to a naturally evolving CVE on a social and functional level.

Another more radical option would be to introduce some kind of immersive or semi-immersive interface, perhaps through the use of a head-mounted display for the former or through the use of magnetic trackers to capture users' gestures in front of a large projected interface for the latter. This possibility was mentioned repeatedly by several users, and could make an interesting contrast to the semi-automated approach to embodiment control.

9.6.2 Navigation

Moving about the virtual environment provided many users with a significant challenge. In particular, difficulties with fine-grained movements, such as aligning to other people, tables, boards and other objects, proved to be a major source of frustration. Several design options might be explored in order to address these problems:

- automatic alignment with designated objects such as meeting tables or other users — sophisticated variants of this strategy might also align the user with other users who were themselves aligned with the object, e.g. on approaching a meeting table one would be steered towards a vacant position that afforded the best possible view of the other participants; such alignment might also operate dynamically as in the example of being automatically aligned to face the current speaker (although such ideas would need careful implementation);
- automatic tracking of moving objects as a further extension to automatic alignment;
- object-centred navigation, as a logical extension of the above, where a user's movements would all be relative to a designated focal point, e.g. after designating a new focus, one would be able to sweep around it and zoom in and out on it without ever losing it from sight.

As a further note, different tasks might require different forms of navigation. For example, movements around a meeting table might suit an object-centred style, whereas navigation of an open space might suit a more traditional Cartesian navigation style. This leads to a final design proposal of context-sensitive navigation, where different styles of navigation (and indeed other forms of interaction), would be triggered within different contexts. For example, on approaching a meeting table, one might automatically swap from Cartesian to object-centred modes of navigation.

9.6.3 Visibility and malleability of spatial model mechanisms

The final issue arises from the use of the spatial model concepts of focus and nimbus in interacting with the board object and in the task where users have to selectively filter a collection of audio sources by moving about a world with a narrow audio focus setting. Users found it difficult to reason about the effects of focus and nimbus or about awareness relationships with other users and objects. For example, it was difficult to tell whether a message board had been successfully engaged or whether another user was aware of oneself. Although this information was available in the text interface and a range of basic settings were provided for focus and nimbus, it is proposed that, where such mechanisms are used, it is important that they are made sufficiently visible and easily malleable, e.g. other users' embodiments might depict the level of their awareness of the observer for a range of different media.

9.7 CONCLUSIONS AND RECENT DEVELOPMENTS

This chapter has presented the results of a series of experiments with the MASSIVE-1 collaborative virtual environment running over the SuperJANET network between BT Laboratories and five UK universities. The main objective of the experiments was to gain an understanding of the kinds of network traffic generated by MASSIVE-1 and, by extension, by the class of CVEs based on a unicast peer-to-peer network architecture. A secondary objective was to explore some of the usability issues associated with this technology.

The consideration of network traffic issues involved an analysis of two kinds of data — multiple system log files generated by different user clients and output from the 'tcpdump' network monitoring command — and proceeded in two stages.

Firstly, a general profile of user activity was developed which provides data on the typical actions of users across a range of tasks (i.e. how often do people move, talk, change worlds, etc) and importantly, the extent to which these different actions are correlated. Secondly, the user profile is combined with information about MASSIVE-1's network protocols in order to develop a model of predicted network traffic for varying numbers of users in different sizes of mutual awareness groupings. Key findings for this work were that:

- on average, users move about 20% of the time and speak (i.e. send network audio data) about 25% of the time, although these figures vary with task;
- when participants are involved in a common task, there is a small but, statistically, highly significant element of correlation between their activities — this implies that some caution may be needed when considering the use

of techniques such as statistical multiplexing of the associated network traffic;

- both world transitions and the frequency and timings of return visits to worlds are important events as they might be expected to generate considerable network traffic, group transitions (i.e. two or more participants using a portal within a short space of time) being a common occurrence as was returning to previously visited worlds — given that these results generalize to other applications, this has important implications for the design of 'group portals' which, at a cost of small delays to individuals, enforce group transitions which can be efficiently dealt with at the network level (e.g. by multicasting world state information) and also for the use of caching;
- the use of more aggressive silence detection algorithms could help reduce audio traffic, although at the risk of missing potentially significant non-speech and background noises.

The predictive model of network traffic for MASSIVE-1 has led to two main findings. Firstly, aura is a useful concept for limiting network traffic in CVEs. For an application involving N simultaneous users, auras would reduce the total traffic from being $O(N^2)$ to being $O(NM)$, where M is the number of users in a mutually aware group of intersecting auras. Secondly, the use of network multicasting would be a significant factor in further scalability. For example, given an average group size of ten participants, there could be a six-fold reduction in required bandwidth rising to a factor of more than twenty times for larger groups.

Although specific to MASSIVE-1 and to the kinds of task undertaken in the experiments, it is believed that these findings will be broadly consistent with the behaviour of unicast peer-to-peer CVEs used for a range of different activities, and also that this general method of analysis should be of interest to those engaged in analysing other classes of CVE or even other kinds of co-operative system.

The exploration of usability issues mainly involved the analysis of questionnaires. Key issues raised by the participants include the need for more expressive and controllable embodiments, problems with navigation (especially managing 'fine-grained' alignment to objects and other participants), and the need to make spatial model concepts such as focus and nimbus more visible and malleable at the user interface. These findings suggest several new design possibilities for MASSIVE-1 including the association of more powerful animations with embodiments and also options for automatic alignment, object tracking, and object-centred and context-sensitive navigation.

In conclusion, two recent developments, which follow on from the ITW project, are mentioned here. The first of these is the development of the

MASSIVE-2 system (also sometimes known as CVE) [8]. MASSIVE-2 represents a considerable extension of MASSIVE-1 both in terms of functionality and network architecture. Key extensions include:

- an extension to the spatial model of interaction-called third parties which support indirect awareness relationships between objects and which can be used to model a range of contextual and environmental effects on communications, with third party objects able to be used to model a diverse range of structures including:

  — nested regions whose boundaries have varying degrees of transparency;

  — mobile crowds which provide aggregate views of the inhabitants;

  — shared objects which manipulate the mutual awareness of their users;

  — group vehicles;

- the mapping of this extended model on to an underlying multicast architecture — essentially, the dynamic and nested structure of a world (e.g. nested regions and mobile crowds) is mapped on to a self-configuring hierarchy of multicast groups, with protocols then being defined to manage group membership as a result of movement within a virtual world;

- support for direct interaction with the world and additional aids to navigation as well as the provision of a general application development environment.

As a final note, at the time of writing, Nottingham University and BT Laboratories are establishing a further collaboration in the form of the eRENA project within ESPRIT IV Long Term Research. eRENA will involve a collaboration between technologists, artists, broadcasters and social scientists to explore how large-scale CVEs such as MASSIVE-2 can be used to support new forms of cultural experience in areas such as art, entertainment, performance and so-called 'inhabited television' — the deployment of very large scale CVEs into the home as a new form of socially participative television.

REFERENCES

1. Benford S D, Bowers J M, Fahlén L E, Mariani J and Rodden T R: 'Supporting co-operative work in virtual environments', The Computer Journal, 37, No 8, Oxford University Press (1994).

2. Rogers A S: 'Virtuosi — virtual reality support for groupworking', BT Technol J, 12, No 3, pp 81-89 (July 1994).

3. Greenhalgh C M and Benford S D: 'MASSIVE: A virtual reality system for teleconferencing', ACM Transactions on Computer Human Interfaces (TOCHI), 2, No 3, pp 239-261, ACM Press (September 1995).

4. Heath C and Luff P: 'Collaborative activity and technological design: task co-ordination in the London Underground control rooms', in Bannon L, Robinson M and Schmidt K (Eds): 'Proceedings ECSCW '91', Kluwer, Dordrecht (1991).

5. Benford S D and Fahlén L E: 'A spatial model of interaction in virtual environments', Proc Third European Conference on Computer Supported Cooperative work (ECSCW'93), Milano, Italy (September 1993).

6. Fahlén L E, Brown C G, Stahl O and Carlsson C: 'A space based model for user interaction in shared synthetic environments', in Proc Inter CHI'93, Amsterdam, ACM Press (1993).

7. Deering S: 'Host extensions for IP multicasting', IETF RFC 1112 (August 1989).

8. Benford S D, Greenhalgh C M and Lloyd D: 'Crowded collaborative virtual environments', Proc ACM Conference on Human Factors in Computing Systems (CHI'97), pp 56-66, Atlanta, Georgia, US (March 1997).

10

VIRTUAL CONFERENCING

A N Mortlock, D Machin, S McConnell and P J Sheppard

10.1 INTRODUCTION

Conferencing systems should enable remotely located groups or individuals to communicate effectively. Current audio- and videoconferencing systems restrict the realism of that communication. In order to overcome some of the barriers, virtual conferencing uses a shared computer-generated space as an environment that more closely resembles a physical meeting. Individuals are represented by realistic 'virtual humans' that move naturally.

Initially this chapter explores the benefits of virtual conferencing, and outlines some of the essential requirements. It then focuses in detail on the underlying technology developed for an early virtual conferencing demonstration (see Fig. 10.1). Finally, standards and future study areas are discussed.

Fig. 10.1 Virtual conference.

10.2 WHY VIRTUAL CONFERENCING?

Current audio- and videoconferencing systems, while enabling effective communication at a distance, limit the naturalness of communication in the following ways:

- audioconferencing allows multiple people to communicate using audio only — for larger conferences it is difficult to tell who is talking and there are no visual cues, such as body language and facial expressions, to complement the interaction;
- videoconferencing adds visual information to the audioconference thereby reintroducing the visual cues for interaction — however, current systems are not able to create spatial positioning for multiple participants and therefore body language and conversations cannot be naturally directed to the relevant people in the conference; in addition, multiple video windows do not scale well both in terms of display and bandwidth.

Virtual conferencing, however, introduces the spatial cues into the communications environment, by representing participants as 'virtual humans' in a computer-generated space (as in Fig. 10.1). Gestures and body language can be directed towards other people. Visual cues, such as who is speaking, are maintained by customizing the visual appearance of each user's 'virtual human'.

Virtual conferencing provides additional advantages:

- the bandwidth required to support the environment is low, as, after initialization, only changes in position of the 'virtual humans' are transmitted;
- a virtual conference can potentially support interactions between much larger groups of people, the terminals do not have to receive and process multiple video streams, and more 'virtual' space is available in the 3-dimensional virtual world than on 2-dimensional video windows;
- the use of 'virtual humans' allows flexibility to tailor the environment to achieve the objective of the conference, the complexity of the virtual representations possibly varying dependent on the requirements of the interaction — if the number of people in the meeting is large, the detail of the users' representations may need to be reduced [1]; on the other hand, with a small number of participants it may be both appropriate and technically feasible to use more complex representations, revealing all body language and facial expressions in the interaction; if the meeting is focused on particular tasks or data in a group-working scenario, the detailed representation of users may prove to be relatively unimportant (see Chapters 5 and 12); in this case available computer resources can be targeted appropriately.

10.3 REQUIREMENTS FOR CREATION AND CONTROL OF A VIRTUAL HUMAN

There are a number of fundamental requirements for the creation and control of virtual humans used in a conferencing environment:

- the virtual humans must be rendered in real time, enabling the users' expressions and movements to be replicated near-instantaneously — delays in simulating the gestures, expressions and body language can affect the meaning of the messages being communicated;
- the representations should be personalized to the users, allowing the participants in the conference to readily recognize each other;
- the control of the virtual humans should be non-intrusive, so each participant can concentrate on the task, rather than on the underlying technology used to deliver the conferencing environment.

10.4 FACE READING

The majority of virtual reality systems use data-gloves or headsets to track the movement of the user's hands and head. These methods for controlling 'presence' (see Chapter 12) within a computer-generated environment are adequate when the user is interacting with data, but are restrictive when the primary objective is communication with other people. Research at BT Laboratories (BTL) is looking at how these problems may be overcome using computer vision-tracking systems.

A vision system [2] has been constructed to analyse facial details without recourse to specialist image-processing hardware. This system consists of a PC containing a simple video input card, which is connected to a video camera aligned to view the user's head and shoulders. Images from this camera are analysed to find the main facial features such as eyes, mouth and eyebrows, and characterize facial expressions and head orientation.

In the following sections, the techniques used to locate the facial features are described.

10.4.1 Locating the eyes

The first stage in the facial feature location is an eye search. Once the eyes have been located, the detection of other facial features is made easier.

The eyes are located by scanning model images of left and right eyes over the top centre region of each video frame until a high correlation is found. After the first few frames, the search time is reduced by constraining the search to the areas immediately surrounding each eye. In the event of failure, the area is widened for each frame until a match is again found.

If multiple matches are found, due to background features with a similarity to an eye template, then a score is derived from the horizontal angle, correlation coefficients and separation. The features with the highest scores are then used.

10.4.2 Head orientation

Knowing the positions of the eyes in relation to the head boundaries in an image dictates where a face is pointing. The skin and hair colours are used to segment the face from the background. Figure 10.2 shows the resultant location of the eyes and the segmentation of the sides and top of the head.

Fig. 10.2 Locating the eyes and the extremities of the head.

During the first few frames, after eye location, a skin image sample is taken from just below the eyes, and hair from above. The colour levels red, green, blue (RGB) of each pixel from these samples are used to calculate an average skin and hair colour. The Euclidean distance from the mean RGB distribution to each pixel to be classified gives the closeness to skin or hair. The sides of the head are determined by finding the ends of the skin class along the horizontal stripe and similarly with the vertical.

The distance of the head from the camera is derived from the ratio of the number of pixels across the head to the assumed head diameter. Simple geometry is used to estimate the face pose angle.

10.4.3 The mouth

Knowing eye position and head width allows a good estimate of mouth location (see Fig. 10.3). Greater accuracy is achieved by passing a horizontal edge detector over the mouth region revealing the mouth as a set of long horizontal lines. This works even in the presence of facial hair which tends to lie vertically.

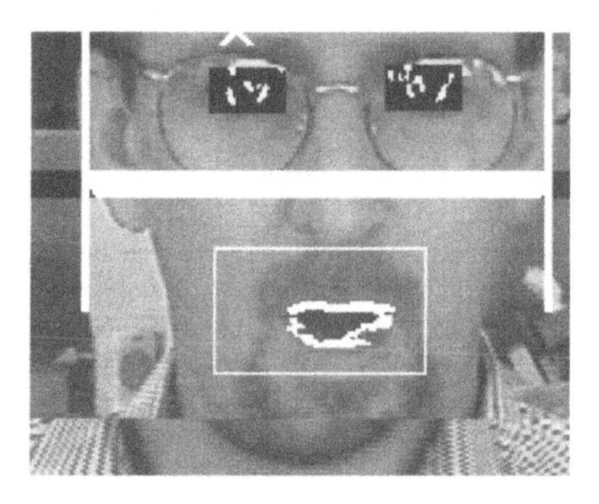

Fig. 10.3 Mouth tracking.

The horizontal edge detector is a simple image processing function where a mask, three pixels square, is scanned over the image. The mask contains the pixel values shown in Fig. 10.4. Each pixel in the mask is multiplied by the pixel in the image which it overlays. If the sum of these values is greater than a predefined threshold, the centre pixel is classified as an edge pixel.

-2	-2	-2
-	-	-
2	2	2

Fig. 10.4 Pixel mask.

The overall width and height of the mouth are normalized and transmitted. These values are used as weighting factors in the mouth animation. For conferencing applications, the mouth shape parameters need to be transmitted at more than 15 frames/sec for a good representation of speech.

The shape of the upper lip is used to determine the degree of smiling to apply to the mouth model. This is calculated using the vertical position of the centre of the upper lip in relation to the vertical position of the corners of the mouth. The resting shape of the mouth is determined when the face reader is initialized. The resting shape is subtracted from all subsequently calculated positions.

10.4.4 Eyebrows

The eyebrows are located in the images using template matching in a similar manner to that for the eye location. The resting positions of the eyebrows are measured and used as an offset to any position change. At present, some of the subtler eyebrow movements have proved difficult to detect, although frowning is detected by counting the number of vertical lines between the eyebrows. A mapping from eyebrow position to muscle action is performed, giving appropriate weighting factors for the frontalis and currugator muscles.

10.5 BODY TRACKING

Upper body gestures and movement are important for effective communication. The ability to gesture in free space enhances the 'naturalness' of the interaction.

Determining the position of the hands and the head in 3-D involves two cameras in the stereoscopic vision system developed by MIT Media Lab, and known as STIVE [3]. The heads and hands are segmented from the background for each of the camera views using skin colour (see Fig. 10.5). Both the 3-D position and the orientation of the head and hands are determined through geometric computation given the information from the right and left camera views.

Fig. 10.5 STIVE body-tracking system.

Person-independent tracking is enabled by training the system to recognize skin colour across a number of users. Images of different users are captured initially and examples of skin colour presented to the calibration system, by clicking on the heads and hands within the images.

10.6 CREATING REALISTIC VIRTUAL HUMANS

The previous section focused on extracting the information required to control a virtual human using non-intrusive tracking systems, and this section goes on to consider the creation of 'virtual humans' that represent the participants in a conference. The virtual human computer model is a three-dimensional mesh of polygons, as illustrated in Fig. 10.6.

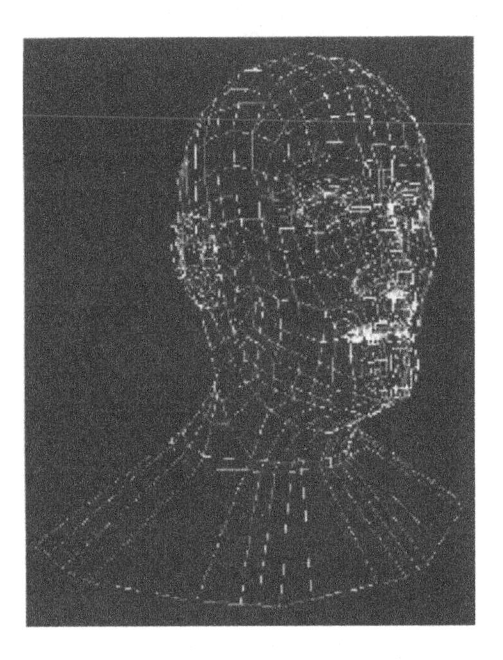

Fig. 10.6 Polygonal head model.

The complexity of the 3-D computer model required to represent a user in a shared environment is dependent on the application and the nature of the information which needs to be communicated. In applications where visual feedback is not required then the complexity of the model can be reduced significantly and may be used merely to indicate the presence of a user. In interactions where the focus of attention is directly on the other users in the interaction rather than on a common task, the facial expressions and body language become paramount for providing realistic and natural interaction. Here the interest is in generating realistic 'human-like' representations of users while maintaining the real-time manipulation of these models to synthesize natural movement.

10.7 HEAD MODELLING

Initially the head model is a generic head, which, in a process referred to as conformance, is modified semi-automatically to match the shape of the user's head. Images of the head are captured from the front, both sides and back. Additional images of closed eyes, teeth and the tongue are also captured to aid the realism in the final model. The range of images captured is shown in Fig. 10.7

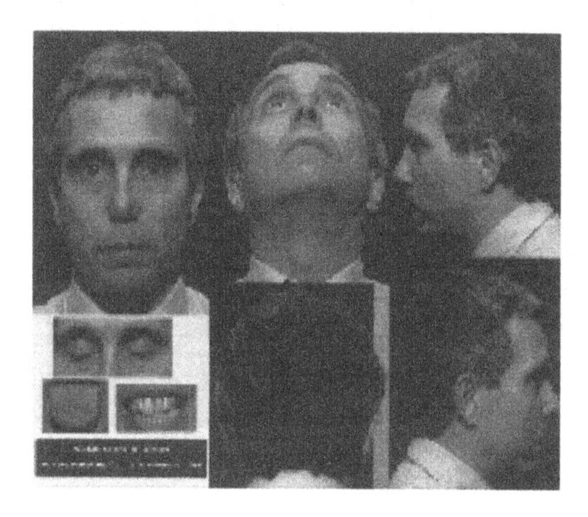

Fig. 10.7 High-quality images of the user.

View-based recognition techniques are used to detect significant features such as the eyes, nose and mouth [4]. The positions of other features are provided within the front image and one of the side images. The polygonal mesh is conformed to match the significant facial features of the user, such as distance between the eyes, width and height of the head and the distance of the eyes to the mouth. Once a model is produced, the head is textured with the captured images without the need to stretch regions of the images, as the underlying model is of the appropriate size and shape.

This is achieved through one of two methods — manually in a modelling package or using an automated texturing system [5].

10.7.1 Manual texturing

The combined texture map together with the head model, conformed to the size and shape of the user's head, is loaded into 'Medit', a 3-D modelling package [6].

The user can then manipulate the texture map by hand so that it maps correctly on to the model. Medit will then export the textured head model in an appropriate format.

The resulting 'normal' textured head is shown in Fig. 10.8.

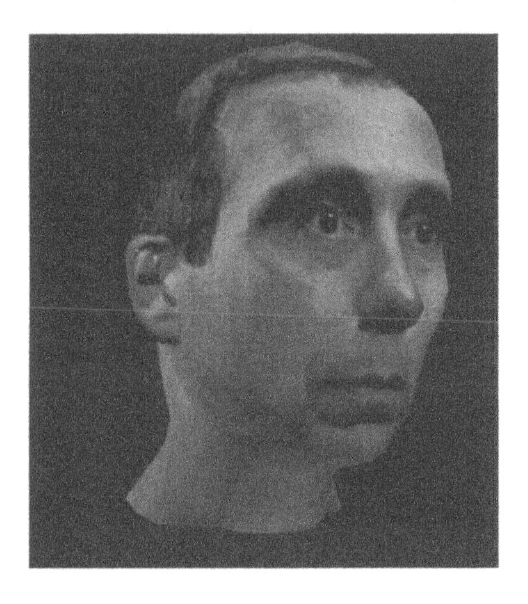

Fig. 10.8 'Normal' head resulting from manual texturing.

10.7.2 Automatic texturing

Automatic texturing applies the photographs of a subject taken from different aspects (Fig. 10.7) to a conformed head. The algorithm applies regions of each photograph to different areas of the head model, so that maximum image resolution is maintained at all times.

For optimum results, a colour equalization and blending of the photographs should be performed before automatic texturing. This will correct any lighting differences when the photographs were taken and ensure the texture blends seamlessly once applied to the head model. A resulting head textured automatically is shown in Fig. 10.9.

10.8 MOUTH MODELLING

The next stage of the modelling process is to define mouth models, which sufficiently describe the extemes of mouth shapes produced during normal speech. Initially the mouth shapes are modelled for a generic head model using video sequences of speech as references. The scaling transformations applied to

Fig. 10.9 Head textured automatically.

the generic model for each user are then applied to the generic set of mouth models to produce the user-specific set of models.

10.9 FACIAL EXPRESSIONS

The synthesis of facial expressions is achieved by modelling the action of a number of muscles within the face. To simplify this problem only the major muscles within the face have been modelled. These muscles are shown in Fig. 10.10.

The effect of each muscle on the polygonal mesh has been modelled by hand for both the relaxed state and at maximum contraction. A weighting factor, representing the contraction of each muscle in the range 0 to 1, is set. An intermediate mesh for each muscle is generated, assuming each muscle acts linearly from relaxation to maximum contraction. The resulting meshes are then combined linearly to produce a single resultant expression. In this way, a whole range of facial expressions can be produced by combining the action of several muscles. Again, to make the facial expressions user-specific, the conformation transformations are applied to each of the muscle models.

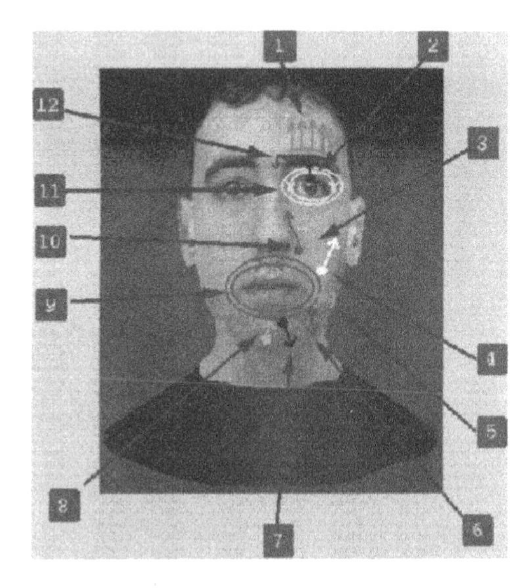

1 Frontalis — raises the eyebrows, as in surprise.
2 Eyelid Muscle — opens and closes the upper eyelids.
3 Upper Lip Muscles — raises the upper lip to produce a sneer.
4 Smiling Muscle — raises the corners of the mouth towards the cheekbones.
5 Risorius — stretches the lower lip horizontally into a grimace.
6 Triangularis — the 'frowning muscle' which lowers the corners of the mouth.
7 Lower Lip Muscle — pulls down on the lower lip, as in pronouncing 's'.
8 Mentalis — the 'pouting muscle,' which raises the chin and lower lip.
9 Lip Purser — constricts the mouth, as in kissing or pronouncing 'oo'.
10 Nose — Upper Lip Muscle — flares the nostrils and pulls the upper lip to produce a wince.
11 Squinting Muscle — compresses the skin around the eyes.
12 Currugator — furrows the brow by pulling it downward.
13 Jaw Muscles — open and close the mouth.

Fig. 10.10 Muscles within the face.

10.10 BODY MODELLING

Up to this point, the chapter has concentrated on the modelling of the face and facial expressions. A realistic representation of a human in a virtual environment requires a body that enables upper-body gestures. To this end, the JACK human body modelling software [7] has been integrated with the facial modelling work.

The JACK body model consists of joints modelled closely on that of a real human, so that the appropriate degrees of freedom and joint limits are expressed in its movement. The software is primarily used in the design industry, in particular by automotive manufacturers for human factors engineering, where they require virtual humans of the appropriate size and shape, which move naturally.

The JACK software environment is a modular system giving users the ability to extend the basic functionality. This feature is used to integrate the facial modelling work with the body model. The standard JACK human model is hierarchical in the sense that the human object is made up of a number of segments, which describe the visual appearance of the human and a number of joints which define how the segments are attached to each other. The head segment of the standard JACK virtual human is replaced with the new head segment modelled previously. As the new segment is placed in the same hierarchical location within the human object description of the JACK virtual human, all the joints defined regarding the connection to the head still apply — only the visual appearance of the head is changing.

The integration of the face model with the body model produces a recognizable virtual human with natural upper body and facial movement as shown in Fig. 10.11.

Fig. 10.11 Virtual human.

10.11 NATURAL MOVEMENT

To produce realistic visual feedback in an application, the user's representation must move naturally, not only with large body motions but also with finer facial movements. In addition, the user's speech signal should be synchronized with the virtual human's mouth and facial movements. To synchronize the audio with the animation of the facial movements frame-based simulation system software [8] is used. This enables the computer system to maintain a fixed simulation frame rate, even if complex changes to the virtual environment are apparent.

10.11.1 Speech synchronization

The animation of the mouth is achieved by morphing between three different mouth shapes — mouth closed, mouth open, as in 'ah', and lips pursed, as in 'oo'. The face-reading system monitors the subject and determines directly the weightings for each shape. Natural movement of the face is limited by the ability to track movement of facial features in real time and with sufficient resolution.

In a complementary approach for computer-generated speech, BT's text-to-speech system, Laureate [9], has been used to control the movement of the mouth with much greater accuracy. This is achieved using a linear interpolation algorithm to morph between the complete range of mouth shapes. The algorithm extracts a stream of mouth shapes and the duration each is active from the text-to-speech system. This is important as different phonemes take differing times to be pronounced.

Given this information the algorithm calculates the appropriate interpolated mouth shape at each frame boundary, given the system frame rate, and renders the appropriate model in real time (see Fig. 10.12). In addition, the appropriate portion of the speech stream is output.

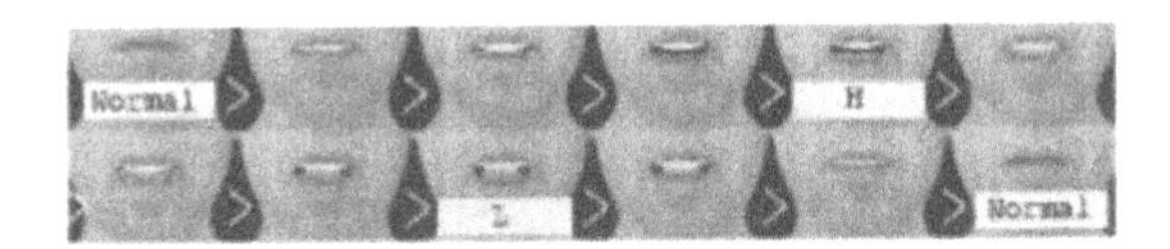

Fig. 10.12 Interpolation between 'H' and 'L' mouth shapes.

All calculations and rendering of the new model need to be performed within one frame duration. The system frame rate is therefore dependent on the processing power of the machine, and a frame rate of 24 Hz is achievable on a graphical workstation.

10.11.2 Head rotation

Head rotations (Fig. 10.13) are achieved by multiplying the vertices in the head by a rotation matrix. This rotation is gradually reduced in the neck region giving the impression that the neck is twisting. The angle of rotation about the X and Y axes is given by the face-reading tracking system, with new rotations being rendered each frame.

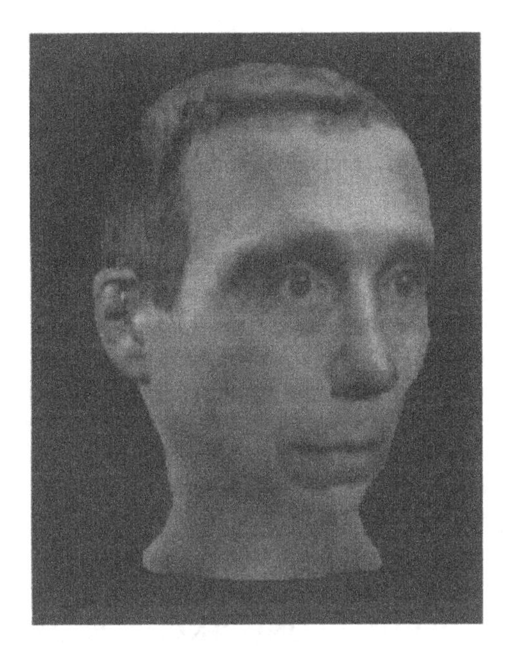

Fig. 10.13 Head rotations.

10.11.3 Facial expressions

Expression changes are currently achieved through the variation of each of the muscle's weighting factors, which are calculated on a frame-by-frame basis. Currently the face-reading system provides weighting factors for a limited number of the muscles. These are the frontalis and currugator muscles, which affect the eyebrows, and the 'smiling' muscle.

10.11.4 Body movement

Upper body movement is achieved using inverse kinematics, which is part of the JACK body-modelling software. Inverse kinematics is the technique used to calculate the position of a limb given its end position. A constraint chain, specifying which joints to position automatically given the position of a constraint, is set up for the left and right arms. The respective starting joints are the left and right shoulders and a hold constraint is specified at the right and left hand palm centre. The *xyz* position of the left and right hand are obtained for each frame through a remote procedure call to the body-tracking system. This information is specified as the global position of the hold constraints. The inverse kinematics algorithm then solves for the positions of the joints within the chain on an iterative basis. The solution provides a realistic simulation of arm movement.

Since there is no direct control of the elbow joint there are a number of joint positions, which solve the criteria of the algorithm. Therefore, it is possible for the elbow to appear in an unusual position. To minimize the problem the wrist joint is frozen, reducing the number of variables within the joint chain.

Movement of the upper body is controlled in a similar fashion, with the joint chain specified from the waist to the top of the neck. The performance of the inverse kinematics algorithm is optimized by reducing the time limit for the algorithm solution. This reduces the accuracy of the final position of the hands and head but increases the frame rate, thereby optimizing the lag between movements in the real and virtual world.

10.12 VIRTUAL CONFERENCING SCENARIO

The preceding sections have described the underlying technology for delivering a virtual conference. Each user appears as a realistic representation of themselves within a computer-generated environment, with computer vision systems to track facial expressions and body movements. As people interact, the other participants see appropriate body postures, gestures and facial expressions. Figure 10.14 summarizes the set-up for an individual user.

The set-up shown in Fig. 10.14 was implemented, as a demonstration, to determine the performance of the system and to gauge initial reaction to the use of 'virtual humans' in conferencing.

The system performed well, achieving a frame rate of 8 frames/sec for the movement of the 'virtual human'. The tracking systems can run at faster rates, as stated earlier, but system performance is limited by the rendering of the 'virtual human'.

The data sent across the network, to control the 'virtual human', was very small. Changes in position to the facial muscles, head movement and body movement, in addition to the speech, were all that were communicated. The data rate to control the 'virtual human' excluding speech was less than 8 kbit/s. No effort was made to compress the data.

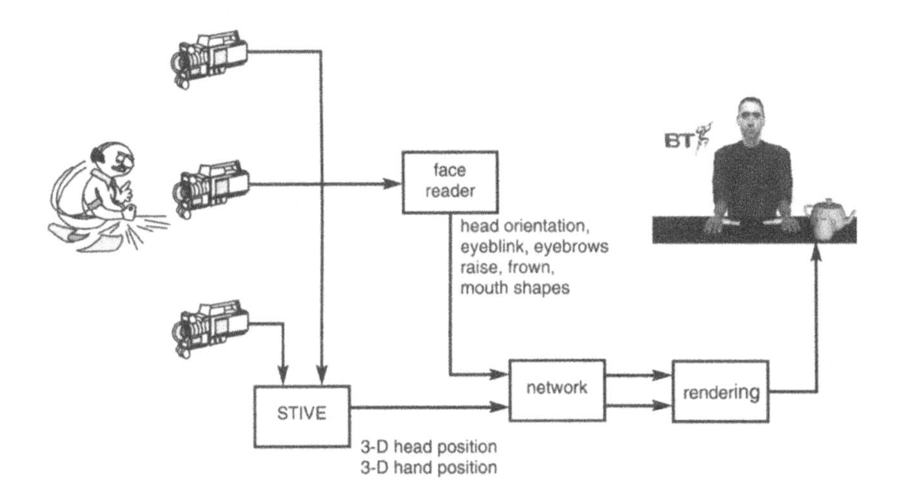

Fig. 10.14 Virtual conferencing system for a single user.

10.13 DISCUSSION

The use of 'virtual humans' as part of a conferencing application is an area of increasing research attention. A number of universities [10, 11] are developing realistic 'virtual humans', while other research groups [12, 13] are developing similar conferencing scenarios which combine computer vision systems with 'virtual humans'. It is therefore important that consideration is given to the standardization of the interfaces to the 'virtual human', so that future conferencing systems can make use of this technology.

The pre-eminent standard looking at this area is the Synthetic/Natural Hybrid Coding (SNHC) section of the MPEG-4 standard [14]. The synthetic coding currently under standardization includes media integration of text and graphics, face and body animation, texture coding and static and dynamic mesh coding.

The face and body animation working group is considering both the appearance and animation of the face and body. The appearance of the virtual human is defined using facial and body definition parameters (FDPs and BDPs), while facial and body animation parameters (FAPs and BAPs) define the movement of the 'virtual human'. The definition parameters are used to customize a generic head and body model to a particular person.

The facial animation parameters are based on the study of minimal facial actions and are closely related to muscle actions. They represent a complete set of basic facial actions and therefore allow the representation of most facial expressions. The body animation parameters describe the movement of the 'virtual human' using joint angles.

A more detailed discussion of the MPEG-4 standard is given in Chapter 8.

The question of customer acceptability of such systems is also under active investigation. The development of the MPEG-4 standard will lead to synthetic data appearing in future conferencing systems, while the work described here will allow the suitability of this new technology to be evaluated. As computer-processing power increases, the realism of the representations will improve, and the boundary between real and virtual data will become blurred. How real do the 'virtual humans' need to become before customers feel as though they are interacting naturally? Reaction to the demonstration was that the head was very realistic, but the simple visual appearance of the body detracted from the realism of the 'virtual human' as a whole.

Early subjective experiments have been carried out by ATR Communication Systems Research Laboratories in Japan, using a similar virtual teleconferencing system [15]. The experiments focused on the required complexity of the 'virtual humans' and the naturalness of the movements. The results show that there is an optimum level of complexity of 'virtual human', above which there is no significant increase in perceived quality. The research team concluded that further improvements in smoothness, naturalness and speed of movement are required before this technique becomes acceptable for conferencing.

These studies have highlighted a number of important areas for future work, including:

- the rendering of the head was much more realistic than that of the body, giving conflicting messages as to whether the virtual human was a cartoon animation or a video-like representation;

- the complexity of the 'virtual human' needs to be balanced with the speed and smoothness of movement;

- occasional gross errors in the body- and face-tracking systems need to be almost eliminated in order to make the gestures appear natural;

- subtle movements in the eye region are very important in the replication of body language — this cannot be animated using the same methods that have been described for the mouth, and a technique such as video texturing may be more appropriate for the eye regions.

In addition to such challenges in image processing and rendering, much work also remains on issues of interface, application design and usability before commercial services can be developed.

10.14 CONCLUSIONS

The ability to model humans and control their movements in a virtual environment without the use of sensors or keyboards has been demonstrated. The BTL system used a single virtual human and an image processing system, and demonstrated several of the essential requirements for the creation and control of virtual humans for real-time conferencing.

Virtual conferencing promises scalability and increased presence in multi-point meetings. It enables appropriate spatial proximity, social cues and gestures to be introduced into multi-location conferences, and with appropriate coding can reduce the bandwidth requirements. In a future of diverse, multimedia telepresence services, the virtual human is certain to be an important contributor.

REFERENCES

1. OnLive Technologies, http://www.OnLive.com/

2. Machin D J: 'Real-time facial motion analysis for virtual teleconferencing', IEEE International Conference on Automatic Face and Gesture Recognition (October 1996).

3. Azarbayejani A, Wren C and Pentland A: 'A real-time 3-D tracking of the human body', Proceedings of IMAGE'COM 96, Bordeaux, France (also M.I.T. Media Laboratory Perceptual Computing Section Technical Report No. 374) (May 1996).

4. Moghaddam B and Pentland A: 'Face recognition using view-based and modular eigenspaces automatic systems for the identification and inspection of humans,' SPIE, 2277 (July 1994).

5. Wallin N: 'Automatic texturing of BT talking head,' MEng Project report, Dept of Electronics, University of York (1997).

6. Medit Modeling software, http://www.medit3d.com/

7. JACK Human Body Modeling Software, Center for Human Modeling and Simulation, University of Pennsylvania, http://www.cis.upenn.edu/~hms/jack.html

8. 'IRIS Performer', Silicon Graphics, http://www.sgi.com/Technology/Performer/

9. Page J H and Breen A P: 'The Laureate text-to-speech system — architecture and applications', BT Technol J, 14, No 1, pp 57-67 (July 1992).

10. MiraLab, University of Geneva, http://miralabwww.unige.ch/

11. University of California, http://mambo.ucsc.edu/psl/fan.html

12. French National Audio-Visual Institute, http://www.ina.fr/INA/Recherche/TV/TV.en.html

13. Ohio State University, http://www.cis.ohio-state.edu/~parent/research/research.html

14. MPEG-4 Synthetic Natural Hybrid Coding, http://www.es.com/mpeg4-snhc/index.html

15. Ohya J, Kitamura Y, Kishino F, Terashima N, Takemura H and Ishii H: 'Virtual space teleconferencing: real-time reproduction of 3D human images', Journal of Visual Communication and Image Representation, 6, No 1 (March 1995).

11

INTERACTIVE COLLABORATIVE MEDIA ENVIRONMENTS

D M Traill, J M Bowskill and P J Lawrence

11.1 INTRODUCTION

Humans have an unrivalled ability for assimilating, understanding and communicating information. This ability when applied to computing is often handicapped by the interface through which we communicate with the information beyond, the ubiquitous interface being the key-board and mouse. Technology is only a solution if the interface is intuitive and closely aligned with the physical and social demands of the human task. Virtual reality (VR) represents a leap in interface technology in which the user is immersed in a graphical and auditory machine representation of a natural (or sometimes abstract) environment.

With the popular present-day perception of VR being reserved almost exclusively for single users playing games and/or simulations, the wider potential for VR within, for example, industry, commerce and health, is being overlooked. This traditional viewpoint of VR is being challenged by the work of numerous researchers around the world working at the leading edge of visualization/ application techniques. VR techniques can be employed within tasks demanding collaboration between individuals or groups of people, where the immersive reality is provided by an interactive physical environment, rather than using head-mounted technologies. The range of display technologies currently being used for VR is diverse (see Fig. 11.1), and a number of 'partial immersion' interfaces are now being developed which will facilitate new ways of working effectively using VR.

Ubiquitous solutions will be based on 3-D graphical interfaces combined with continuous presence video and spatialized audio (see Chapter 4). The synthesis of many such techniques is being explored within 'media environments' that are networked for collaborative applications.

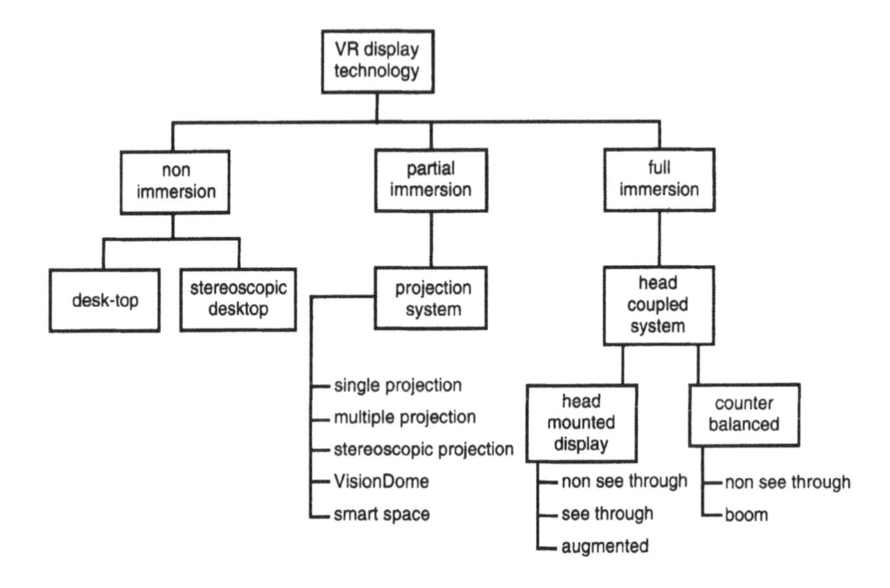

Fig 11.1 Break-down of current virtual display technologies.

The discussion covers the reasons why media environments are being developed, with emphasis on the research covering immersive and desktop systems.

11.2 VIRTUAL ENVIRONMENTS

Both modern business, and society in general, is underpinned by a need for effective communications. As humans, we are equipped with complex biological mechanisms which allow us to communicate within an equally complex social etiquette. For many decades telephone technology has been able to satisfy our basic need to communicate within ever expanding communities and organizations. However, demands and expectations change and remote collaboration with larger communities of people, discussing diverse forms of information, requires new technologies and solutions, in order to communicate effectively.

Kalawsky [1] defines virtual environments as synthetic sensory experiences that communicate physical and abstract components to a participant. An illustration of the interrelationships in a virtual environment is given in Fig. 11.2, a partitioned virtual environment. The shaded region shows a fully interactive virtual environment embodying visual, auditory and touch environments; this region is commonly termed 'virtual reality'. The ability for technology to stimulate most of our senses including vision, hearing and touch, enables human immersion in artificial virtual environments.

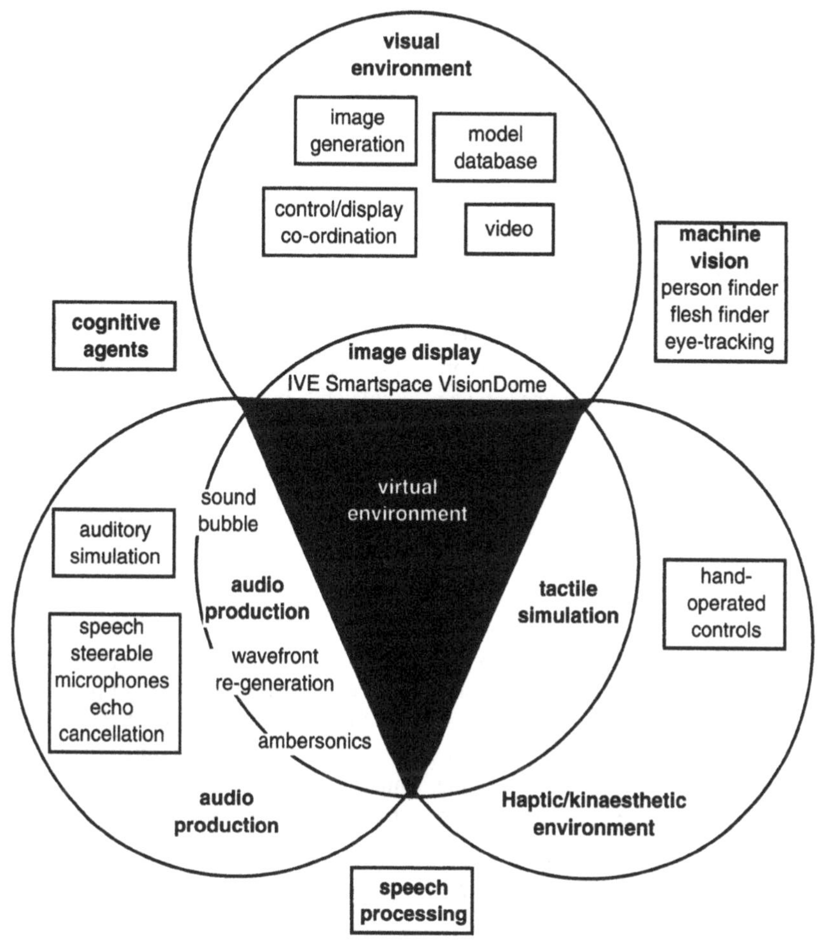

Fig. 11.2 Partitioned virtual environment.

In terms of collaborative working, virtual environments can produce the mutual sense of presence that is an important trait within any collaborative activity. Shared virtual environments such as MASSIVE [2] allow users to meet and communicate in different worlds (see also Chapter 9). However, present display and interface peripherals limit the potential of virtual environments within many activities or scenarios. Mainstream immersive VR is centred on head mounted display (HMD-) based solutions in which the user is isolated from their surrounding environment. The occlusion of real-world interaction within such systems imposes unnatural social and physical constraints on the user. Consider the difficulty in reacting to interruptions such as the telephone or carrying out 'normal' background activities such as note taking. Desktop VR breaks the user from the immersive environment and thus detracts from many applications in which interaction or 'life scale' is important. Being in a world and looking into a world are two distinct experiences.

11.3 MEDIA ENVIRONMENTS

The media environments [3] project at BT is looking at ways of creating more natural immersive spaces. Media environments are spaces in which the physical world is enhanced with 2-D or 3-D computer graphics, which can be complemented with spatialized audio. These elements when combined produce an immersive experience within a more natural human environment. An office, for example, could include back-projected or flat-screen technologies to create areas in which virtual artefacts can be manipulated using gestures.

Many human tasks are based on groups of people collaborating and not just on individuals. Activities, such as design reviews, rely on collaboration between people not all of whom are physically remote. Therefore, spaces in which people can meet, discuss ideas in person, and be immersed in a wider collaborative activity, offer significant benefit over many interactive situations. Telepresence technology is biased towards spatially distributed individuals with, typically, one user workstation per location. This can be characterized as 'networked personal immersion'. Successful solutions to many tasks rely on technology which supports not only interaction between sites but also allows each point of presence to be inhabited by more than one person, i.e. 'networked group immersion'. The areas of enhanced reality, interactive video environments, desktop immersion and group immersion represent complementary examples of media environments and technologies, the integration of which forms the basis of current research.

11.3.1 Enhancing Reality

While much effort is being focused on the realization of virtual worlds, a potentially important grouping of technologies is emerging. There is an intermediate stage between the worlds of fact and fiction, a form of 'enhanced reality' [4]. In an enhanced reality environment the user sees, and interacts with, a view of reality that has been modified to augment the operator's perception and handling of the associated task. The key to this technology is the idea of visual annotation, whereby video images of an operator's surroundings are enhanced by computer generated graphics.

The conventional view of a VR environment is one in which the participant is totally immersed in, and able to interact with, a completely synthetic world. Such a world may mimic the properties of some real-world environment, either existing or fictional; however, it can also exceed the spatial bounds of physical laws ordinarily governing reality. Virtual reality has been developed from a convergence of video technology and computer hardware, the underpinning being computer graphics. Enhanced reality, in contrast, represents a convergence of the aims and technology of virtual reality with the domains of image processing and machine vision. Underpinning is provided by image processing techniques, the extraction of information from the real scene, as opposed to the graphical construction of a virtual scene. An aim which enhanced reality supports is the notion of intelligence amplification as envisaged by Brookes [5] in which the technology augments the decision-making criteria of human operators by enhancing their visual perception. For a discussion on the forms of graphical annotation and reality enhancement, refer to Bowskill and Downie [6].

Milgram and Kishino developed a taxonomy of mixed, or enhanced, reality in an attempt to define the concept of a 'virtuality continuum' between real and virtual environments [7], as represented in Fig. 11.3. Milgram also emphasizes that the next-generation telecommunications environment must provide an '*... ideal virtual space with [sufficient] reality essential for communication*'. Both 'virtual space' and 'reality' are available within the same visual display environment. Real environments are defined as environments consisting solely of 'real objects' and include, for example, what is observed via a conventional video display of a real-world scene. Virtual environments consist solely of 'virtual objects', an example of which would be a conventional computer graphic simulation. The most straightforward way to view a mixed reality environment is one in which real-world and virtual-world objects are presented together within a single display. As indicated in Fig. 11.3, this is anywhere between the extrema of the virtuality continuum.

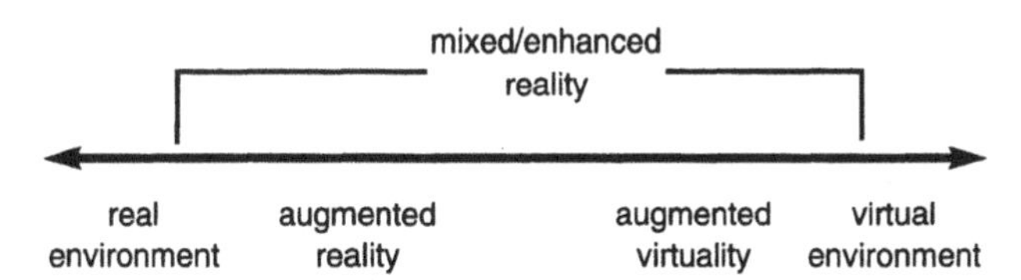

Fig. 11.3 The Milgram 'virtuality continuum'.

VR is inherently suitable for modelling and interacting with abstract environments that are beyond our present capabilities. However, in applications where a real environment must be modelled in order to create an acceptable level of immersive realism, some form of an enhanced reality is called for. After all, why generate a computer model of reality when reality is all around? Is it not more appropriate to annotate reality with enhancements to improve the task in which the user is engaged? For collaborative virtual environments (CVEs), the nature of the interface space is determined by the users' real-world constraints and the social constraints of the task in which they are involved. For example, surgeons could benefit from being able to participate in collaborative visualization and diagnosis sessions. However, the physical constraints of their working environment would dictate an augmented reality interface to the CVE in which the virtual participants could be seen to co-exist within the real environment. If immersion in a single real or virtual space is prohibited, perhaps some form of 'metaworld' is required in which participation with a number of real and virtual spaces or users is possible.

Enhanced reality can also include audio rendering, adding spatialized audio to objects to act as navigation aids, to aid the performance of a task. Audio cues are already used as alert tones on personal computers, but with advances in spatialized and 'rendered' audio, auditory feedback will become increasingly feasible. In an enhanced space these can act as a navigational guide, either when the visual scene may be occluded or as an audio icon for additional directional information.

11.3.2 Interactive video environment (IVE)

While exploring the unique artistic potential of the computer, Krueger can be credited with many significant experiments in which the possibilities of mixing video, computer graphics, and gesture/position sensing technologies were demonstrated [8]. METAPLAY (circa 1970) first demonstrated a 'responsive environment', as Krueger termed it, in which a participant viewed and interacted with a back-projected image of themselves annotated with computer graphics. Although graphics were generated and mediated initially by a remote human facilitator, the evolution of this system extended the role and capabilities of computer-based mediation. VIDEOPLACE, an open-ended laboratory in which image processing is used to identify user attributes, has been under development since 1975 as a form of 'shared video space'. Krueger perceived the potential as a telecommunications medium to be significant, as illustrated by the following quote: *'Even in its fetal stage, VIDEOPLACE is far more flexible than the telephone is after one hundred years of development'* [9].

MIT Media Lab, in collaboration with BT Laboratories, developed ALIVE (Artificial Life Interactive Video Environment) [10], a system in which a human can interact with a virtual agent within an unconstrained environment. The system is based on a magic mirror metaphor, as is VIDEOPLACE, in which a person in the ALIVE space sees their own image as in a mirror. Real-time video of the user in the reflected world is augmented with 'Silas' the dog, an autonomous computer animated graphic (virtual agent). This is displayed on a back-projected screen, which forms one wall within the room in which the user is standing.

Interaction between real and virtual participants of the reflected world is facilitated via the 'Pfinder' vision-based tracking system [11]. This extracts the user's head, hands, and feet position, as well as the gesture information, from the real-time video. Pfinder's gesture tags and feature positions are used by the artificial character to make decisions about how to interact or respond to the user. Pfinder also allows graphics to be placed correctly in the 3-D environment, such that video of the person must be able to occlude, or be occluded by, the graphics. A subsequent application of the Pfinder system has been to enable body position and poise of a user to directly control navigation within the SURVIVE (Simulated Urban Recreational Violence IVE) 3-D virtual game environment. This system demonstrates the ability to use gesture recognition to interact with the computer without using a keyboard or mouse.

The potential also exists for IVE systems to create convincing shared virtual spaces and this has been demonstrated in the form of a collaborative conferencing tool (described in Chapter 10). In this a remote participant can be viewed as an annotated figure (avatar) within the user's local environment, as illustrated in Fig. 11.4. Information about the users, via Pfinder, is shared between geographically separate locations. At the remote end, information about the user's head, hands, and feet position is used to drive a video avatar that represents, or perhaps purposely misrepresents, the user in the scene. Such an approach is inherently scalable, with the potential for large numbers of people to collaborate and communicate in a single shared space. Firstly, network bandwidth is efficient, as it is possible to create convincing telepresence without transmitting video to the remote site. An IVE conference between MIT and the Human Interaction Centre at BT Laboratories has, for example, been demonstrated via a 64 kbit/s ISDN network connection.

Fig. 11.4 ALIVE, including a real person, a virtual avatar and Silas, the virtual dog (agent).

The visual perception of a synthesis of real and life-size virtual artefacts, with corresponding auditory cues from a spatialized audio system, creates an effective form of telepresence and offers great potential for teleconferencing and CVE activities. A practical advantage is the suitability for IVEs to be integrated within 'real-world' spaces, for example, meeting rooms or lecture theatres. A metaphor which current research is exploring is the 'morphic table', as termed by the authors. A real table is complemented with an adjoining display, allowing the physical surface to extend into a virtual environment, as illustrated within Fig. 11.5. In terms of the future development of IVE-based interfaces to collaborative spaces, two primary areas of activity can be identified — multi-participant support (including appropriate mechanisms for interaction and control arbitration), and comprehensive mechanisms for annotating and interacting with imported virtual artefacts.

Fig. 11.5 Morphic table.

11.3.3 Desktop immersion

All computer users are familiar with the non-immersive desktop interface. Desktop immersion has been largely restricted to HMD peripherals. For example, Kalawsky identifies Desktop VR as: *'CAD systems with the added advantage of some form of animation, so that ... objects can be dynamically controlled'*, and introduces immersive virtual environments as: *'... systems that employ a helmet-mounted display or a BOOM-like display to present a visually coupled image'* [1]. Interacting with a CVE via a non-immersive desktop interface can be adequate if the surrounding physical environment accommodates the user. However, it has been observed that a 'degree of presence' problem exists when users become distracted by local events within their physical space, leaving their virtual embodiment unoccupied [2]. In this respect increased desktop immersion, while maintaining a degree of real-world awareness, is desirable.

Two approaches exist for desktop immersion — either create a personal 'enhanced-reality desk space' using a look-through display, or create a physical desk built around advanced interface technologies. An example of the former enhanced reality is provided by 'Windows on the World' [12]. Perhaps it is obvious to suggest that wherever we use a personal computer interactively within a task there may be times when the display would be better placed **within** our view rather than on a monitor **in** our view. In 'Windows on the World' the user wears a look-through display, which is tracked, and the display indexes into an XWindows 'virtual desktop' bitmap. As the user moves, the display is updated to a different part of the XWindows workspace. This effectively places the user inside a display space that is mapped on to part of a surrounding virtual sphere. Application windows can either be displayed at fixed positions within the virtual desk space, fixed within the user's field of view or attached to real-world objects. Not only can annotations be attached to stationary objects, but, if the object is tracked, the annotation window can also be made to follow objects as they move.

BT SmartSpace is an example of a physical advanced interface — a novel integration of technologies which gives the user enhanced visual and auditory immersion. SmartSpace, as shown in Fig. 11.6, is designed as a chair-based workstation, which is able to replace the conventional combination of a desk, and associated personal computer.

The current SmartSpace prototype provides a user interface to a performance personal computer, which is mounted remotely from the chair and connected via an umbilical cable.

Fig. 11.6 The BT SmartSpace.

Visual display is provided by two distinct areas. The primary display and control surface is a high resolution (XGA) Pixel Vision touchscreen which is mounted above the user's lap, a position which is highly intuitive. A screen-based keyboard, or OCR software, allows text to be entered. A secondary display is provided by a wrap-around screen, which extends in an arc 120° within the user's field of view. The screen is a laminate of glass and liquid crystal which is voltage switchable, allowing the normally transparent 'window' to be made opaque when the display is projected. Projection is provided by two video projectors mounted in the headrest.

The sense of visual immersion produced by such a wide screen is complemented via a transaural audio system, as described in Chapter 4, which allows sound sources to be positioned relative to the user. Interaction within applications can be via a conventional tracker ball, touch-screen mouse emulation, 3-D space mouse, or voice. While user trials have largely still to be undertaken, the SmartSpace prototype has been demonstrated for high-quality (life-size, eye-to-eye) videoconferencing, video telepresence (with physical chair movements driving the position of a remote camera) and the navigation of virtual environments.

11.3.4 VisionDome™

As described previously most telepresence technologies are aimed at single users and ignore the fact that tasks are often undertaken by groups of people. The 'VisionDome'™ [1] [13], developed by Alternate Realities Corporation (ARC), attempts to address this problem by creating an immersive environment for a group of people. The main advantage of the dome is that viewers do not have to wear inhibiting hardware.

BT has been collaborating with ARC for the last two years on the development of the VisionDome. Much of the work has been concentrated on gaining a greater understanding of how to create applications that give the viewer a sense of immersion.

Current research is concentrating on how this type of environment can be used to provide an 'enhanced' videoconferencing environment. The large screen enables live video images and real-time computer graphics to be displayed at the same time. This will enable face-to-face videoconferencing with the users being able to view and manipulate a shared computer-generated model.

---

[1] VisionDome is a registered trademark of Alternate Realities Corporation.

11.3.4.1 Technical background

The idea of viewing images within a dome is not new. Cyclorama panoramic paintings were first created in the 1700s, to depict historic events and modern planetariums were opened in the 1920s. The VisionDome attempts to take this one step further by giving the audience the ability to interact with a 3-D world, creating a walk-in VR experience.

The VisionDome, as illustrated in Fig. 11.7, is a hemispherical projected digital display system. A central projector unit projects on to the tilted hemisphere of the dome creating a 360° by 180° image. The projector uses a hemispherical lens that matches the curvature of the dome surface. When coupled with spherically rendered images, the projected scene appears undistorted to the viewer. The system is capable of projecting real-time computer-generated graphics, high-definition TV (HDTV) video images, and live video camera images. The advantage of the dome is that it gives the participants the experience of life-size 3-D models without having to wear inhibiting hardware. Linked with compelling audio, the audience can be temporarily transported into a virtual world.

Fig. 11.7 Demonstration of an architectural review meeting inside the VisionDome.

User immersion is achieved by projecting an image that is greater than the viewer's field of view. In doing so the viewer loses their normal depth queues, such as framing of the image by the edge of the screen, and, if the content is correct, can see beyond the surface of the screen. This illusion gives the sense of

3-D to the objects projected on to the screen. The dome itself allows freedom of angular motion (head motion), so that the viewer can turn their head slightly, up and down or left and right and still have their field of view occupied by the image [14].

11.3.4.2 Content creation

Little support currently exists for spherical rendering and hemispherical projection. Projecting a flat plane results in the image being 'stretched' across the surface and this appears distorted and unnatural to the eye. To create the correct spherical perspective, that matches our spatial perception, images have to be distorted before projection.

For this development, content has been created using 3-D computer animation tools. Each frame of a scene is rendered as sequential images. These images are transferred on to videotape that can then be played back on the dome. To give sufficient resolution on the VisionDome such that the image does not appear blurred, HDTV equipment is used. HDTV has a resolution of 1920 vertical by 1035 horizontal lines giving an aspect ratio of 16:9. The VisionDome screen is effectively a circle with an aspect ratio of 1:1. Spherical images are created, as in Fig. 11.8, for the dome with an aspect ratio of 1035 by 1035 lines.

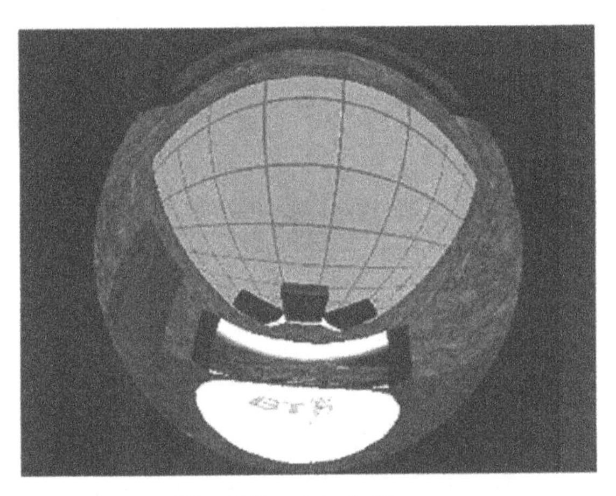

Fig. 11.8 Spherical image.

To create the spherical image an 'anamorphic' lens is placed in front of the camera. The anamorphic lens allows a greater field of view to be captured in a single frame. To create the anamorphic lens in a 3-D computer animation package, a reflective hemisphere matching the shape of the projection lens is placed in front of the virtual camera. Using a technique called ray tracing, the hemisphere reflects the scene into the camera. Ray tracing, is computationally intensive — coupled with HDTV resolution, single images can take several hours to render. The three-minute animation shown at Innovation '97 at BT Laboratories took approximately four months to render on two Silicon Graphics (SGi) high-end workstations.

Real-time graphics enables the audience to interact with an application giving a pseudo-virtual environment. The graphics are distorted before viewing on the VisionDome, by using a 3-D transform on the vertices in the model. This transform has the same result as an anamorphic distortion. ARC have produced a modified OpenGL library for an SGi workstation that performs the transformations. OpenGL is SGi's standard graphical library, which is utilized by a number of current VR applications, which could therefore be used on the dome.

The second main problem with the real-time graphics is referred to as 'object subdivision'. This causes distortion of large objects that have a low number of vertex points. Consider the simple example in Fig. 11.9, where a straight line is to be transformed. In the left image just the end-points have been transformed, with the result that when the straight line is rendered on to the dome it will be badly distorted. The longer the line, the worse the distortion becomes. To correct this distortion, the object can be broken up into smaller shapes ensuring a more faithful transformation of each component part. This is shown for the straight line in the right image.

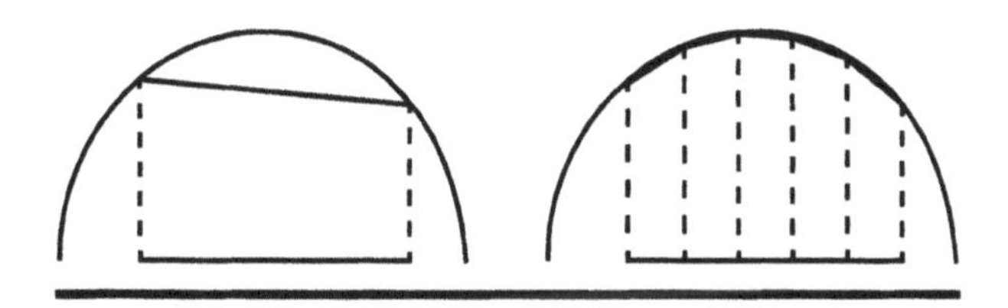

Fig. 11.9 Distortion of large objects (left) and with object sub-division (right).

Distorting the graphics in software results in a degradation in performance of the machine. This has limited the complexity of real-time models on the VisionDome. Current developments by ARC have concentrated on optimizing their libraries to utilize the parallel processing available on an SGi workstation.

As these types of projection systems become more common, content creation will become easier, with support for spherical projection being moved from software into hardware.

11.4 VIRTUAL COLLABORATIVE WORKING

Effective collaboration in a computer-mediated environment demands an intuitive interface with the ability of having a sense of perceptive coupling with members of the shared environment (see Chapter 12). The technologies described above open up the possibility for collaborative working. Semi-immersive display systems enable several video windows to be displayed at the same time, giving face-to-face videoconferencing with the ability to visualize data. However, in multi-user environments such as the VisionDome, issues over mediation and control between local participants need to be addressed.

Creating an effective human interface to ever-increasing volumes of complex information is a critical challenge. People will be able to interact with data in a way that is natural and implicit, without needing special procedures or tools, thus generating the need for computers to understand and naturally interact with emotion (see Chapter 13).

Creating the right environments and infrastructure for these possibilities to be realized will take the integration of features from a number of technologies, forming a broadband network infrastructure supporting transfer of data from a number of different sources — live video, video servers, and data servers.

Ranges of activities have been envisaged that may be carried out in such environments. Imagine, for example, a shared space [15] in which:

- the public, planners and politicians could walk through a proposed urban development, experiencing the environmental impact;
- engineers from around the globe could meet to review a virtual product design, exploring options in form and function;
- military commanders or commercial managers could be immersed in scenarios and information, collaborating on complex, time-critical decisions;
- scientists could travel through their data, interacting in real time with the underlying instruments or experiments;
- students could be transported in time and space, collaborating with tutors and colleagues in an unfamiliar environment.

Undoubtedly collaborative spaces offer significant opportunities in product design, development and testing to display life-size images with an appropriate sense of perspective. Developers could use the environment linked to real-time computer-generated graphics to develop and understand new products. By projecting live or prerecorded video into the space, particularly into a dome or large wrap-around screen, a compelling sense of immersion within the environment can be created. Telepresence applications could include remote real-time projection of sporting events and concerts, immersive videoconferencing, and re-creation of remote or environmentally hostile locations.

Virtual environments will have a significant impact on both education and entertainment — but these two areas may well migrate together into 'edutainment'. Children could visit anywhere on earth, the present, the past and maybe even representations of the future. A class could be immersed in an informative and entertaining manner that they would be unlikely to forget. Not only would the events or images be viewed at the correct scale, but such environments offer the interaction essential to effective learning.

11.4.1 Differences between HMDs and projected display systems

Head-mounted displays give a high degree of immersion within a virtual environment, but are only accessible to a single user. An immersive spatialized display is for a group of people experiencing the same information at the same time while having some feeling of immersion. A major advantage of a projected immersive display is that it can be more convenient and easier to use.

A simple analogy would be the difference between a motorbike and a car. The motorbike offers a single viewer an exhilarating experience through a limited view of the world, i.e. their helmet visor. The car can contain several people with a single driver, who is responsible for the direction, etc. Each person in the car has a large viewing area, the windscreen, which can be obscured by others. One does not need any specialized training to be a passenger in a car, whereas a complete novice cannot be expected to get on to a bike and be able to drive it safely or effectively. It could be argued that the car is safer and more comfortable, whereas a motorbike is more exhilarating and closer to the elements, giving better feedback to the user.

11.4.2 Network futures

Underpinning collaborative virtual environments is the interconnecting network. The network has to support both connectionless data traffic and connection-oriented speech and video. At BT Laboratories an experimental network, 'The Futures Testbed' [16], has enabled the testing of collaborative environments unhindered by bandwidth constraints.

Other advances in network technology [17] have seen the development of asymmetric connections over copper wire. Asymmetric connections enable a high bandwidth connection in one direction with lower bandwidth control information being sent in the other. This type of connection is suitable for video-on-demand applications and may be appropriate for CVEs with a virtual world rendered on a central server distributed to remote users. Asymmetric digital subscriber loop (ADSL) technology is expected to deliver approximately an 8-Mbit/s channel. A hybrid version VADSL using copper/fibre connection is predicted to offer channels in the range of 25 Mbit/s to 51 Mbit/s.

11.5 CONCLUSIONS

This chapter has introduced and defined media environments, a collection of interfaces in which real and virtual spaces are mixed. Media environments can be classified as one form of enhanced reality, based around immersive physical spaces enhanced for effective collaborative activities. Current research is directed at three forms of enhanced spaces — immersive projected displays, interactive video environments, and immersive desktop. While HMD and desktop VR facilitate many collaborative tasks, the synthesis of real and virtual realities within a life-size environment offers distinct advantages within other applications. The rationale is to develop immersive environments, such as the VisionDome or IVE, which support more than a single user. Developments to date indicate that media environments provide an effective interactive interface within teleconferencing and collaborative visualizations. In the future it is envisaged that networked media environments will allow remote groups of people, and individuals, to communicate in many novel and intuitive ways.

REFERENCES

1. Kalawsky R S: 'The science of virtual reality and virtual environments', p 331, Addison-Wesley (1993).

2. Greenhalgh C M and Benford S D: 'MASSIVE: A virtual reality system for teleconferencing', ACM Transactions on Computer Human Interfaces (TOCHI), 2, No 3, pp 239-261, ACM Press (September 1995).

3. Bowskill J M and Traill D M: 'Interactive collaborative media environment', Collaborative Virtual Environments 96 conference proceedings, Nottingham University (1996).

4. Bowskill J M and Downie J D: 'Extending the capabilities of the human visual system: an introduction into enhanced reality', Computer Graphics, 29, No 2, pp 61-65 (May 1995).

5. Rheingold H: 'Virtual reality', p 25, Mandarin, London (1992).

6. Bowskill J M and Downie J D: 'A taxonomy of reality enhancement', Technical Report TR6502, Department of Electrical and Electronic Engineering, University of Brighton (1995).

7. Milgram P and Kishino F: 'A taxonomy of mixed reality visual displays', IEICE Transactions on Information Systems, E77-D, No 12 (December 1994).

8. Krueger M W: 'Artificial reality', Addison-Wesley (1991).

9. Krueger M W: 'Responsive environments', Proceedings of the National Computing Conference, pp 423-433 (1977).

10. Pentland A P: 'Smart Rooms', Scientific American, pp 54-60 (April 1996).

11. Wren C, Azarbayejani A, Darrell T and Pentland A: 'Pfinder: real-time tracking of the human body', published in SPIE Conference on Integration Issues in Large Commercial Media Delivery Systems, 2615 (October 1995).

12. Feiner S, MacIntyre B, Haupt M and Solomon E: 'Windows on the world: 2D windows for 3D augmented reality', Proc UIST 93, Atlanta GA, pp 145-155 (November 1993).

13. Walker G, Traill D, Hings M, Coe A and Polaine M: 'VisionDome: a collaborative Virtual environment', British Telecommunications Eng J, 15, Part 3, pp 217-223 (October 1996).

14. Lantz E: 'Introduction — Spherical image representation and display: a new paradigm for computer graphics', Graphics Design and Production for Hemispheric Projection Course Notes for SIGGRAPH95, pp A-8 to A-22 (1995).

15. Bradley L, Walker G and McGrath A: 'Shared spaces', British Telecommunications Eng J, 15, Part 2 (July 1996).

16. Barnes J W, Chalmers J, Cochrane P, Ginsburg D, Henning I D, Newson D J and Pratt D J: 'An ATM network futures test bed', BT Technol J, 13, No 3, pp 102-109 (July 1995).

17. Young G, Foster K T and Cook J W: 'Broadband multimedia delivery over copper', BT Technol J, 13, No 4, pp 78-96 (October 1995).

12

REAL PEOPLE IN SURREAL ENVIRONMENTS

C K Sidhu and P A Bowman

12.1 INTRODUCTION

"It is not the technology ... what matters is what happens in the hearts and minds of the users."

(Judith Robin, Worlds Away, E2A '96)

The emergence of virtual reality (VR) is dramatically changing the way we communicate. It is redefining the role of computers and telephony to provide compelling virtual environments (VEs) that help mediate collaboration, conversation and social interaction between people.

Advances in technology over the past decade have led to geographical boundaries coming down as we start moving towards an age of networked communities. The concepts of telepresence, teleoperation, virtual conferencing and shared environments are becoming a reality in supporting these communities.

Preliminary research has generated a wealth of information and insight into this area [1-3]. However, it is realized that as application areas for VR become more refined and the choice for the user increases, it will be the provision of intuitive, compelling, and easy-to-use interfaces that will become the key differentiator. This chapter looks at some of the issues surrounding these new ways of communicating. An overview is provided of the human perception of the physical aspects of VR, highlighting the need to understand human perception and the interaction between the senses, and describing both the human cognitive element in VR, and the approaches and techniques used for designing and developing compelling and usable VR applications.

Understanding of the user issues related to VEs delivered over the Internet has increased significantly over the last year through field trials and concept testing, and specific areas for further research have been identified. Researching the human element of collaborative VEs is still at an early stage. The aim is to understand and address these issues across the different media spaces available for delivering VR.

In section 12.4, it is stressed how user perception of a VE consists of complex interaction between the various elements which contribute to a successful interaction.

Finally, an example of an Internet VE, The Mirror [4], is provided to demonstrate the assessment methodology used to address technology and the interface factors required for delivering successful VEs.

12.2 PHYSICAL ASPECTS OF VIRTUAL ENVIRONMENTS

Virtual reality can be applied to a wide range of application domains including training, business, simulation, education, design, manufacturing, medicine, and entertainment.

12.2.1 Presence

For there to be benefit and added value over conventional means of communicating, interacting within virtual environments must be compelling, efficient, intuitive and responsive. However, virtual environments offer more than images and sound, they can offer presence (or, synonymously, telepresence, in the case of distributed or shared environments).

Presence is not new and can mean different things to different people. For a long time playwrights, artists, film and TV producers have used devices to engage audiences through their particular medium. Presence is hard to define and even harder to measure. However, most definitions of presence include the notion of 'being there' and the challenge is to engineer systems and devices to fool the senses into maximizing this notion and thus produce systems that are sufficiently immersive for the application. The range of specification or granularity for virtual environments is wide, including:

- high-end flight simulators incorporating high specification visual, auditory, motion and tactile display and feedback;
- head-mounted stereoscopic displays with motion tracking;
- SmartSpace [5] with a widescreen display and acoustic bubble;
- PC or workstation with a physically narrow field of view compensated by appropriate content, e.g. networked games can offer some of the criteria described above using quite modest resources.

One of the principal ingredients that impacts on this sense of presence is the ability to perceive spatial information enabling navigation and interaction with virtual surroundings. The essential location information is in the form of the depth, volume and space cues which can be provided by a number of possible techniques including:

- occlusion or interposition whereby more distant objects can be hidden by nearer objects;
- stereopsis as a result of binocular disparity which is particularly significant when 3-D objects are close to a viewer;
- perspective;
- the use of shadow and motion parallax.

These visual cues can also be complemented by audio cues associated with the visual elements such as Doppler shift or appropriate reverberation profiles.

12.2.2 The visual environment

In a virtual environment the visual information covers a range of complexity:

- simple wire-frame graphics;
- solid objects;
- rendered objects;
- photo-realistic models.

There might be a temptation to assume that high levels of graphical detail in VR environments is desirable or essential. However, there are times when scenes can become too intricate or cluttered resulting in the key elements being less effective. At times, more is better, less is more.

12.2.3 The audio environment

The audio environment is often underestimated and can be exploited to good effect for most applications if there is:

- adequate scope for accurate manipulation and synthesis of the soundfield;
- good temporal and spatial correlation, such as synchronization, between the audio and visual geometric cues or video components.

The general audio quality in terms of level, noise and distortion performance should be commensurate with any corresponding video quality. However, when virtual environments are shared with other people and conversational speech is a major component, it is also necessary to ensure that there is adequate echo cancellation and minimum transmission delay. Likewise, if voice switching is employed, there should be good immunity to any extraneous background noise and it should not clip the speech to any significant degree.

12.2.4 Contrasting the contribution of audio and video

There can be a danger in considering audio and video perception as separate entities. The relative significance or importance of the audio and visual channels are factors that need to be carefully considered to gain the optimum effect.

Evidence of subtle cross-modal interaction between these channels has been demonstrated in a number of cases, for example:

- perception of TV picture quality is affected by the presence of audio [6];
- the McGurk effect [7] illustrates how the ears can be tricked into hearing different sounds when the eyes are open than when they are closed;
- a highly detailed, photo-realistic, VR fly-through of a building will have a 'dead' feel to it if just the visual channel is used — the addition of even low-level acoustic background sounds or incidental noises can heighten the feeling of being part of the environment being visualized and contribute to our sense of presence.

The audio and video components of a virtual environment have fundamentally different properties, as different views on to a common virtual world can be seen on individuals' displays at the same time and in the same room with negligible interference. This is not the case with audio as sources from several directions, including behind the listener, can be heard simultaneously. As a result, consideration needs to be given to locations where different virtual environments are physically close to each other in order to minimize any acoustic interference between them.

There is a further potential enhancement that can be gained from the inclusion of an appropriate audio environment. In some applications it is possible to maintain a degree of continuity of presence so that when the head is turned away from a display system the audio can still be heard. If, however, the user was restricted to a visual stimulus and turned away, then any sense of presence that was being experienced would be lost.

12.2.5 Network considerations

Many interactive VR environments have been developed for the simulation and computer aided design (CAD) industries; these are generally implemented on local high-performance workstations, involve a single user, and are normally limited by the local hardware and software. When virtual environments are to be shared among users who are not collocated, the performance of different network technologies and topologies needs to be considered.

Delivery of interactive VEs can span from public Internet connection with bandwidths constrained by comparatively low-speed modems and copper cables, through to high-speed wide-bandwidth photonic networks.

Apart from the fundamental limitations of bandwidth that might impact on audio quality and video information, transmission delay is a major consideration when real-time responses and interactions are anticipated.

12.3 RESEARCHING THE HUMAN ELEMENT

> "You [the user] not only give your avatar[1] life, but you give other avatars life via your presence. Pixels do not matter unless you have people behind them."
>
> (Randy Farmer, Electric Communities, E2A '96)

> "It seems like the people really make the worlds."
>
> (User comment from The Mirror trial)

People are the main content of CVEs (collaborative virtual environments). It is through their actions and experiences that the environment comes to life. Technology is the enabler to achieving user goals and tasks effectively and efficiently. As a result, technology must be harnessed to support the user experience and must not be the sole driver.

With collaborative VEs, we are moving into a new domain where the user becomes an integral part of the technology. Researching the human element requires a modification of more traditional approaches in order to cater for the fact that, unlike most other technologies, with VR we are dealing with a dynamic media where the environment is generated in real time in response to user actions and experiences. User experiences vary depending on how they decide to interact with objects in the VE and also who is sharing the space at the time.

---

[1] Individuals are represented in virtual environments through their avatars — a body image generated by the computer and controlled by the user.

There is a tendency for human/computer interaction (HCI) to reuse and reinvent techniques drawn from its own narrow sphere. Successful design, development and evaluation depends on collecting approaches from different domains and tailoring these to the different life-style stages in the product development life cycle. To succeed, novel approaches need to be developed by combining techniques such as storyboarding and ethnography with existing traditionally used ones such as user trials. In addition, multidisciplinary teams are essential for the development of successful VR applications, including a combination of 3-D designers, software designers, human factors specialists and service providers.

Below are some examples of approaches to increase understanding when designing for the user.

- Requirements capture through visualizations

  Techniques using early visualizations of complex concepts have been developed which can be taken to users early on in the development phase. This was applied to the information place (IP) [1], a 3-D application involving the use of an astral metaphor for organizing and representing information in a collaborative virtual environment. At a top level, the user sees a constellation of planets which represents the different groupings of information. The user can navigate around the space, visiting planets, meeting people, and interacting with objects and information.

  The information place is an example where technology and design principles were used to create an innovative demonstrator. Early visualizations, in the form of story-boards were taken to a potential segment of users (business users) to gauge their requirements for such a virtual environment.

  Again, the requirements focused on the key components of CVEs, such as navigation, avatar representation, and communication. Additional issues, including security, business versus personal use, were also addressed.

  The use of storyboards as a technique for visualizing concepts proved very effective in demonstrating the concept and eliciting key requirements for the IP. Figure 12.1 shows an example of a clip from the IP storyboard. The image shows the astral metaphor and the scenario sets the scene.

  This technique can be implemented cost effectively and very quickly with maximum impact. Feedback from such exercises can be used to implement a range of potential prototypes based on an element of user attitude and innovative design. These prototypes can then be tested by potential users to further refine understanding from the user perspective.

When Rob has finished he notices Jane's avatar. Jane has been discussing her work with Peter and when they see Rob enter they invite him to join the conversation ......

Fig. 12.1 Sample from the information place storyboard.

Other techniques that can be used to capture requirements include the use of rich pictures to obtain a diagrammatical representation of the users' thoughts through their own drawings, and careful observation of the real-world situations which are to be supported in the VE.

- Understanding VR through users' actions

The walk-through technique has been used to understand:

— the usability of particular features from the user perspective;

— thought processes in relation to problem solving and task interpretation;

— how users deal with errors encountered when using an interface.

Based on the 'think aloud' technique, users speak out their thoughts and actions, describing what they like and dislike about the particular interface they are using. This approach can be used in combination with other techniques, such as part of the requirements capture process, in user trials.

This strategy was used early on in VR research to identify components and issues that are important to users and which contribute to 'usable' virtual environments. The data was used to inform designers and contribute to the development of a questionnaire for assessing VEs.

- Controlled user trials

  User trials allow researchers to assess user attitude and performance in a controlled environment. Tasks can be designed to assess particular aspects of the system under trial. For example, in The Mirror trial, users were asked to complete a set task in one of the worlds where they had to collaborate with other avatars to complete the task.

  Depending on the design of the user trials, they can be used to compare different implementations, e.g. a comparison of immersive environments with non-immersive environments to investigate ease of use [8].

- Field trials

  The key advantage a field trial has over a user trial is that it allows researchers to investigate usage over time and in the real context of use. In the absence of controlled tasks, field trials of virtual environments pose new challenges for human factors (HF) researchers. With the main Mirror trial (see section 12.5), there was very little opportunity to control who went into the world, and at what times. As a result a range of data-gathering techniques, existing and new, were employed to collate objective and subjective data.

Additional approaches to assessing VEs currently being considered include:

- ethnography [9], an observational technique which can be used to analyse user interaction/communication within real environments to establish salient attributes that can be supported within VEs;
- modelling techniques, to understand behavioural changes and social implications of CVEs;
- spatial syntax theory [10], to understand the relationship between space, movement and interaction in VEs.

12.4 WHAT CONTRIBUTES TOWARDS A USABLE VIRTUAL ENVIRONMENT?

The scope of research required in this area is vast and some of the questions that need to be addressed include the following.

- Which tasks are most suitable for users to perform in VEs?
- What are the physiological effects on the users? Will users get sick or be adversely affected by exposure to VEs? How much sensory feedback from the system can the user process?

- How can VE technology be improved to better meet the user's needs?
- Do design metaphors enhance a user's performance in VEs?

Over the past year BT has undertaken preliminary research to address the design, technology and HF issues involved in delivering virtual environments over the Internet. The objectives of the research were to assess the technical performance of a CVE, and to investigate user perceptions related to inhabiting 3-D virtual environments.

An early study in BT consisted of a series of user walk-throughs carried out with novice users. The study addressed the fundamental goals of interacting with virtual worlds, e.g. communication and navigation, in order to establish issues that are important to potential users.

The main aims of the study were to assess the usability of particular components that make up a virtual world from the user perspective, and to identify particular aspects of the worlds that users like and dislike. People were asked to use three virtual worlds which were accessible in real time over the Internet. They were instructed to interact with each world in order to get an overall feel for what each had to offer. As a secondary task, they were asked to communicate with other avatars in the space.

The study identified the following components that contribute to users' experiences of a VE available over the Internet:

- entry into the VE, and initial impressions;
- communication and collaboration in virtual environments;
- avatar representation, customization and animation;
- navigation;
- overall look of the world (i.e. the use of graphics);
- use of audio/visual cues;
- sense of presence.

The above is a top-level list which can be broken down further. For example, avatar representation includes methods for selecting and changing appearance, facial/body movement, expression of emotions, how to present the information to users, etc.

Some of the main conclusions from the study are now presented. There is no one component that makes a successful virtual world. Navigation, communication, and system feedback are all intrinsically related. This was captured in a user's comment:

> "They are sort of equally important. Being able to choose your avatar is very important. Conscious of how you look and how others are perceiving

> you. At the same time, what's the point of being able to navigate wonderfully if you can't communicate with somebody. What's the point of communicating with somebody if you can't see them; you might as well be on the phone. You kind of need it all."
>
> User comment

The goal is to understand the relationship between the three different components.

12.4.1 Communication

Communication was seen as the key component within any CVE. Users need to communicate and interact with other people, information and application objects. All other components contribute towards making communication easier for the user — navigation, avatar representation, choice of text or speech, use of sound and motion, and the general look of a world.

The study highlighted some of the key steps for supporting communication within the CVE.

- Making the decision to communicate

  The virtual environment must support users in their decision to communicate; this decision could result from an interest in the appearance of another avatar, a particular motion indicating interest, information held on avatar profiles, or hearing other conversations. Such information must be both easy to convey and access.

- Deciding on the type of communication

  Users may require a range of communication 'types', e.g. one-to-one, one-to-many, group chat or formal meeting, public chat, private chat as part of a larger group chat, etc. Again these should be supported within the shared environment.

- Initiating the communication

  There are numerous techniques used to initiate communication in the real world. These range from a simple informal one-to-one chat resulting from chance eye contact, through to a pre-arranged formal meeting set up in advance over the telephone, by e-mail or by a third person. Virtual worlds need to provide the necessary tools to support such requirements, e.g. the use of gestures or specific movement can be used to attract someone's attention. A noticeboard or an on-line e-mail facility can be used to set up a larger chat.

- Carrying out the communication

  The virtual world needs to support user requirements during the communication, e.g. use of gestures (facial expressions and body movement, availability of people through their representations), use of position and location in the space to provide feedback cues to other avatars, easy access to information, and availability of virtual equipment, such as whiteboard, document transfer facility, the ability to invite others to join a chat, etc.

The above is by no means an exhaustive list and the requirements will vary depending on the user population.

Communication will be greatly influenced by the use of text or speech, each of which has its good and bad aspects (see Table 12.1).

Table 12.1 Advantages and disadvantages of speech and text.

Method of communication	*Advantages*	*Disadvantages*
Text	Gives users time to compose messages.	Open public chats using the same window can be potentially confusing, as a number of separate conversations are presented which can increase the amount of processing required by user.
	Users have a record of past conversations.	May detract from realism. If text is in a separate window, it can draw attention from main window.
	Helps communication if text can be easily attributed to avatars within the environment.	Typing introduces significant delays into the communication process.
Speech	Users perceive speech as the natural form of communication between people.	Technical constraints can lead to distortion.
	Has the potential of adding to the sense of presence.	Potentially confusing if all conversations are broadcast.

The advantages and disadvantages are dependent on the system implementations. A combination of technical advances and design can help overcome the current limitations.

Other issues that impact on ease of communication include the speed of the system, implementation of the aura, and use of audio and visual stimuli. Each needs to be implemented differently depending on the specific application. For example, for entertainment, it may be necessary to go against user expectations, whereas for business use, it is more important to map user expectations.

12.4.2 Navigation

Navigation impacts on a user's perception of the quality of the interaction and sense of presence. As such, navigating in a virtual environment should be as intuitive as possible requiring minimum effort from the user. Users need to be able to navigate easily so that moving around in the VE does not detract from the main goal of communicating.

The key elements that contribute towards 'intuitive' navigation include the response time, contextual information and system feedback in terms of the visual information. This becomes more important in a text-based world where the text window is separate from the main graphics window. Without feedback the graphics window becomes redundant as all the 'action' is in the text window.

Different implementations of navigation within VEs are currently being investigated to reduce the amount of effort required on behalf of the user.

12.4.3 Use of audio

Audio plays an important role by helping to maintain continuity in the VE. It enhances the sense of presence, providing users with realistic stimuli in the form of ambient cues and collision cues. Not only does this add to the sense of presence, but it also aids navigation and communication.

In summary, all these aspects of CVEs need to be implemented to reduce the load on the user. They need to become part of the automatic background functionality.

12.5 ASSESSING VIRTUAL ENVIRONMENTS — THE MIRROR CASE STUDY

> "A rapidly expanding system of networks, collectively known as the Internet, links millions of people in new spaces that are changing the way we think, the nature of our sexuality, the form of our communities, our very identities."
>
> (Sherry Turkle, *Life in the Screen,* 1995)

In addition to researching specific aspects of VEs, it is essential for developers and designers to learn about the whole process of delivering virtual environments to end users — the customer experience. This involves consideration of the technology, system performance, and content design, together with the user interface issues. The Mirror project provided an ideal opportunity to investigate these areas.

The Mirror was a research project created by BT Laboratories, Sony Corporation, the BBC and Illuminations. The project brought together key disciplines of 3-D design, media, engineering and human factors to produce six compelling and stimulating 3-D virtual interactive worlds. For a detailed description of The Mirror, see Walker [4].

The worlds in The Mirror reflected the themes of the six broadcast programmes on the BBC2 technology series, The Net. They were designed to experiment with specific aspects of VR content, to explore what would be most appealing to both new and experienced participants. The worlds were accessed via the entry portal (Fig. 12.2) which highlighted the 'World of the Week' corresponding with the programme.

Fig. 12.2 Entry portal for The Mirror.

When designing the world, one of the goals was to develop a self-sustaining environment in which users contribute to the content of the VE not only through communication, but also through involvement in the special events, providing material for the art show in the 'Creation' world, leaving memories in 'Memory' world, and interacting with the application objects within the worlds.

The assessment of The Mirror, which consisted of two phases, looked at the success of these features.

- Usability trial

  This was a controlled user trial which ran over a two-week period. The overall aim was to test the process for designing and delivering a virtual environment to end users in preparation for the 'live' trial with the BBC. The main trial consisted of four one-hour sessions in which the participants were asked to complete a task.

  Additional assessments were carried out to evaluate the Sony browser usability, prepare the guidance material for the main trial, and design the cover for the CD.

- The Mirror field trial

  The main trial ran for 7 weeks between 13 January and 28 February 1997. The main aim was to investigate and understand the technology and interface issues required for delivering interactive 3-D environments over the Internet. This extended trial allowed for changes in usage to be noted over time. It was also possible to find out what users had expected from the technology.

12.5.1 Measuring usage and usability in The Mirror CVE

It is believed that a comparison of actual (objective) behaviour and perceived (subjective) feedback is required to provide an overall assessment of usage and usability. Subjective assessment examines user attitude while objective assessment involves logging platform statistics to establish actual usage and behaviour.

12.5.1.1 Objective analysis (quantitative)

Usage data, which took the form of performance statistics of interaction between users and the VE, included individual data and average data for all users. Types of measurements included frequency of access, total time spent in each session, etc.

Recorded interactions between users and the VE included the analysis of the text files of communication between users. Where possible, recordings of visual content were also used.

12.5.1.2 Subjective analysis (qualitative)

On-line questionnaires/focus groups were used to target a sample of users to obtain feedback related to all aspects of the trial. Focus groups were carried out both face to face and in the actual Mirror VE.

Evaluators accessed and observed behaviour within the worlds, both as passive and active participants.

Helpline data included a breakdown of problems and issues raised via the 'Helpline'. These contained both technical and usability-related issues.

All correspondence between the users and the assessment organizers was carried out via the Internet.

12.5.2 Results from The Mirror experiment

Some of the high-level results are presented below:

- over 2300 people registered to become citizens of The Mirror spending a total of 4408 hours logged on to the server (see Fig. 12.3);
- in terms of the content of the worlds, the graphics and the interactive applications provided an initial focus of interest which was essential in attracting people and ensuring return visits;
- the special events played a key role in bringing together new people to share novel experiences with attendance totalling over 500 hours across all the worlds — such events contributed to an evolving 3-D space where users' experiences and interactions change the environment;
- users who became regular visitors to The Mirror developed a real sense of community over the trial period and feedback indicated a genuine emotional involvement rarely associated with new technologies;
- the main reasons for return visits were to meet new people, look around in the worlds, and interact with objects (in that order) — over time, communication between users was the most important reason;
- avatars were perceived as essential in allowing individuals to project their own personalities — The Mirror included a choice of four avatars which participants could customize (see Fig. 12.4), with users highlighting the need to include additional functionality, e.g. gestures, use of emotions, functionality to enhance self-awareness in the VE.

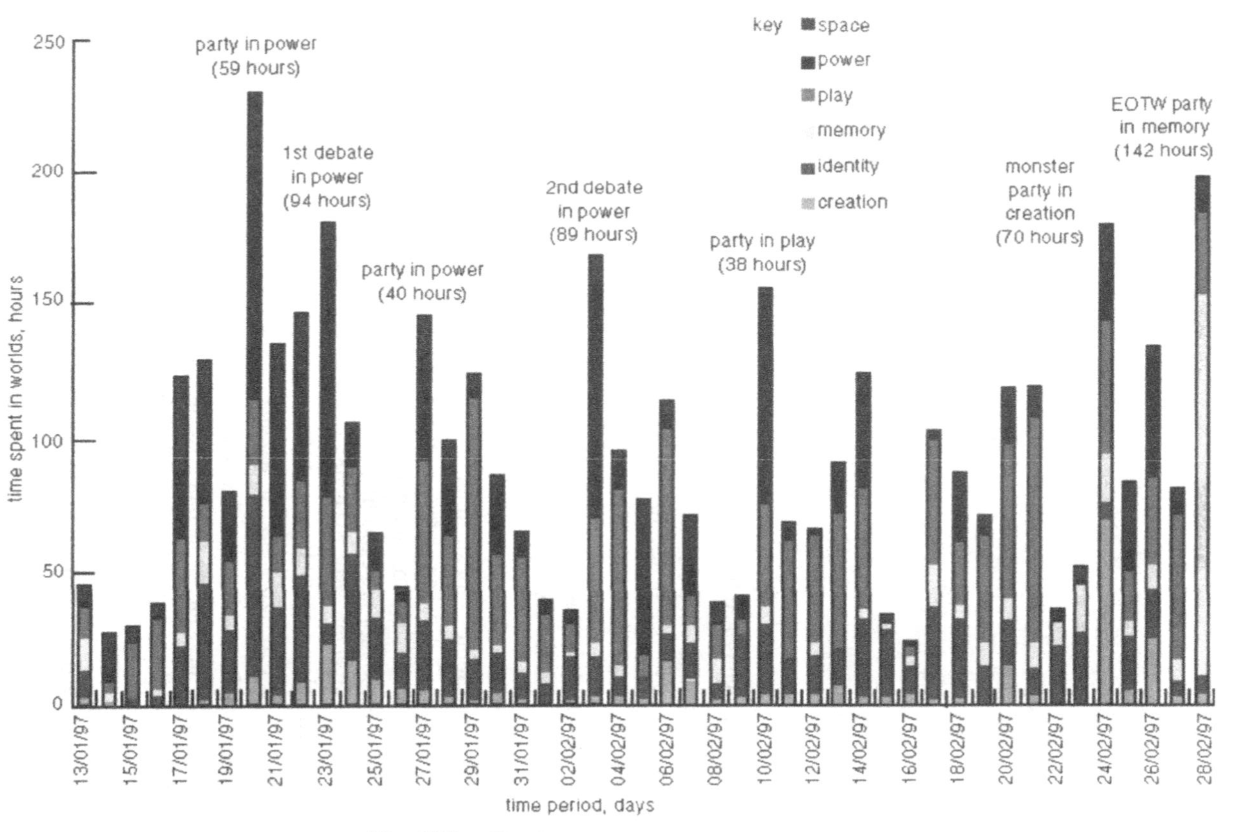

Fig. 12.3 User hours spent in The Mirror.

Fig. 12.4 The Mirror avatar changing room.

The trial highlighted four main critical issues:

- initial contact with The Mirror — the early contact with customers (the initial advertising, registration process, packaging, and delivery of CD and installation process) are key to providing a positive first impression;
- speed of loading worlds/interaction speed — long loading/response times resulted in some users abandoning the trial at the onset;
- the existence of a personal aura was seen as a major barrier to communication — for small numbers of users, this does not cause too much confusion, but when large numbers are gathered in a small space, restrictions arising from the aura mean that the users may not be able to see other people they are expecting to see, e.g. if the aura size is ten, it may be that the eleventh person expecting to meet friends may seem to be alone or with complete strangers;
- achieving a critical mass of people within the worlds — the primary motivator for visiting or returning to a particular world was the number of people in it, with users reporting that many worlds were empty, thus reducing motivation to return there in future.

12.6 DISCUSSION — FUTURE DIRECTION

This chapter has provided an overview of some of the factors that will influence a user's decision both to explore and, in the longer term, to inhabit virtual media environments.

In order to develop CVEs which are effective and well received by their users, the following areas have to be better understood:

- perceptual aspects of VEs, e.g. an understanding of human perception and the interaction between the senses;
- design of the content for compelling, persistent VEs;
- usability of CVEs, e.g. impact of navigation, sense of presence, communication, etc, on the user experience;
- physiological impacts of VEs on the user;
- social/ethical implications of VE technology;
- platform performance threshold required to deliver successful VEs across different media environments.

Clearly, there is a diverse range of possible applications and implementations of virtual environments and as a result there is a need for developers to be able to measure, benchmark, quantify and model systems. The basic properties of hearing and vision are fairly well understood and there are evolving recommendations from bodies such as the ITU for the assessment of audio and visual quality. However, when users are immersed in a virtual environment there is a resultant notion of presence setting new challenges requiring new metrics and assessment techniques.

In addition to understanding the perceptual issues associated with physical VE interfaces, the techniques for the capture of user requirements, design and implementation of key functions such as communication, navigation and the selection and manipulation of avatars are being developed to provide appropriate and effective implementations. The HCI field has developed models, techniques and methodologies for understanding the 'human element' across different products and services. However, with virtual reality there is a need to revisit the way in which these techniques are used.

Unlike many other available technologies, virtual reality is a totally new type of interface which is dynamically changing and evolving in response to the actions of users. Researchers and developers need to understand the nature of this medium, both from a technical and from a user-centred point of view. The challenge is to develop novel ways of combining and using new and existing tools and techniques that will help to effectively understand, design, develop and evaluate this fast-moving medium.

REFERENCES

1. Crossley M, Davies N J, Taylor-Hendry R J, McGrath A J: 'Three-dimensional Internet developments', BT Technol J, 15, No 2, pp 179-193 (July 1997).

2. Myatt E, Alder A, Ito M and O'Day V: 'Design for networked communities', Proceedings of CHI '97 (1997).

3. Benford S D et al: 'Networked virtual reality and co-operative work', in 'Presence: teleoperators and virtual environments', MIT Press (1995).

4. Walker G R: 'The Mirror — reflections on inhabited TV', British Telecommunications Eng J, 16, Part 1 (April 1997).

5. Bowskill J and Traill D: 'Interactive collaborative media environments', CVE '96 Workshop proceedings (1996).

6. Rihs S: 'The influence of audio on perceived picture quality and subjective audio-video delay tolerance', Proc MOSAIC Workshop Advanced Methods for the Evaluation of Television Picture Quality, Institute of Perception Research, Eindhoven (September 1995).

7. McGurk H and MacDonald J: 'Hearing lips and seeing voices', Nature, 64 (December 1976).

8. Boyd C: 'Does immersion make virtual environments more usable?' CHI 97, p 325 (1997).

9. Agar M: 'The professional stranger: an informal introduction to ethnology', Academic Press (1980).

10. Hillier W: 'Space is the machine', Cambridge Press (1996).

13

AFFECTIVE INTELLIGENCE — THE MISSING LINK?

R W Picard and G Cosier

13.1 INTRODUCTION

The topic of emotion is one that many engineers and scientists would prefer to eschew. After all, emotional behaviour is regarded as irrational — deficient in reasoning — and counter to the entire scientific method. Too much emotion, poorly managed, is clearly undesirable. However, there is another side to emotion that is only recently beginning to be understood, a side that is essential for rational human behaviour. Scientists have discovered that not only is too much emotion detrimental to decision making and rational behaviour, but so too is too little emotion. It is time to take a closer look at the communication and management of emotion in human/computer interaction, what is here called 'affective intelligence', and to consider how this will have an impact on developments in technology and machine intelligence.

In this chapter, the words 'emotional' and 'affective' are used interchangeably as adjectives describing both physical and cognitive aspects of emotion, although 'affective' will sometimes be used in a broader sense than 'emotional', particularly when including computers and other communications technologies. Note that the word 'feeling' does not always imply emotion; squeezing scrambled eggs between your fingers feels squishy, and is a tactile feeling, not an emotional one. Feelings such as hunger and pain are also usually not considered emotion. On the other hand, having a good feeling or a bad feeling about something — so-called 'gut feelings' — are known to be important valenced emotional indicators.

Many years ago, Gardner revolutionized thinking about human intelligence with his theory of 'multiple intelligences' [1] which includes not only the traditional mathematical and verbal abilities, but also other abilities, especially social intelligence, which consists of interpersonal and intrapersonal skills. Later, Salovey and Mayer identified these skills as 'emotional intelligence' which they define as 'the ability to monitor one's own and others' feelings and emotions, to discriminate between them and to use this information to guide one's thinking and actions' [2]. The importance of these skills has been underscored by Goleman [3] in his book which argues that affective abilities are more important than traditional IQ for predicting success in life.

Emotional intelligence in people involves many factors such as the ability to accurately recognize emotions in others, to be aware of them in oneself, to express them in an appropriate and controlled way, and to manage emotions, for self-motivation, persistence, and social deftness. But, what does all of this have to do with technology?

Today it is easy to find people who spend more time interacting with a computer than with other humans. Every minute people enter the on-line communities of the Internet where they communicate with each other through computers or other mediating technologies. Daily interaction between humans and computers has billions of dollars of economic impact, not to mention psychological impact, which is harder to quantify. Furthermore, when humans interact with technology, they do so in a very interesting way — naturally and socially, almost as if it was no different than interacting directly with a person.

The claim of social and natural interaction is based on the findings of Reeves and Nass at Stanford University, who have conducted dozens of classical tests of human social interaction by substituting computers into a role usually occupied by humans. Hence, a test that traditionally studies a human/human interaction was used to study a human/computer interaction. For example, one experiment examined how what is said by human *A* about human *B*'s performance changes when *A* gives the evaluation face-to-face with *B*, as opposed to when *A* gives the evaluation about *B* to another (presumably neutral) person. Studies of human social interaction indicate that, in general, humans are nicer face-to-face. In a variation on the traditional test, human *B* is replaced with computer *B*. Human *A* now has to evaluate the computer's performance after the computer has given *A*, say, a short lesson. Human *A* gives *B* its evaluation 'face-to-face', and then is asked by a different computer for an evaluation of how *B* did. The classic human/human results still held; for example, the tendency to be nicer face-to-face remained. Numerous similar experiments were done by Reeves and Nass, revealing that the classic results of human/human studies were maintained in human/computer studies. The findings held true even for people who 'know better', such as computer science students who know that computers do not have emotions. After accounting for potential biasing factors, the researchers concluded that individuals' interactions with computers are inherently natural and social. For details of the experiments described here, see Nass, Steuer and Tauber [4], Nass and Sundar [5], and Reeves and Nass [6].

When you see a person expressing anger at their computer, this behaviour is an example of a natural and social interaction. They know that the computer does not recognize their anger, and that it does not have nor understand anger, but they still express anger at it nonetheless. Emotion is a natural and social part of human/human communication, and people naturally use it when they interact with computers and other forms of media technology. Consequently, the critical role of emotional intelligence in human/human interaction applies to human/computer interaction, and to interaction between humans which is mediated by computers — human/computer/human interaction (see Fig. 13.1). The latter is particularly important in telepresence environments, where the critical factor in connecting geographically remote people is the natural link — human to human.

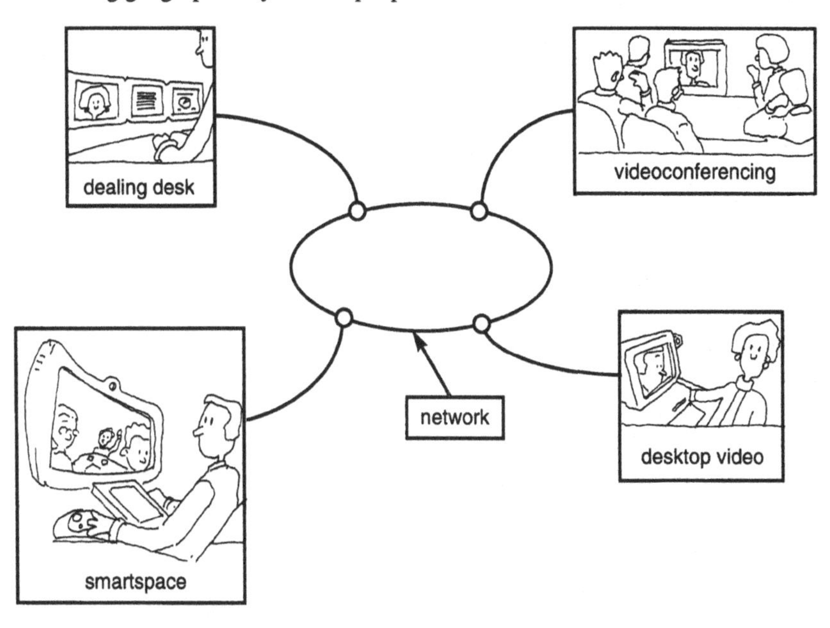

Fig. 13.1 Human/human computer-mediated communication.

13.2 COMMUNICATING EMOTION — FROM E-MAIL TO TELEPRESENCE

Telecommunications is playing an ever-increasing role in today's society, within business, education and at home. It has developed from its historical past into being a medium that not only transports voice, but is now a critical business component. However, the communications interface has developed less rapidly and while videoconferencing currently offers the richest form of telepresence today, examples of on-line communities [7] are now challenging this position. Stimulating our senses with experience or illusion is a key element in delivering a more natural communications environment. Turin, of London University, compares our sensory system to the most wonderful personal assistant — always there, beavering away for us, never complaining, and constantly underrated.

Scientists working in artificial intelligence have found out just how amazing our perceptual abilities are in trying to replicate human perception with machines.

Stimulating this amazing machine has been a vision of leading writers such as Kubrick and Clarke for many years, and the ability to appear collocated in multiple locations or worlds, which these authors have implied, has become a major business driver for the telecommunications industry today. The quote 'Beam me up, Scottie' is probably familiar to all, but can one actually appear in an alternative location, with all of the benefits and feelings of a real-world interaction? Videoconferencing has opened up the possibilities for this vision with face-to-face meetings now an everyday occurrence within the business sector. However, to date the quality of presence is somewhat limited and research shows that the interaction is poor if the participants have not previously met in person [8]. Research at BT Laboratories (BTL) has started to look into what differences are experienced between the real-world face-to-face encounter and the same situation over a networked connection. Preliminary results show that the feeling of presence is considerably reduced when communicated from a remote location [8]. Clearly this statement needs to be qualified in that the application, task, latency, cognitive load, intimacy of the participants, and pre-conditioning will all affect the perception of the interaction.

Today's videoconferencing terminals are opening up new ways of working for many business applications. However, as business becomes more like a molecular structure where each node represents a virtual partner participating from the most appropriate corner of the globe, then high-quality eye-to-eye contact becomes increasingly critical to the new business environment. This will become increasingly important as virtual business partnerships develop in this way. High-quality videoconferencing conveys facial expression effectively, but it still fails to deliver a strong sense of presence. For example, we can detect the wink of an eye or squint, but may be totally unaware that this was due to a video camera light disturbing the participant. The shared experience of meeting face-to-face in the real world is somehow lost when conventional videoconferencing is used. Experience is more than just eye-to-eye contact, it relies on the assimilation of a wide range of human gestures and on building an appreciation of the surrounding physical environment. In particular, some of the key pieces that are missing are the physical handshake, and subtle cues of affective timing and expression. Affective communication is generally impaired by current forms of mediated communication, and when the bandwidth of affective communication is reduced, then problems can occur.

E-mail, for example, is one of the most affect-limited forms of communication. Very few people have the skill to masterfully communicate feelings using text only. Almost all people that rely on e-mail have wasted one or more days trying to straighten out a misunderstanding generated by a hastily produced message that offended the recipient. It was not intended as obnoxious, and had the same words been delivered as a casual spoken comment — even had they been said on the telephone — the misunderstanding may well have been avoided. Instead, the recipient detected a nasty tone in the text and took offence. E-mail is affect-limited, because it carries very little intonation or information about the affective state of the sender. Affect is so important to the communication process that when it is constrained people find some way, however limited, to add this facet to their communication. Actors and animators exaggerate affective expression, because it is crucial that emotion be communicated accurately [9]. Users of e-mail have developed limited means for expressing their emotions. Symbol combinations like :-) and :-(are now in common use. However, these 'emoticons' are a woefully inadequate way to express the range of natural human emotions. Figure 13.2 shows how 'emoticons' are being used in an on-line chat world. This strategy places a burden on the user that is unnatural.

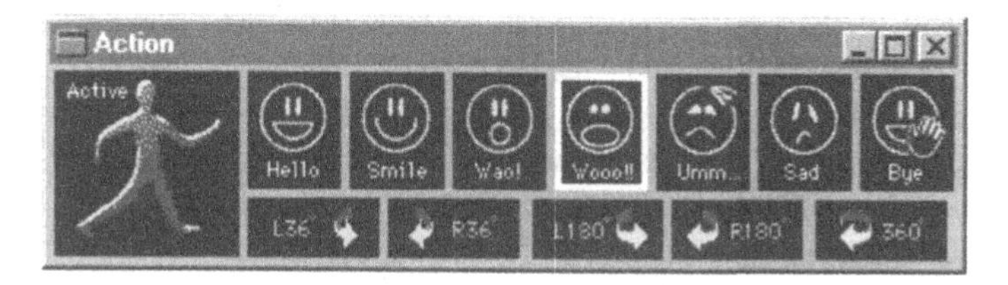

Fig. 13.2 Emoticons.

It is interesting to note that although affect might not be communicated by a piece of technology, people still tend to perceive it — to fill it in when it is missing. Even a neutral message may be perceived as emotional. Studies have shown that the mood of the perceiver influences perception of ambiguous stimuli. If healthy subjects are asked to quickly jot words they hear, then they are more inclined to spell 'presents' than 'presence' if they are happy, and to spell 'banned' than 'band' if they are sad. Subjects resolve lexical ambiguity in homophones in a mood-congruent fashion [10]. Similar results occur when subjects look at ambiguous facial expressions. Depressed subjects judge the faces as having more rejection and sadness [11]. Moods also bias perception of the likelihood of events — an individual in a negative mood perceives negative events as more likely and positive events as less likely, and the reverse holds true for people in positive moods. A variety of studies have indicated that not only does affect play a critical role in perception and judgement, but it also plays a critical role in creativity and other aspects of cognition [12].

Affective intelligence is a new area of research, being explored by BT Laboratories and the MIT Media Lab, which focuses on understanding how emotion is communicated and managed in human/computer interaction, and in human/human interaction when that interaction is mediated by technology. In particular, affective intelligence involves communication of affect — recognizing human affective expression, and transmitting it in a meaningful way.

Examples are a synthetic avatar representing a real human and mimicking the expressive gestures and intonation, and a software agent that is trying to adapt to its user, and recognizes its user's expression of dislike and like, in order to figure out when it needs to modify its behaviour. In these cases and others like them, the role of affect is a key link in facilitating communication. The focus of the present research is to establish some means of recognizing affect, and to understand how this new information can contribute to more natural human/computer interfaces and telepresence applications. In so doing, the so-called hidden elements of human communication can be treated scientifically, and the technology can be adjusted to overcome some of its present limitations.

Understanding and accommodating affective communication is a key part of affective intelligence. In particular, affective communication is usually subtle, and it can be important to preserve its subtlety. Consider, for example, the problem of trying to run a brainstorming meeting when all the participants are not physically in the same place. If you try to do this by e-mail, it will be easy to collect a lot of ideas, but it will be hard to 'sense' when one of the ideas appeals to everyone. This is critical, because people tend to build on the ideas that they sense people like. The way that 'liking' tends to be communicated is by subtle affective cues — did everyone perk up and look interested when that idea was spoken? Did people lean forward or raise their eyebrows? Did they respond quickly, or with hesitation? Did people slump back, exhale, or frown? These subtle affective cues cannot be replaced by more overt expressions, such as asking people to rank every idea as it is expressed, which would be unnatural and upset the creative process. However, these things are not communicated by e-mail, and only a few of them are communicated by present videoconferencing technology. The new enhanced reality environments being developed at BT [13, 14] aim to address these deficiencies.

Consider the application of affective intelligence to such new scalable collaborative environments. Traditional videoconferencing is not scalable; as more remote parties join the conference, each appearing to look out of the screen, it becomes increasingly difficult for participants to appreciate the shared experience. In scalable collaborative environments, participants share space in a 3-D virtual world, enabling communities to interact in a way that is more analogous to typical human interaction. However, the behaviours or rules of engagement within these shared environments need to be governed so as to mediate the interaction between the real-world surroundings within which the user is located and the computer-generated world within which they are engaged. For example, if a user is interrupted within your real-world space, then the affective interpretation of their presence within the shared environment will need to be addressed. If one person responds quickly, leaning forward with interest, then this needs to be communicated. Even subtle differences in timing — instantaneous versus hesitating — are important for judging affective response. These environments need enough affective intelligence to recognize what cues are important, no matter how subtle, and to preserve the communication of these cues.

The blurring of the real world with the computer-generated world is of particular importance in emerging communication environments. The concept of immersion is relevant with respect to traditional television, which in its current form, can be viewed as a socially isolating medium, although paradoxically broadcasts are essentially a shared media involving millions of people at the same time. Extending the experience of this media could imply modifying our interface to television into one whereby the viewer relaxes in their own home complete with virtual friends and family occupying virtual chairs all watching (and sharing) the same telepresence experience, and perhaps also sharing in the joint excitement or disappointment of an interaction. Imagine if, when you are on the road travelling, your hotel room allowed you to virtually join your family in a computer-generated living room — laughing at the same jokes, seeing the happiness in a child's eyes while you play a game, reading a bedtime story with expressive voices and faces, all the time allowing awareness of each other's feelings — facilitating the kind of full-bandwidth communication that brings people closer, in spite of physical distance.

13.3 BACKGROUND — BASICS OF EMOTION

Nearly a hundred definitions of emotion have been collected from the literature [15], and will not be repeated here. Internal human states like fear, anger, sadness, and joy, which psychologists call 'affective states', are examples of emotion upon which there is wide agreement. But what about interest, boredom, hope, frustration, and lust? Emotion theorists do not agree on precisely what is and is not an emotion. Fortunately, this lack of a solid definition does not mean that we cannot proceed to conduct solid research. To quote John McCarthy: 'We can't define Mount Everest precisely — whether or not a particular rock or piece of ice is or is not part of it; but it is true, without qualification, that Edmund Hillary and Tenzing Norgay climbed it in 1953. In other words, we can base solid facts and knowledge on structures that are themselves imprecisely defined.' Most people's intuitive concept of emotion is well-established, and serves the purposes of this chapter well, for beginning to understand how a broad range of affective phenomena significantly influence mixed media telepresence.

Many theorists have suggested that there are 'basic emotions' and have defined these to satisfy different criteria — for example, to have a basic facial expression, such as a joyful smile. However, by this definition, 'love' is not a basic emotion because it does not have a basic facial expression. Alternatively, some researchers emphasize not the physical aspects of emotions, but the cognitive aspects. A person might cognitively appraise a situation as favourable to their goals, and thereby have a joyful emotion. To recognize such an event, a computer would need to be able to perceive a situation and reason about the emotions that it might generate. Additionally, computers would benefit from being able to perceive what emotions are actually expressed in a situation. Towards this goal, there is much research describing the physical changes in a person which accompany an emotion, such as a furrowed brow when confused, raised vocal pitch when annoyed, increased heart rate and temperature when angry, and so forth. We believe that computers need to understand both the physical and cognitive aspects of emotion.

Computers might use many representations for emotion — not just discrete categories as implied by lists of emotions such as fear, anger, joy, and sadness, but also continuously dimensioned spaces. The two most common dimensions are 'valence' and 'arousal'. In general, two dimensions cannot be used to distinguish all emotions; for example, intense fear and anger lie in the same region of high arousal and negative valence. However, these two dimensions do account for many important affective phenomena (see Fig. 13.3). Depending on the application, a representation using continuous dimensions or discrete categories may be preferred. For example, the dimension of arousal is known to be highly correlated with attention and memory [6]. If an application involves describing viewer interest, as for assessing TV broadcasts, then the dimensions of arousal and valence are very useful.

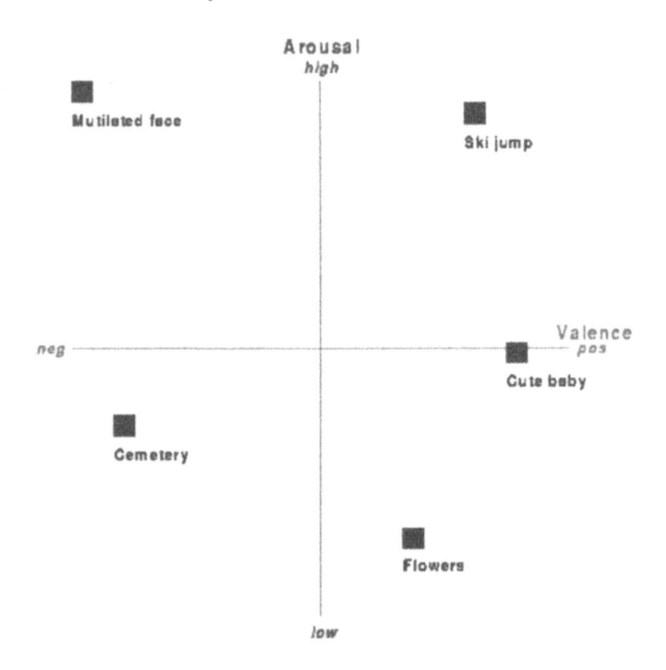

Fig. 13.3 The dimensions of 'valence' and 'arousal' [16].

The details of what is and is not an emotion are not as important for affective intelligence as are the abilities to accurately communicate a broad range of affective phenomena. The communication of emotion, even if subtle, can powerfully influence actions. An example of the impact of emotional expression occurs in the film *Schindler's List* when a one-armed Jewish machinest goes to thank Schindler for giving him a job. His face lights up when he sees Schindler and he leans towards him with dignity and gratitude: 'I want to thank you, sir, for giving me the opportunity to work ... the SS beat me up, they would have killed me. But, I'm essential to the war effort thanks to you ... God bless you, sir. You are a good man ... God bless you.' The sincere joyful appreciation on the man's face clearly affects Schindler's feelings, and those of many viewers. The emotions communicated in this scene become a turning point in the film, after which Schindler's actions begin to change.

Even without trying to communicate emotion, emotions can be contagious. People may recall a favourite teacher who was so enthusiastic about a subject that they became interested in it too. Alternatively, you may have watched an actor express emotion so effectively that its impact, not only on the other characters in the drama, but also on the audience, was electrifying. In order to achieve this heightened level of communication, actors study posture, gestures, facial expressions, and vocal intonation and use a wide range of these attributes to communicate their character's feelings. An effective actor is a master of affect.

Emotions can be communicated in many ways. The most visible form is through facial expressions. However, this form is also the easiest to control, and therefore can be the least sincere. Vocal inflection is less easy to control, and is perhaps the second most common way to communicate emotion. Other visible forms include gesture, posture, behaviour, and more subtle cues such as pupilary dilation. Pupils tend to dilate with interest and with mental workload, as well as with decreased illumination. Less visible bodily changes also accompany emotional responses — changes in galvanic skin conductivity, heart rate, blood pressure, skin temperature, respiration, muscle tension, and so on. These changes may not be noticed in casual social interaction, although some of these can be picked up from a handshake or from sitting close to a person. All of these cues, even very subtle ones, can be used to communicate important information, such as whether or not a person liked the suggestion you just made in your virtual meeting.

13.4 EXAMPLE — THE AFFECTIVE MIRROR

Imagine the most important interview of your life, with the head of the company that you have always wanted to work for. You are asked tough questions about problems you have solved, challenges you have faced, and why you want to leave your present job. At the end of this gruelling meeting, he tells you that you were too nervous-sounding, had unusually short pauses in your speech, were evasive with eye contact, and had cold clammy hands. This was not the real thing, fortunately, but a practice session in front of your trusted computer. Your computer interviewing agent, displaying the face of the chief executive, asked you questions while listening to changes in your voice and discourse parameters. It watched your facial expressions and body language, sensing changes in physiological parameters such as your skin conductivity and temperature. It watched your affective responses to see where they differed from what it usually senses from you in day-to-day interaction.

This scenario is still in the future, but most pieces of it are present technology. Inspired by the high school boy practising asking a girl out for a date, rehearsing in the mirror, the 'affective mirror' would be an agent that interacts with a person, helping individuals to see how they appear to others in various situations. With a camera, microphone, and various sensors, the system is far more advanced than the mirror in front of which most people practice, even though it is potentially not as good as practising in front of a skilled human listener. Nonetheless, there are many times when a computer's availability, patience, and non-partisan judgement cannot be beaten. Being able to 'try out' that important talk in front of your computer has a certain convenience and privacy that some find encouraging and comfortable. This use of affect recognition is one of many ways in which the research described here might be used in a virtual environment, not just for advancing one's own interactive skills, but, in general, for improving communication that is mediated by technology.

13.5 DEVELOPING TECHNOLOGY THAT COMMUNICATES EMOTION

Present technology tends to block the transmission of affective information. We think of e-mail as having less affective bandwidth than a phone call, a phone call as having less affective bandwidth than a videoconference, and a videoconference as having less affective bandwidth than physically 'being there.' One of the goals of telepresence is to provide a much closer sense of being there — especially the subtle and significant aspects of emotion communication.

How might tomorrow's enhanced reality systems facilitate truly natural and intuitive human communication? Research on developing basic techniques to assess perceived performance employs models of the human senses in order to include aspects of perception in engineering metrics. Models of hearing are used as a basis for assessing speech quality performance [17, 18], while further research is combining audio and video models for a multi-sensory assessment [19] (see also Chapter 14). Such an approach represents an important breakthrough since it allows performance assessment to be based on only those aspects of signal streams which are available to human consciousness. In the future such models may be embedded in the delivery technology, exploiting algorithmic models of perception to dynamically maximize subjective performance. To date, however, the emphasis in these research areas has been predominantly on tasks such as objective prediction of perceived quality, and performance optimization in view of human sensory performance. Affective cues, such as intonation in speech, are usually treated as noise. Sometimes it is not what is said that is most important, but how it is said. The expression on a speaker's face, the way they are leaning forward, and the subtle intonation changes in their speech can communicate powerful messages — sometimes even the opposite of what is spoken.

Tools that enable affective communication are useful not just in enhanced reality systems, but also for helping people with various kinds of disabilities. Stephen Hawking, who relies on text-to-speech synthesis, is one of millions who is without effective natural speech communication. One of the problems with present text-to-speech systems is that they say everything with the same tone of voice. This makes it particularly difficult to communicate feelings — to interrupt angrily, to express anxious concern, to soften your voice in approval, or to indicate empathy and other expressions which illustrate a speaker's emotional intelligence.

Research is being conducted by Murray at Dundee University in Scotland [20], Alm, Murray, Arnott, and Newell [21], and by Cahn at the MIT Media Laboratory in Cambridge, USA [22] to determine parameters of speech which these systems can modify to enable the computer to speak with suitable intonation. However, there is an interface problem, in that people talk much faster than they can type. Speech is typically around 180—250 words per minute (wpm), while a good typist may type 50—60 wpm or a stenographer with a chording keyboard may type 180 wpm. Some disabled people using standard keyboards persevere at rates of under 10 wpm. People are already limited by how quickly they can give input to a text-to-speech device. To add affective parameters suggests more control knobs for them to set, which is undesirable given that they are already busy typing as fast as possible. In some cases, when the emotion is more important than the semantics, a person might opt for hitting a button producing 'an angry interrupting sound' or some other 'audio emoticon' that conveys the emotion of the response without words. Computer recognition of affect could help in a more direct way — by recognizing directly from a person's physiological changes how their affective state is changing, and how using this can potentially help speaking disabled persons to be more expressive, without increasing their typing burden.

Within the telepresence experience that will deliver truly natural communication, we see a vision of affective audiovisual transmission. Not only should facial, gestural, and verbal communication deliver 'affective bits', but the system could potentially be intelligent about understanding these bits, and helping people express them, as in the affective mirror scenario above.

13.6 RESEARCH ON AFFECT COMMUNICATION

Research in computer recognition and expression of emotion is in its infancy. Current research efforts at BTL and the MIT Media Lab focus on recognition of facial expression, voice affect synthesis, and recognition of emotions from changes in physiological state parameters. Note that you cannot directly recognize somebody's emotions — you can only recognize affective expressions and try to infer an underlying emotional state. Where this chapter refers to 'emotion recognition', this should be interpreted as 'combining recognition of expressions with reasoning about emotion generation to infer an underlying emotional state.'

Computers, like people, can use cognitive reasoning, a form of common sense, to understand people's goals and predict their affective state when they are disrupted. One way to recognize an expression is to record facial movements, digitize the sequence, then apply pattern recognition. Recognition from a moving sequence is generally more accurate than recognition from a still image. If, for example, a person's 'neutral' expression is a pout, only deviations from the pout (captured by video as movement) will be significant for recognizing affect.

Pattern recognition can utilize a variety of techniques such as analysing individual muscle actuation parameters or (more coarsely) characterizing an overall facial-movement pattern. In a test at the MIT Media Lab involving eight people, recognition rates were as high as 98 per cent for facial expressions of four emotions (see Fig. 13.4) [23]. The emotions shown in this example are clearly exaggerated, or simulated emotions. This is analogous to controlled enunciated

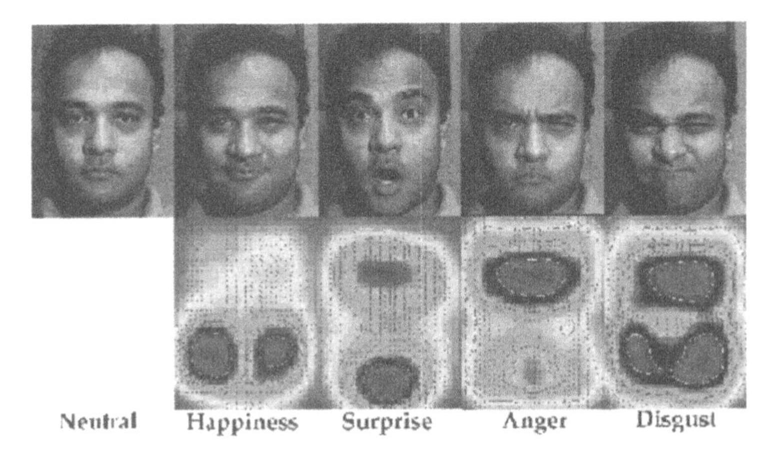

Fig. 13.4 Facial expression recognition and motion energy maps. Top row: shots of the neutral expression and four others — happiness, surprise, anger and disgust. Bottom row: templates of coded energy of each facial movement, in contrast to the still image of the neutral expression. [Courtesy of Irfan Essa]

'yes' and 'no' in speech recognition but clearly demonstrates how this technique can develop. Studies are under way to determine how the recognition rate changes when there are more experimental subjects. As yet, this method of recognition does not work in real time; it takes a few seconds to recognize each expression. However, advances in hardware and pattern recognition should make recognition essentially instantaneous in the near future at least for familiar expressions.

Although facial features are one of the most visible signs of underlying emotional states, and can of course be transmitted with conventional videoconferencing techniques, they are also easy to control in order to hide emotion and can be masked with video-coding techniques. One strand of this research is looking into how we perceive these hidden cues in the straight face-to-face interaction.

Emotional expression is clearly not limited to facial movement. Vocal intonation is the other most common way to communicate strong feelings. Several features of speech are modulated by emotion; they are divided into such categories as voice quality (e.g. breathy or resonant, depending on individual vocal tract and breathing), timing of utterance (e.g. speaking faster for fear, slower for disgust), and utterance pitch contour (e.g. showing greater frequency range and abruptness for annoyance, a smaller range and downward contours for sadness), as illustrated in Fig. 13.5. As these features vary, the emotional expression of the voice changes. The problem of precisely how to vary the vocal features to synthesize realistic intonation remains part of the research programme.

Preliminary experiments indicate that computers can synthesize affect sufficiently well for human listeners to recognize six different categories of emotional expression with 50-90% accuracy, significantly higher than the chance levels of 17% [22]. Humans, on average, can recognize affect with about 60% reliability from other humans [24] when tested on neutral speech or on speech where the meaning has been obscured.

The inverse problem — computer intonation analysis, or recognizing how something is said — is quite difficult. Research to date has limited the speaker to a small number of sentences, and the results are still closely dependent on the particular words spoken. A method of precisely separating what is said from how it is said has not yet been developed.

New research has also begun at the MIT Media Lab in recognizing affect from physiological changes in a person, especially from signals which can easily be sensed by off-the-shelf sensors, such as a comfortable elastic heart monitor, commonly worn during athletic exercise. Four such signals are shown in Fig. 13.6, measured from an actress while she consciously expressed anger (left) and grief (right). From top to bottom in Fig. 13.6, the physiological signals are:

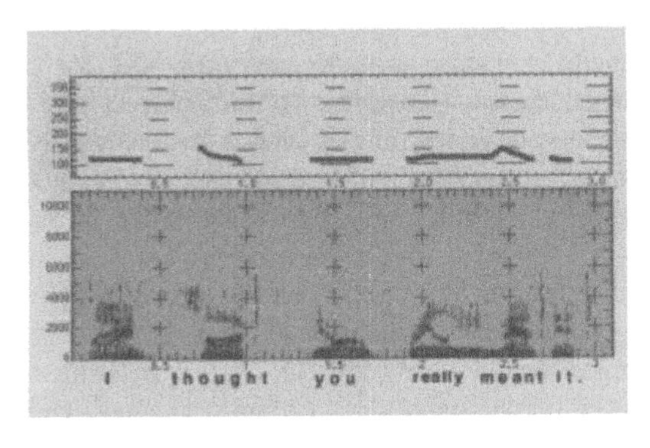

(a)

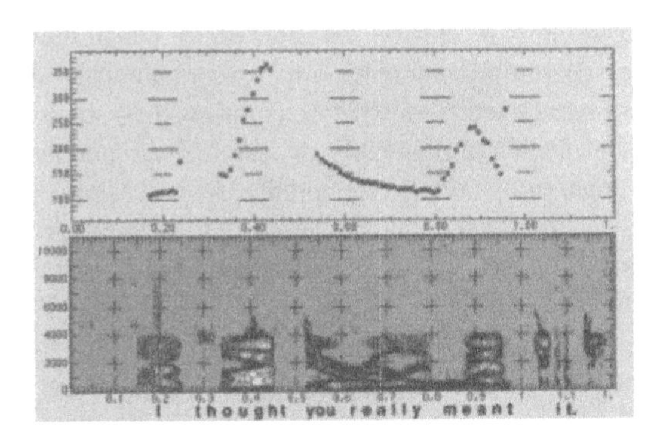

(b)

Fig. 13. 5 Voice inflection synthesis. The same sentence ("I thought you really meant it") is spoken with (a) sad effect and (b) annoyed effect. The pitch track (top) and spectrogram (bottom) for each case are shown. It should be noted that the pitch range for the annoyed affect is greater than the relatively compressed range for the sad affect. The spectrograms also record differences in the speed, hesitation, and enunciation of the two cases.
[Courtesy of Janet Cahn and Lisa Stifelman]

- electromyogram (microvolts);
- blood volume pulse (per cent reflectance);
- galvanic skin conductivity (microSiemens);
- respiration (per cent maximum expansion).

All of these signals can be gathered from sensors on the surface of the skin, without any pain or discomfort to the person. Each box shows 100 seconds of response. Preliminary pattern recognition results, run on various features of these signals, indicates significant recognition rates of five emotion categories, at rates 2-3 times higher than chance [25].

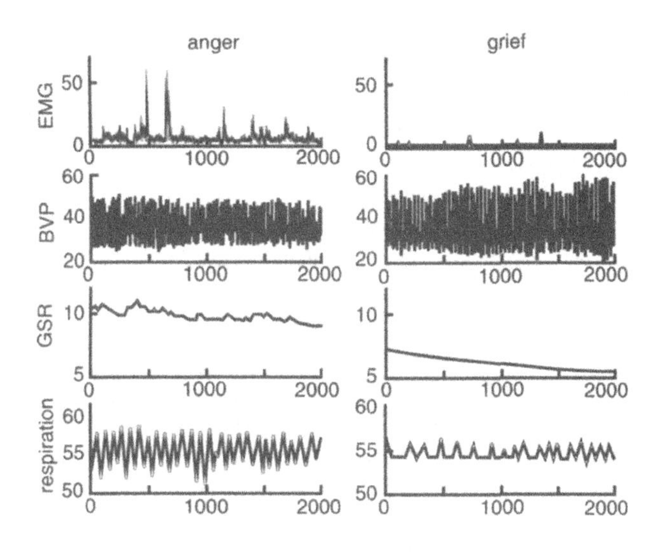

Fig. 13.6 Example of physiological signals.

There are several other efforts under way, both at BTL and MIT Media Lab, to teach computers how to recognize and intelligently respond to a broad range of affective expressions — including frustration, like versus dislike, confusion, and distress. These kinds of emotions are naturally expressed in both human/human interaction mediated by technology, and in human/computer interaction. It is therefore important not to limit them in communication.

It is to be noted that no one method — whether recognition of facial expression, voice intonation, or physiological changes — is likely to produce reliable recognition of emotion. In this sense, affect recognition is similar to other recognition problems like speech recognition and lipreading. It is probable that a personalized combination taking into account both perceptual cues (e.g. from vision and audition) and cognitive analysis of the situation is most likely to succeed. These cues will undoubtedly work best when considered in context — is it a poker game, where bluffing is the norm, or a marriage proposal, where sincerity is expected?

Although emotion researchers have worked hard to uncover universal aspects of emotional behaviour, few have explored emotions from the perspective of a person who knows another person well and can often guess their affective state. The key here is 'get to know.' Working closely with a handful of people, it can be discovered how each one of them expresses — or does not express — emotions under various circumstances. By becoming familiar with several individuals, commonalities can be detected in the ways they express emotions and perhaps used to improve recognition of affect in strangers.

There is an analogy between affect recognition research and speech recognition research in terms of the kinds of progress that it is reasonable to expect. Initially, speech-recognition systems could only recognize speech when trained on an individual speaker. Similar progress in affect recognition can be expected — initially, the best results will occur when the system is trained on the individual whose affect is being recognized. As the research and technology advance, however, the systems should improve their ability to recognize person-independent affective expressions, as has been the case in speech recognition.

There are many factors that complicate emotion recognition. The generation of emotions is influenced by goals, personality, diet, attitude, expectation, perception, and culture. Rules about the social display of emotion, e.g. the inappropriateness of showing anger, also influence its expression and make it difficult for researchers to link emotional states with forms of expression. There is presently no widely accepted comprehensive theory of emotions although much has been learned. Furthermore, nobody is 100% accurate at recognizing emotion. It is clear, however, that emotions can be communicated, and that even their imperfect communication is extremely important in intelligent human interaction.

Even with the limitations of today's technology we are able to see the benefits of gesture and presence. In November 1995 BTL and MIT Media Lab demonstrated the world's first transatlantic virtual conference. This consisted of two active telepresence environments connected via an ISDN network connection. The representation of the participants is seen at the opposite end of the connection as an avatar which mirrors the movements of the host. This technique, referred to here as host and ghost, benefits conventional videoconferencing in that participants of the connection have the social etiquette to which we are accustomed in real-world interactions. Figure 13.7 shows an example of real and virtual representations embodied within the telepresence environment. Figure 13.8 also shows an example of a virtual human [26].

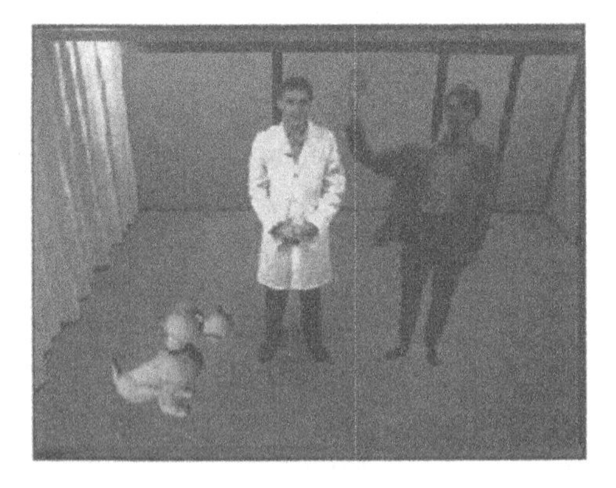

Fig. 13.7 Example of a virtual conference, mixing real video with avatars.

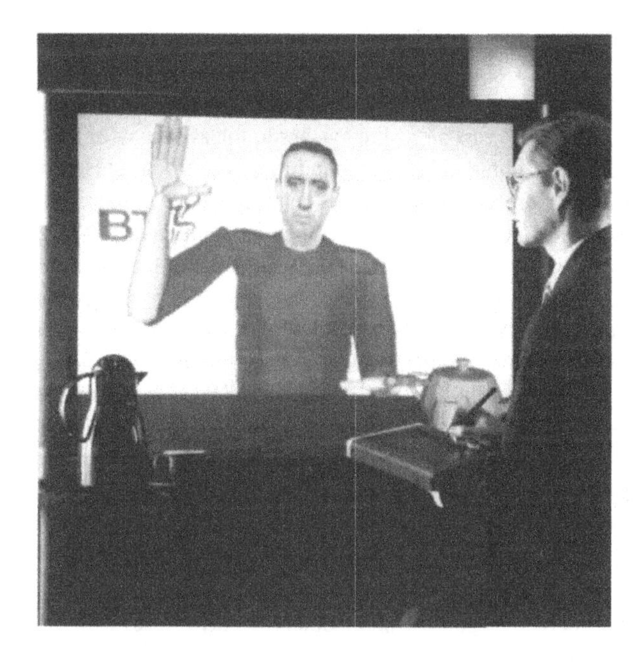

Fig. 13.8 Example of a virtual human.

13.7 TOWARDS TRULY PERSONAL COMPUTERS

Whether through a mouse, keys, joy stick, or touch screen, many people have more physical contact with computers than they do with other people. Moreover, you can now wear computers in your shoes, shirt pocket, or belt, as clothing or accessories. Wearable computers, especially when they become as common as watches, will have the opportunity to get to know their wearer in a variety of situations. They could have access not only to the signals shown above, but also to subtle changes in temperature, blood pressure, and possibly even internal biochemistry such as blood sugar levels. Instead of being restricted to perceiving only your visible and vocal forms of affect expression, they could get to know you intimately — or as well as you will permit them to. At this point, they will also, like underwear, probably cease to be shared and will become truly personal computers. The MIT Media Lab has developed a prototype affective wearable computer, which can gather the signals mentioned above [27]. As this work proceeds, it will be important to highlight the difference between recognizing affective symptoms which are available in face-to-face situations, and those that can be found by slightly more invasive sensors, needed in this preliminary research.

Affective information could also be communicated in unconventional ways. Imagine that a child's wearable computer could detect the interest or excitement in the classroom and broadcast it to their teacher or virtual tutor. The result would be a sort of 'mood ring' that alerts the teacher to the affective (or receptive) state. Consider the same technique being applied to a boardroom presentation, or to a large audience-participation game. A real need for such communication becomes apparent if you have ever given a lecture by videoconference to a large audience. There, in particular, you usually cannot see the details of the faces and posture of the people in the audience, whether they are expressing confusion or interest, enthusiasm or boredom. A barometer which portrays audience arousal would be useful feedback for the remote speaker, so they could adjust their presentation more effectively to the audience's response — changing their speaking style and telling more stories or jokes if they are falling asleep, or maintaining the present pace if the audience is eagerly listening. This information from the audience might be detected by sensors, which are wearable or embedded into things that they are naturally in physical contact with, such as the arm of their chair, their shoes or eyeglasses, or a ring or writing instrument.

Applications of affective recognition could extend to entertainment as well; for example, interactive games might detect a player's level of fear and give bonus points for courage. When the responses of a student playing the computer game Doom were measured, the electromyogram of jaw clenching was expected to peak during high-action events such as when a new deadly enemy starts an attack. Indeed, we found many small peaks during stressful events of the game. However, we were surprised to find that the biggest peak — significantly higher than all the others — occurred not during a point of game action, but occurred when the student had trouble with the software (see Fig. 13.9).

Computers coupled with suitable sensors and pattern-recognition algorithms should soon be able to recognize the basic affective states of a willing individual in a typical context. The emphasis on 'willing' participant here is important. Measurements of affective states obtained in an underhanded manner are not likely to be accurate. People who want to deceive such systems will probably succeed, and this is the right approach. For example, one subject managed to fool a polygraph (which gathers physiological signals similar to those used in affect recognition) by putting a thumbtack in his shoe under his big toe; he stepped on it every time he was questioned in a particular way. Affective computing will be most accurate, and useful, when it is willingly communicated, presumably for the mutual benefit of everyone involved. There are potentially harmful abuses of this technology, as with most technology, but safeguards can be evolved to prevent most of these from occurring [25]. The effect of mood on perception shows that facial expression as a communicator of affect is not a simple process. The possibilities for positive and negative feedback between affect perception and expression must be very significant — maybe our whole feeling about a

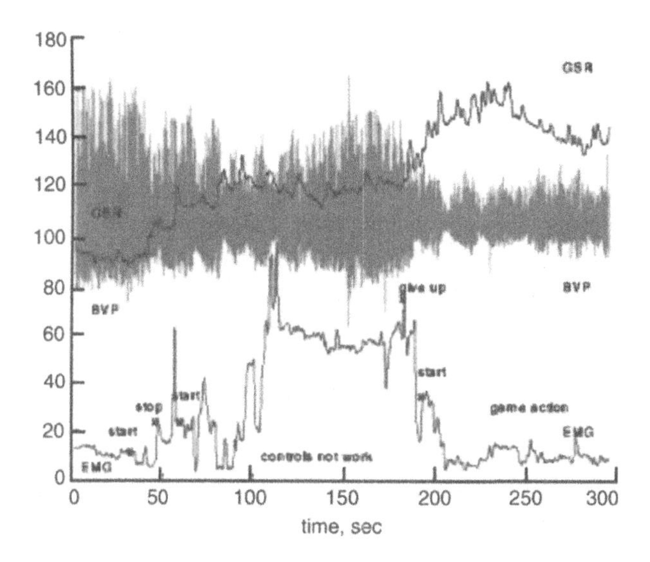

Fig. 13.9 'Doom' electromyogram.

conversation can be determined by a few seconds of affect exchange, like modems talking to each other.

There are many other applications of affect recognition for making smarter technology. Much research on software agents focuses on giving them the ability to learn their user's preferences.

The way a human friend learns your preferences is largely by passively watching what you do and do not like. Computers could potentially be given this ability, so that they could learn to adjust their own behaviour in accord with your affective expressions of like and dislike.

A simple example is learning when to interrupt you. Intelligent people do not interrupt you willy-nilly, especially if you look annoyed when they barge in. If they repeated their interruptions despite the disapproving glances you gave, then you would consider them to be unintelligent, to say the least. In contrast, the telephone and computer interrupt us all the time, but do not have the ability to see if we were pleased by the interruption or angered.

These devices, like agents, could learn to adjust their behaviour based on the cues that humans naturally use for communicating preferences — affective expressions. One of the key pieces they are missing, however, is the ability to recognize human affective expression.

13.8 WHERE DO WE GO FROM HERE?

> "Sometimes the truth of a thing is not so much in the think of it, but in the feel of it."
>
> P Stanley Kubrick, *2001: Filming the Future*

Scientific evidence indicates that emotions are not a luxury; they are essential for 'reason' to function normally, even in rational decision-making, and they are a natural and significant part of human interaction and face-to-face communication. Whether it is used to indicate like/dislike, interest/disinterest, or excitement/disappointment, emotion plays a key role in human communication. Affective intelligence is a new area of research which will permit technology-mediated communication to proceed more naturally among people. By giving computers the ability to recognize affect, and by considering how this can contribute to more natural human/computer interfaces and telepresence applications, the whole nature of the interface can be changed. As technology matures and the focus shifts toward human-centred issues, it is absolutely essential to pay more careful attention to the affective responses of people — the customers of the new technology. Consider, for example, a personal numbering system which allowed the affective state of the caller to anonymously become a part of the call routeing process, taking into account such issues as the potential impact on the caller of routeing their call to a voice messaging service rather than direct to the intended recipient.

Several potential applications to telepresence environments have been described — an affective mirror for virtual role-playing, a barometer for videoconferenced lectures, affective communication in shared spaces so that avatars react in an appropriate manner, and virtual presence games that sense affective responses. Additional applications for computers that can interpret and understand affect include better learning systems (computer recognizes interest, frustration, or pleasure of pupil/subject or inhabitant), and even wearables that sense chronic anger or distress, helping to provide personalized feedback for those interested in reducing stress, to improve their overall health.

In many ways, research in affective intelligence is basic to developing understanding about humans. Human emotion theory has been held back in many ways by the inability to run experiments in natural and social situations, outside artificial laboratory environments. By developing pattern-recognition and signal-processing tools for affective information, it is hoped to make a substantial contribution to research on basic human emotions. This is a key element of understanding the process of communications between people and the factors that influence both the communications process and the perception of that process by the participants. Furthermore, by developing such an understanding it should be possible to make better informed decisions regarding the styles or forms of communication which will be in demand in the future, and thence the nature of the technologies needed to underpin these new forms of communication.

By giving computers affective intelligence, we give them the ability to sense the affective responses of people, to help people in communicating affect, and to begin to intelligently adapt their behaviour to people. The ultimate goal is one of human-centred technology, where computers adapt to people and facilitate what comes naturally to people, instead of vice versa. Telepresence environments in particular need to facilitate the kind of interaction that comes naturally to humans. Emotions should be allowed to be communicated seamlessly, and subtly, as is natural among people sharing the same physical space.

REFERENCES

1. Gardner H: 'Frames of mind', BasicBooks, New York (1983).

2. Salovey P and Mayer J D: 'Emotional intelligence', Cognition and Personality, 9, No 3, pp 185-211 (1990).

3. Goleman D: 'Emotional intelligence', Bantam Books, New York (1995).

4. Nass C I, Steuer J S and Tauber E: 'Computers are social actors', CHI '94 Proceedings, Boston, MA, pp 72-78 (April 1994).

5. Nass C I and Sundar S S: 'Is human-computer interaction social or parasocial?', Tech Rep 100, Stanford SRCT (1994).

6. Reeves B and Nass C I: 'The media equation', Centre for the Study of Language and Information (1996).

7. Morningstar C and Farmer F R: 'The lessons of Lucasfilm's Habitat', in Benedikt M (Ed): 'Cyberspace: First Steps', MIT Press, Cambridge, MA (and http://www.communities.com/habitat.html) (1990).

8. 'Comparison of face-to-face and video mediated interaction', Interacting with Computers, 8, No 2, pp 177-192 (1996).

9. Thomas F and Johnson O: 'Disney animation — the illusion of life', Walt Disney Productions (1981).

10. Halberstadt P B, Niedenthal P M and Kushner J: 'Resolution of lexical ambiguity by emotional state', Psychological Science, 6, pp 278-282 (September 1995).

11. Bouhuys A, Bloem G M and Groothuis T G: 'Induction of depressed and elated mood by music influences the perception of facial emotion expression in healthy subjects', Journal of Affective Disorders', 33, pp 215-226 (1995).

12. Mayer J D and Salovey P: 'The intelligence of emotional intelligence', Intelligence, 7, pp 433-442 (1993).

13. Bowskill J, Morphett J and Bownie J: 'A taxonomy for enhanced reality', International Symposium on Wearable Computers (WEARCON '97), Cambridge, MA (October 1997).

14. Walker G: 'The Mirror — reflections on inhabited TV', British Telecommunications Eng J, 16 pp 29—38 (and http://virtual business.labs.bt.com/msss/IBTE_Mirror/index.htm) (April 1997)

15. Kleinginna P R Jr and Kleinginna A M: 'A categorized list of emotion definitions, with suggestions for a consensual definition', Motivation and Emotion, 5, No 4, pp 345-379 (1981).

16. Lang P J: 'The emotion probe: studies of motivation and attention', American Psychologist, 50, No 5, pp 372-385 (1995).

17. Hollier M P, Hawksford M O and Guard D R: 'Objective perceptual analysis: comparing the audible performance of data reduction systems', Presented to the 96th Audio Eng Soc Convention (March 1994).

18. Hollier M P, Hawksford M O and Guard D R: 'Error activity and error entropy as a measure of psychoacoustic significance in the perceptual domain', IEE Proc Vis Image Signal Process, 141, No 3 (June 1994).

19. Hollier M P and Voelcker R: 'Objective performance assessment: video quality as an influence on audio perception', Presented to the 103rd Audio Eng Soc Convention (September 1997).

20. Murray I R and Arnott J L: 'Toward the simulation of emotion in synthetic speech: a review of the literature on human vocal emotion', J Acoust Soc Am, 93, pp 1097—1108 (February 1993).

21. Alm N, Murray I R, Arnott J L and Newell A F: 'Pragmatics and affect in a communication system for non-speakers', Journal of the American Voice Society, 13, pp 1-15 (special issue — People with Disabilities) (March 1993).

22. Cahn J E: 'The generation of affect in synthesized speech', Journal of the American Voice Society, 8, pp 1-19 (July 1990).

23. Essa I and Pentland A: 'Coding, analysis, interpretation and recognition of facial expressions', IEEE Transactions on Pattern Analysis and Machine Intelligence, 19, pp 757-763 (July 1997).

24. Scherer K R: 'Speech and emotion states', in Darby J K (Ed): 'Speech Evaluation in Psychiatry', Chapter 10, pp 189-220, Grunne and Stratton Inc (1981).

25. Picard R W: 'Affective computing', Cambridge, MA, MIT Press (September 1997).

26. Mortlock A, Sheppard P, Machin D and McConnell S: 'Virtual Conferencing', IEE Colloquium on Teleconferencing Futures (June 1997).

27. Picard R W and Healey J: 'Affective wearables', Proceedings of the First International Symposium on Wearable Computers, Cambridge, MA (October 1997).

14

TOWARDS A MULTI-MODAL PERCEPTUAL MODEL

M P Hollier and R Voelcker

14.1 INTRODUCTION

Reliable objective assessment of perceived performance is required for optimal design, commissioning, and monitoring of quality. Nonlinear processes such as data reduction have necessitated the development of a new generation of objective assessment algorithms. Conventional engineering metrics do not take adequate account of the properties of the human receiver, and new perceptually motivated techniques have started to emerge.

Perceptually motivated audio assessment is relatively advanced and numerous models have been proposed for objectively assessing both high-quality audio, e.g. Paillard et al [1], and telephone speech quality, e.g. Hollier et al [2], and Beerends and Stemerdink [3]. Visual perceptual models which reproduce the gross psychophysics of vision are also described in the literature, usually specializing in particular aspects of visual perception, e.g. Watson and Solomon [4], and Ran and Farvadin [5]. Models of the individual senses do not reproduce the full complexity of human perception but rather allow signals to be transformed, taking account of the dominant psychophysics, to provide a more perceptually relevant representation of the signal. In this way parts of the signal which are imperceptible are given a lower weighting and parts of the signal which coincide with peaks in sensory sensitivity are given a higher weighting. The value of this approach is highlighted by the success of perceptually motivated codecs that achieve data reduction by exploiting the perceptual significance of signal components [6]. Although higher level perceptual issues have a bearing on opinions expressed by subjects, the capacity to exploit basic sensory performance remains. This is the aim for multisensory analysis where low-level cross-modal effects can be usefully measured and modelled with high-level cognitive issues dealt with by constraining predictions to particular application scenarios.

Advanced telepresence, virtual reality and multimedia applications and services rely on more than one of the human senses. As such services become widely used, a multisensory perceptual model will be required to account for the interaction between the senses in order to predict, and optimize, perceived performance. This chapter reviews recent advances in single-sense perceptual models, reports on an early experimental investigation into cross-modal dependency, and proposes an algorithmic basis for a multisensory perceptual model.

Future trends in perceptual modelling are discussed including the need to develop more complete descriptions of the influence of task on perception. Accurate measurement of perception and the influence of task will require immersion of test subjects in highly realistic environments in order to provide algorithmic models of perception which are sufficiently robust to be exploited in the delivery technology.

14.2 PERCEPTUAL MODELS

To measure the subjective performance of complex communications systems it is necessary to take account of the properties of the human senses. For example, a conventional measurement of audio signal-to-noise performance may indicate that a system has poor performance. However, the noise component may be highly audible or quite inaudible depending on whether the noise was masked by the desired signal. A performance measurement that includes a model of the masking that occurs in hearing is capable of distinguishing between these two cases.

Using models of the human senses to provide improved understanding of subjective performance is known as 'perceptual modelling'.

14.2.1 Auditory perceptual models

To determine the subjective relevance of errors in audio systems, and particularly speech systems, assessment algorithms have been developed based on models of human hearing. The prediction of audible differences between a degraded and reference signal can be thought of as the sensory layer of a perceptual analysis, while the subsequent categorization of audible errors can be thought of as the perceptual layer. Models for assessing high-quality audio, such as PERCEVAL [1], have tended only to predict the probability of detection of audible errors since any audible error is deemed to be unacceptable, while early speech models have tended to predict the presence of audible errors and then employ simple distance measures to categorize their subjective importance [2, 3, 7]. It has been previously shown [8] that a more sophisticated description of the audible error provides an improved correlation with subjective performance. In particular, the amount of error, the distribution of error, and the correlation of error with original signal have been shown to provide an improved prediction of error subjectivity.

Figure 14.1 shows a hypothetical fragment of an error surface. The error descriptors used to predict the subjectivity of this error are necessarily multi-dimensional — no simple single-dimensional metric can map between the error surface and the corresponding subjective opinion. The error descriptors, E_d, are in the form:

$$E_{d1} = fn_1\{e(i,j)\}$$

where fn_1 is a function of the error surface element values for descriptor 1. For example, the error descriptor for the distribution of the error, error-entropy (Ee), proposed in Hollier et al [8], was given by:

$$Ee = \sum_{i=1}^{n} \sum_{j=1}^{m} a(i,j) \ \ln \ a(i,j)$$

where $a(i, j) = |e(i, j)|/Ea$, and Ea is the sum of $|e(i, j)|$ with respect to time and pitch;

$$\text{opinion prediction} = fn_2 \ \{E_{d1}, E_{d2}, ..., E_{dn}\}$$

where fn_2 is the mapping function between the n error descriptors and the opinion scale of interest.

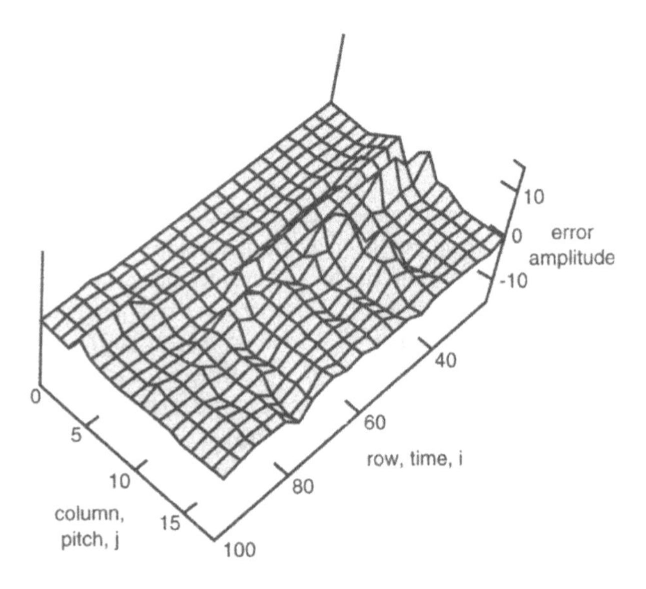

Fig. 14.1 Fragment of audible error surface.

It has been shown that a judicious choice of error descriptors can be mapped to a number of different subjective opinion scales [9]. This is an important result since the error descriptors can be mapped to different opinion scales that are dominated by different aspects of error subjectivity. This result, together with laboratory experience, is taken to indicate that it is possible to weight a set of error descriptors to describe a range of error subjectivity, since different features of the error are dominant for 'quality' and 'effort' opinion scales. The general approach of dividing the model architecture into sensory and perceptual layers, and generating error descriptors that are sensitive to different aspects of error subjectivity, is validated by these results.

14.2.2 Visual perceptual models

An approach similar to that of the auditory perceptual model has been adopted at BT Laboratories (BTL) for a visual perceptual model. A sensory layer reproduces the gross psychophysics of the sensory mechanisms:

- spatio-temporal sensitivity known as the human visual filter;
- masking due to spatial frequency, orientation and temporal frequency.

A number of visual perceptual models are known to be under development and several have been proposed in the literature. For example, Watson and Solomon [4] propose the use of Gabor functions to account for the inhibitory and excitatory influences of orientation between masker and maskee, and Ran and Farvadin [5] use a simple image decomposition into edges, textures and backgrounds. However, most of the published algorithms only succeed in optimizing individual aspects of model behaviour; Watson and Solomon [4] provide a good model of masking, and Ran and Farvadin [5] a first approximation to describing the subjective importance of errors.

Following the sensory layer the image is decomposed to allow calculation of error subjectivity, by the perceptual layer, according to the importance of errors in relation to structures within the image, as shown in Fig. 14.2.

If the visible error coincides with a critical feature of the image, such as an edge, then it is more subjectively disturbing. The basic image elements, which allow a human observer to perceive the image content, can be thought of as a set of abstracted boundaries. These boundaries can be formed by colour differences, texture changes and movement as well as edges. Even some Gestalt effects [10], which cause a boundary to be perceived, can be algorithmically predicted to allow appropriate weighting.

The boundaries are required to perceive image content and so visible errors that degrade these boundaries have greater subjective significance than those which do not.

Fig. 14.2 Image decomposition for error subjectivity prediction.

It is important to note that degradation of these boundaries can be deemed perceptually important without identifying what the high-level cognitive content of the image might be. For example, degradation of a boundary will be subjectively important regardless of whether the image is of a teapot or a spaceship. False contours in plain areas will be recognized as highly visible errors, but will not be classed as defining boundaries since they did not appear in the original image. The output from the perceptual layer is a set of context-sensitive error descriptors that can be weighted differently to map to a variety of opinion criteria.

14.2.3 Multisensory model

To assess a multimedia system it is necessary to combine the output from each sensory model and account for the interactions between the senses (which depend on the task or activity undertaken). It is possible to provide familiar examples of intersensory dependency, and these are useful as a starting point for discussion, despite the more sophisticated examples that soon emerge. Strong multisensory rules are already known and exploited by content providers, especially film makers. Consistent audio/video trajectories between scene cuts, and the constructive benefit of combined audio and video cues are obvious examples. Exploitation of this type of multi-modal relationship for human/computer interface design is discussed in May and Barnard [11]. Less familiar examples include the misperception of speech when audio and video cues are mismatched [12], and modification of error subjectivity with sequencing effects in the other modality [13].

The interaction between the senses can be complex and the significance of transmission errors and choice of bandwidth utilization for multimedia services and telepresence is correspondingly difficult to determine. This difficulty highlights the need for objective measures of the perceived performance of multimedia systems. Fortunately, to produce useful engineering tools, it is not necessary to model the full extent of human perception and cognition, but rather to establish and model the gross underlying (low-level) intersensory dependencies.

Figure 14.3 shows a diagrammatic representation of a proposed multisensory perceptual model developing the concept of sensory and perceptual layers for the individual senses and introducing cross-modal dependencies and the influence of task. The main components are:

- sensory model;
- cross-modal model;
- scenario-specific task model, with an average across the target user group used for higher level perceptual issues such as personal preference and user expectation.

The main cross-modal effects are:

- timing:
  - — sequencing;
  - — synchronization;
- quality balance between modalities.

Error subjectivity also depends on task:

- high-level cognitive preconceptions associated with task;
- attention split;

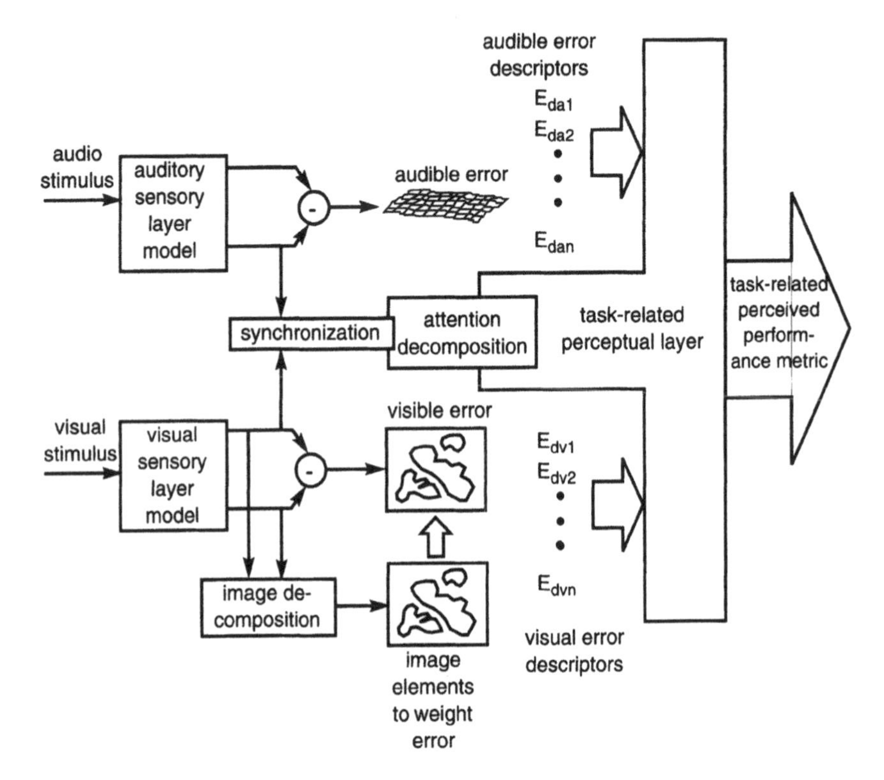

Fig. 14.3 Overview of proposed multisensory perceptual model.

- degree of stress introduced by the task;
- experience of user (novice versus expert).

A mathematical structure for the model can be summarized:

E_{da1}, E_{da2}, ..., E_{dan} are the audio error descriptors, and

E_{dv1}, E_{dv2}, ..., E_{dvn} are the video error descriptors.

Then, for a given task:

fn_{aws} is the weighted function to calculated audio error,

fn_{vws} is the weighted function to calculated video error subjectivity,

fn_{pm} is the cross-modal combining function.

The task-dependent perceived performance metric, TDPM, is then:

$$\text{TDPM} = fn_{pm}\ \{fn_{aws}\ \{E_{da1}, E_{da2}, ..., E_{dan}\}, fn_{vws}\ \{E_{dv1}, E_{dv2}, ..., E_{dvn}\ \}\}$$

Alternatively a task term, which varies according to task, could be made explicit giving a prediction of the performance metric (PM) as:

$$\text{PM} = fn_{pm}\ \{fn_{aws}\ \{E_{da1}, E_{da2}, ..., E_{dan}, T_{wa}\}, fn_{vws}\ \{E_{dv1}, E_{dv2}, ..., E_{dvn}, T_{wv}\}\}$$

where T_{wa} is the task weighting for audio, and T_{wv} is the task weighting for video.

The perceptual models described here rely on features that can be inferred from the audio and video signals. To weight the significance of errors, perceptual weightings must be derived from the signal without reference to higher level cognitive knowledge of content. This approach can be referred to as an implicational model. A more complete description of error subjectivity would be possible if the higher level content was known. Use of higher level knowledge about cognitive content can be referred to as a propositional model. For example, an implicational model would predict that a high-quality image of a table with three legs contained no subjectively important errors (because the image is undegraded), while a propositional model would identify the table, noting that it should have various properties including four legs, and predict that there was an error.

The signals arriving at the current generation of perceptual models do not contain any higher level description of the content and therefore it is not possible to form and test propositions. Current work is concentrating on the extent to which perceptual importance can be deduced by implication. This approach will result in perceptual weightings being applied to regions of the image that, due to the image content or propositional considerations, are not subjectively important. However, it is reasoned that weighting all perceptually important features will provide an evaluation of the image superior to any where such a weighting has not been applied.

Future perceptual models may be able to exploit the raising and testing of propositions by utilizing:

- either content descriptors supplied with the signal, for example, as proposed in MPEG7, which may be available to facilitate intelligent searches and indexing;
- or knowledge of a virtual environment within which a known range of objects and properties can exist.

The inclusion of proposition information will be useful in order to provide additional weighting on error significance. However, a deliberate attempt to develop implicational models is under way, since it is expected to be difficult to exploit proposition information in certain classes of future system. For example, a normal proposition to be tested would be that an object in free space will fall. In a virtual environment this will not always be true since it would be possible, and potentially advantageous, to define some objects which remain where they are placed in space and not subject to gravity. Equally, future, advanced telepresence systems may be very unfamiliar and lead to unexpected behaviour and expectations. In such circumstances, it may be better to rely on errors which can be predicted by an implicational model rather than risk an inappropriate prediction of a propositional effect.

To develop our understanding of multi-modal perception and to provide the data required for the construction and verification of a multisensory model, it is necessary to perform a series of subjective tests. These tests must be conducted with sufficient scientific rigour to provide hard data for numerical modelling, while at the same time be conducted efficiently so that the many tests required can be conducted with feasible resource.

The technical execution of the multi-modal tests required falls beyond the scope of subjective tests previously carried out. The next section discusses a multi-modal subjective experiment, along with its results, and the future requirements for multi-modal testing.

14.3 EXPERIMENTAL INVESTIGATION INTO MULTI-MODAL PERCEPTION

The first stage in the development of objective techniques to measure the subjective performance of multimedia systems is the collection of data on how these systems are perceived, and in particular the interactions which occur between the senses.

To ensure that this data, which will form the basis of multi-modal models, is statistically valid, formal subjective assessments should be performed which conform to recognized and controlled design constraints. Because of the complex nature of the interactions which occur between the senses, and the influence of task and environmental factors it is necessary to ensure that subjects are immersed in highly realistic simulations of the application/task under test.

When characterizing multimedia systems, the potential area of investigation is vast. The scope of most early experiments therefore is to focus on specific aspects such as the influence of video quality on audio perception. These specific aspects must be explored in order to determine gross intersensory dependencies that can be used as a foundation for the initial models and as a base for further experimentation.

14.3.1 Report on experiment

For this experiment, the major goal was to test and quantify the hypothesis that video quality has an influence on the perception of audio quality, as well as obtaining general information on the impact on the quality of interactions between the two. To maximize the amount of information obtained from this experiment, a non-interactive viewing-only experiment was used, because employing an interactive task (e.g. one involving human-to-human communication over a multimedia system) would have required a reduced scope to control the additional degrees of freedom. The subjects watched a set of presentations and were asked to give discrete opinions of the quality of each presentation on its completion.

14.3.1.1 Source material

The material viewed in the experiment consisted of eight video clips, each of approximately 10 s in length, with supporting audio commentaries. The visual element consisted of a series of virtual-reality fly-throughs of a hypothetical building. Unique audio commentaries were digitally recorded for each of these clips to provide two narratives, one describing the software used to create the graphics and the other describing the building simulated.

One male and one female talker were used giving a total of 32 combinations (8 video clips × 2 audio themes × 2 talkers).

14.3.1.2 Degradations

The video degradations used were a sub-set of the degradations available in the ITU VIRIS software toolkit [14], as this provided algorithmically defined, well-characterized degradations. Three video degradations were selected from the toolkit:

- edge business — a distortion concentrated at the edges of objects, characterized by temporally varying sharpness or spatially varying noise, which adds a flickering type of degradation to strong edges in the image;
- blurring — makes the image blurred and unfocused;
- white noise — produces a snowstorm-like effect.

Six video degradations (three types, each at two levels), together with an undegraded condition and one with no-video gave eight video degradations.

The four audio conditions used were:

- no degradation;
- band-limited — telephone bandwidth (300-3500 Hz);
- MNRU (Q = 20 dB) (modulated noise reference [15][1]) — this adds noise proportional to the amplitude of its input signal, is a widely used degradation in audio experiments, and is designed to simulate the effect of tandeming PCM systems;
- band-limited — a combination of the above, where the band-limiting and MNRU were performed and then the band-limited signal was passed through the MNRU.

14.3.1.3 Experiment design

To provide statistically verifiable results, a formal balanced design was used, using a pair of Graeco-Latin squares. This ensures that, over the experiment as a whole, all possible combinations of source material and degradations are used once and only once. In total, 32 non-expert subjects were used, each completing the experiment twice, once being asked a question on audio quality only, and once on overall quality.

14.3.1.4 Experiment procedure

The subjects were seated in an acoustically isolated room directly in front of the screen, at a distance of 5H (5 times the vertical height of the screen). A pair of speakers, one either side of the screen were used to play back the audio commentaries.

The wording for audio or overall quality was:

Please give your opinion of the overall quality of the clip.

Please give your opinion of the audio quality of the clip.

For both of these questions, the permitted answers were those used in the ITU Recommendation P.800 [16] for the quality scale:

---

[1] The algorithm used was taken from the ITU Software tools library and applied directly to the commentaries sampled at 44.1 kHz. Strictly speaking, this is outside the design limits of the software, which has been designed for use with speech sampled at 8 or 16 kHz. However, it still provides an algorithmically defined degradation which was all that was required for this experiment.

Excellent
Good
Fair
Poor
Bad

These opinion categories are ascribed a numerical value, typically 0, 1, 2, 3, 4, which is known as an opinion score. This score may be averaged across a number of subjects to yield a mean opinion score (MOS), which is not necessarily an integer value.

14.3.1.5 Experiment results

The first stage in reviewing the results of a subjective experiment is to review their statistical validity. The balanced design methodology permitted the use of analysis of variance (ANOVA) to estimate the experimental error and determine which of the factors within the experiment (such as subjects, degradations, presentation orders, etc) have had a significant effect on the results.

Once the analysis of variance has been completed, the scores can be averaged across the factors to yield MOSs within any constraints dictated by the ANOVA process.

14.3.1.6 Analysis of variance

An analysis of variance was performed on the results, the key points from which were:

- subject variations were significant, an expected effect, due to the random sampling from a larger population — scores can still be averaged across subjects who are considered to be representative of the larger population;
- the order in which the subjects saw the material and the degradations did not have a significant bearing on the results;
- the variations caused by the degradations significantly affected the results;
- the question asked (audio or video quality) had a significant effect on the resulting scores;
- the ordering of the questions (whether a subject did the audio or the overall question first) also significantly affects the results;
- the material to which the degradations were applied also significantly affected the scores — this was an expected effect, as the video material was computer generated and some of the clips contained more apparent distortion artefacts than others; the use of two different talkers also accounts

for some of this variation, this effect being small compared to the variations due to the degradations, and separate conclusions need not be drawn for each piece of source material.

14.3.1.7 Discussion of results

Noise was rated as the worst video degradation while the subjective responses for the other two degradations were found to be broadly similar. For the audio degradations, the order of preference (best to worst) was undegraded, band-limited, MNRU, followed by the combination of both these degradations. It is important to note that these findings do not imply that certain types of degradation are generally considered worse than others, as selecting a different severity of each of the degradations would have produced different orderings.

When no video is present, the perceived audio quality is always worse than if video is present. However, as no guidance was given to subjects on how this should be rated, interpretation of the no-video case is uncertain.

For the audio question, the effect of whether the question was asked in the first or second session has little effect on the results. However, for the overall question, the scores are higher if the overall question is asked in the first session. The notable exception to this is when no video is present, where the scores are lower.

Figure 14.4 summarizes the effect of varying the video quality on the perception of the audio quality. The graph is averaged over all the audio qualities for clarity. The variation in audio MOS is shown for each video distortion. To illustrate the effect of audio quality, the same data is broken down for each audio degradation in Fig. 14.5.

The differences shown are small, but the trend in the data, as summarized in Fig. 14.5, is consistent across the majority of the video degradations. It is therefore concluded that within the constraints of this experiment, a decrease in the visual quality had a detrimental effect on the perceived audio quality.

14.3.1.8 Experiment review

In the experiment reported, each opinion question (either overall or audio) was asked in separate sessions; the subjects therefore knew, while viewing the clip, which aspect of performance they would be asked to judge. This method was chosen since the object was to determine audio perception when audio was known to be the stimulus of interest. It should be noted, however, that randomizing the questions and not presenting them to the subjects until they had finished viewing the clips would be expected to alter the audio/video dependency measured, since the subjects would pay attention in a more uniform way if they did not know the nature of the question.

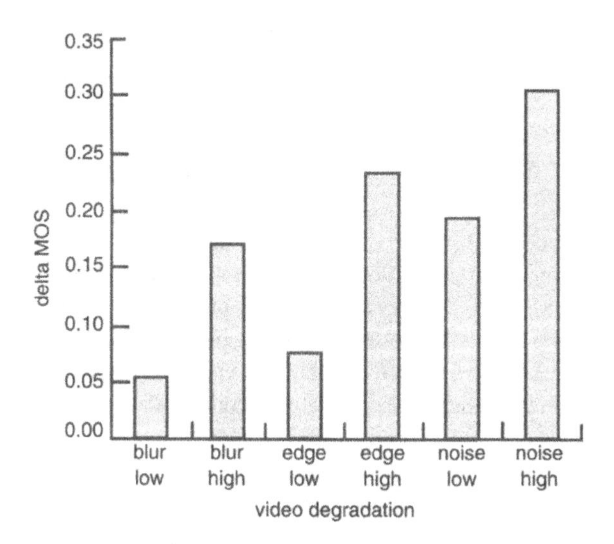

Fig. 14.4 Differences between MOS for degraded video and for undegraded video, averaged over all audio conditions. The higher the bar, the greater this difference.

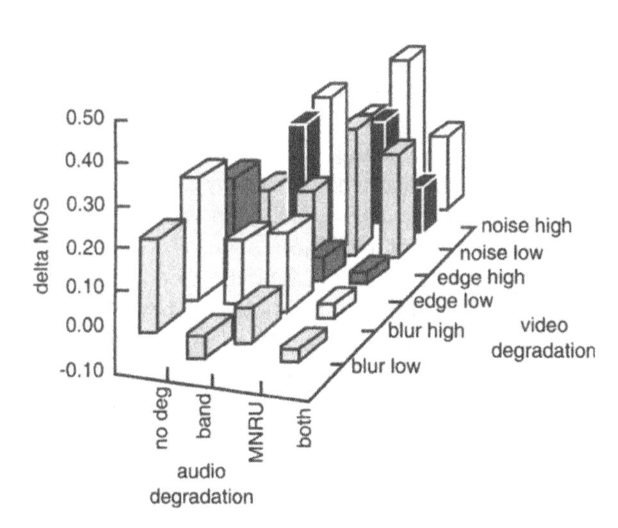

Fig. 14.5 Differences between MOS for degraded video and for undegraded video, shown separately for each audio degradation. The higher the bar, the greater this difference.

Using a less passive, more involving task, where the subject is not able to consciously consider the audio and video performance until they are required to vote, would also be likely to alter the judgements of the subjects.

Further experimentation, including questions relating to the perceived video quality, are planned to enable more information on the relationships between overall perceptions and those of the individual senses to be established.

14.3.2 Future multi-modal testing

It was noted above that for multi-modal perception there is a strong interdependency between perception, task and environment. Further, as with any subjective test, each user will have a variety of preconceptions, experience and aptitude. It appears impractical to separate all the potential dependencies between perception and task using a large number of individual tests and so it is proposed to immerse subjects in highly realistic simulations of the specific scenarios of interest.

If a subject is immersed in a task/application of interest it is not always practical to collect their reactions and opinions by asking explicit questions.

In order to recreate a realistic experience for test subjects, and to allow response collection without distracting from the task in hand, a new approach is required for some subjective tests. This new approach may be thought of as 'immersive subjective testing'. For example, a subject immersed in a virtual environment could be asked to perform one or more tasks. While these tasks are performed, the subject is fully involved with both the task and the experimental environment, ensuring that preconceptions and other experiential influences are appropriately exercised.

During the subject's activity within the virtual environment different aspects of performance, and preferences, can be assessed by recording reaction times and behaviour. In this way it is possible to collect a variety of multi-modal data without the explicit knowledge of the subject.

It is possible to imagine collecting data on the results of degradation of the audio/video quality, the impact of usability issues, and even factors influencing trust. During a hypothetical information collection task a subject might encounter parts of the audiovisual environment which are degraded. The subject's reaction time and following decisions will be influenced by these distortions and using the data from many subjects the influence of the degradation, including cross-modal effects, can be determined. Direct questioning of the subject can provide the opportunity to determine the relationship between overall opinion and the responses collected during the task.

An experimental virtual environment, which was developed at BTL, is available for this type of testing and can be customized by the experimenter to suit particular requirements. A view of this experimental virtual environment is shown as Fig. 14.7 in the next section.

14.4 APPLICATION OF MULTISENSORY PERCEPTUAL MODELS

As sufficient information about cross-modal behaviour is collected it will be possible to build algorithmic models of multisensory perception. These models will find application in the same way that auditory perceptual models are already being exploited — for the assessment of performance, and within the delivery technology in order to reduce bandwidth and optimize performance.

14.4.1 Assessment

To assess the perceived performance of a multimedia system it is necessary to model both the human senses and to simulate the 'sensory interface'. The sensory interface includes the acoustic air-path between the system under test and the user, as well as the properties of the user's ear. Similarly, it is necessary to include the viewing distance between the display equipment and user's eyes. In order to assess a complete multimedia system, the most complete way to ensure that the human sensory interface is accurately simulated is to use a mannequin equipped with artificial eyes, ears and mouth. Such a system can be referred to as a 'sensory interface simulator' (SIS) and would form the front end of the multisensory perceptual model. Figure 14.6 shows an early prototype SIS that has been built by adding 'eye' cameras to a modified Bruel and Kjaer 4128 Head and Torso simulator. Figure 14.7 shows the prototype placed in an immersive interface to an experimental distributed virtual space.

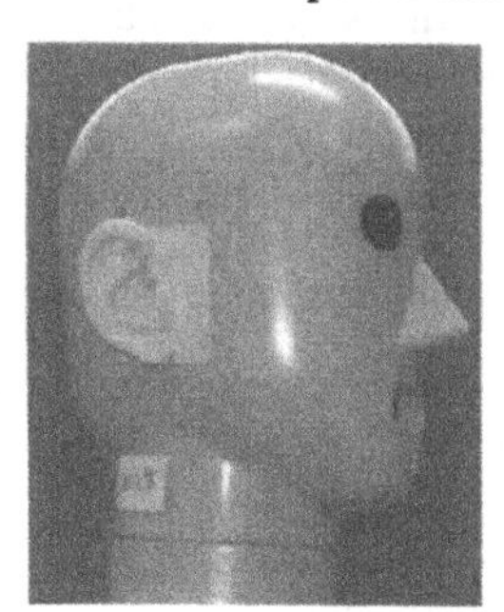

Fig. 14.6 Close up of early prototype SIS with eyes, ears and mouth; and snap-shot of system in position in front of a distributed virtual space application.

14.4.2 Delivery technology

As multisensory models become more advanced they will be able to model cross-modal dependencies in real time and will therefore be suitable for use in the

Fig. 14.7 Early SIS prototype positioned within an immersive interface to shared virtual space.

delivery technology for multimedia and advanced telepresence applications. Auditory models are already used in a similar way for coding and decoding audio signals to maximize the perceived performance of systems with limited bandwidth.

In an advanced telepresence system there will be two principal uses for perceptual models.

- Maximization of perceived performance within available bandwidth

  Modelling cross-modal dependencies in real time will allow best use of available data rates, maximising the perceived performance of a multimedia system. This optimization for data rate will be different for different users who may be joining a shared space via radically different bandwidths or by using different types of application.

- Maximization of perceived performance within available processing power

  Even with powerful desktop personal computers it is not possible to recreate life-like virtual spaces. The quality of the percept created can be improved by employing the available processing power to produce events and environments that are supported by cross-modal consistency. In this way, the perceived performance with the available processing power can be optimized by trading off quality audio and video according to predictions from a cross-modal model which ensures that the required percept is effectively created. Early examples of this are likely to be relatively simple, for example:

  — 'cross-modal masking' would allow aspects of the reproduction which will receive little attention to be rendered with less accuracy;

— 'sequencing', such as a temporal pattern, in one modality will be expected to increase sensitivity to the other modality in relation to the sequence.

The need to optimize multimedia systems for limited bandwidth and processing power is expected to remain an important consideration for the foreseeable future. Even when very high bandwidth connections and supercomputer processing are available to some users, it is likely that many people will wish to join virtual meetings in other ways. For example, many users will be on the move perhaps gaining access via limited radio spectrum, while the processing capabilities of portable and wearable computers is likely to be restricted for some time. Indeed, if a compelling percept can be achieved with limited computing and bandwidth via appropriate multi-modal optimization, this would fuel the drive towards miniaturization rather than increased power.

14.5 CONCLUSIONS

The requirement for a multisensory perceptual analysis for the objective assessment of multimedia systems has been introduced. An essential part of the requirement for such an analysis is knowledge of the gross dependencies that exist between audio and video components and how to grade their impact. The important influence of higher level perceptual factors such as task and utility is also recognized.

Existing work on single-sense perceptual analysis has been summarized and attention drawn to the advantageous nature of the BTL perceptual models that provide error descriptors which may be mapped to a variety of opinion scales. Knowledge of the relationship between audio and video perception, as well as specific task-related influence on error subjectivity, is required to complete the multi-sensory model. A subjective experiment to investigate aspects of intersensory dependency has been reported and results shown which reveal a falling perception of audio quality with deteriorating video quality in certain circumstances.

A strategy has been proposed to render the overall modelling task tractable. Immersion of subjects in realistic simulations is proposed as one means to collect the required data. Gross intersensory dependency can be modelled and included in a multi-modal model with aspects of higher level task dependency included as scenario-specific weightings. An outline model structure incorporating these concepts has been shown.

The architecture of BTL's perceptual models lends itself to multisensory combination due to explicit multi-dimensional description of error subjectivity. An outline of a multisensory model is proposed and further development is planned. It is apparent that a majority of applications in the communications and entertainment industries will require multisensory evaluation.

In the future, multi-modal models will be applied both for assessment and within the delivery technology for multimedia and telepresence systems.

REFERENCES

1. Paillard B, Mabilleau P, Morissette S and Soumagne J: 'PERCEVAL: perceptual evaluation of the quality of audio systems', J Audio Eng Soc, 40, No 1/2 (January/ February 1992).

2. Hollier M P, Hawksford M O and Guard D R: 'Characterization of communications systems using a speech-like test stimulus', J Audio Eng Soc, 41, No 12 (December 1993).

3. Beerends J and Stemerdink J: 'A perceptual audio quality measure based on a psychoacoustic sound representation', J Audio Eng Soc, 40, No 12 (December 1992).

4. Watson A B and Solomon J A: 'Contrast gain control model fits masking data', ARVO (1995).

5. Ran X and Farvadin N: 'A perceptually motivated three-component image model-Part I: description of the model', IEEE Transactions on Image Processing, 4, No 4 (April 1995).

6. Hollier M P, Hawksford M O and Guard D R: 'Objective perceptual analysis: comparing the audible performance of data reduction schemes', Presented to the 96th Audio Eng Soc Convention in Amsterdam, Preprint No 3879 (February 1994).

7. Wang S, Sekey A and Gersho A: 'An objective measure for predicting subjective quality of speech coders', IEEE J on Selected Areas in Communications, 10, No 5 (June 1992).

8. Hollier M P, Hawksford M O and Guard D R: 'Error-activity and error entropy as a measure of psychoacoustic significance in the perceptual domain', IEE Proc Vis Image Signal Process, 141, No 3 (June 1994).

9. Hollier M P and Sheppard P J: 'Objective speech quality assessment: towards an engineering metric', Presented at the 100th Audio Eng Soc Convention in Copenhagen, Preprint No 4242 (May 1996).

10. Gordon I E: 'Theories of visual perception', John Wiley and Sons (1989) (summarized at: http://hyperg.uni-paderborn.de/0x83ea6001_ 0x0001cc42).

11. May J and Barnard P: 'Cinematography and interface design', in Norbdy K et al (Eds): 'Human Computer Interaction', Interact '95, pp 26-31 (1995).

12. McGurk H and MacDonald J: 'Hearing lips and seeing voices', Nature, 264, pp 510-518 (1976).

13. O'Leary A and Rhodes G: 'Cross-modal effects on visual and auditory perception', Perception and Psychophysics, 35, pp 565-569 (1984).

14. ITU-T Recommendation P.930: 'Principles of a reference impairment system for video', (August 1996).

15. ITU-T Recommendation P.810: 'Modulated noise reference unit (MNRU)', (February 1996).

16. ITU-T Recommendation P.800: 'Methods for subjective determination of transmission quality', (note that this Recommendation was previously numbered as P.80) (August 1996).

15

DISTRIBUTED ENTERTAINMENT ENVIRONMENT

S J Powers, M R Hinds and J Morphett

15.1 OVERVIEW

The development of interactive computer games can be traced back to about 1960, with the development of Spacewar at MIT. However, it was not until 1972 that computer gaming penetrated the mass market, in the shape of Pong. Developed by Nolan Bushnell, founder of the first videogame company (Atari, 1968), it was an overnight sensation. Throughout the 1970s Atari continued to develop and release arcade-based gaming systems, and in 1976 it took videogames into consumers' homes, with the VCS 2600. Since that time, although the technology has drastically improved with the standard PC now a major player in the games market, the fundamental style of gaming has not changed.

While the latest Nintendo 64 console may boast the sort of graphics power previously only available on £50 000+ graphical workstations, the games themselves follow a very familiar pattern. The game designer sets certain goals (highscore, reaching a certain level, etc) together with certain tasks (avoid the monsters, collect the treasure, overtake the other cars, etc). Players follow an essentially directed path through the tasks, progressing from a starting point to a termination point, specified by the game designer. For each new game the player restores the game state to its initial conditions, and restarts along the same path.

This traditional style of gaming is now under threat, with the convergence of several games-related technologies, and a game genre known as MUDs (multi-user dungeons or multi-user dimensions).

MUDs offer the exception to the single player game described above [1]. Firstly, they are inherently multi-player, running over networks such as the Internet and on bulletin boards (BBs). Players can join or leave the game at any time, and usually are able to play the same on-line character each time they play. Secondly, they are far less goal directed, offering a more organic, persistent playing environment, that can grow or shrink over time as it is modified by the users. Finally, while also offering tasks for players to complete, a large part of the game is made up of the interaction between players. This may manifest itself in either competition or co-operation, dependent on the game goals and the inclination of the players.

While these features of MUDs make them a highly compelling and interesting gaming environment, their user base has thus far been small. One of the key reasons for this is their outdated, crude and simplistic client interface. Players interact through a text channel, receiving textual descriptions of their current location, and typing simple commands (e.g. 'take gold', 'go north', etc) to manipulate the environment.

With almost no professionally authored content, and a cliquey user base, the MUD concept, as it currently stands, is doomed to a niche existence. However, this may be about to change, due to development in two key technical areas.

Firstly, increasing computer power has allowed complex 3-D virtual environments to be modelled and displayed. These environments allow players to both move through a virtual world and to interact with the entities found within it. Games such as Doom and Quake, utilizing this style, have proved astonishingly popular, selling millions of copies and penetrating the cultural mainstream.

Secondly, networking facilities on both PCs and game consoles have become commonplace. With the Internet, corporate LANs and dedicated gaming networks widely available, the network game has come of age. Most newly released games offer some form of network support, and some gaming companies are even making networking the primary form of play (e.g. LucasArts with X Wing versus Tie Fighter).

Combining these technologies with some of the ideas behind MUDs offers the potential for an exciting new game genre, the shared virtual environment MUD (VEMUD).

Like a MUD, this is based around the idea of a persistent world, which exists and operates irrespective of the presence of players. Also, like a MUD, it utilizes the idea of open-ended gameplay, with competition and co-operation between players to complete game goals. However, unlike a MUD, it draws on the most popular features of modern games — complex 3-D graphics and sophisticated sound effects. This allows for very intuitive interaction between player and environment, making the gaming experience both involving and of widespread appeal.

15.2 VIRTUAL GAMING ENVIRONMENT REQUIREMENTS

In delivering any shared VE, be it a game or business tool, the key technical issues are [2]:

- consistency;
- scalability;
- persistence.

A game-oriented VE also requires that the following three areas are addressed [3]:

- integrity;
- balance;
- extensibility.

15.2.1 Consistency

In a single player game, modelling a virtual environment is comparatively simple. The game maintains an internal model of the world which is repeatedly updated, based on the internal game logic (triggering puzzles/traps, moving autonomous creatures, etc) and the input of the player (moving the viewpoint, selecting objects with which to interact, etc). After each update of the world a 2-D image of this 3-D model is rendered and presented to the player. This loop is repeated until the player wins, loses or gets bored (see Fig. 15.1).

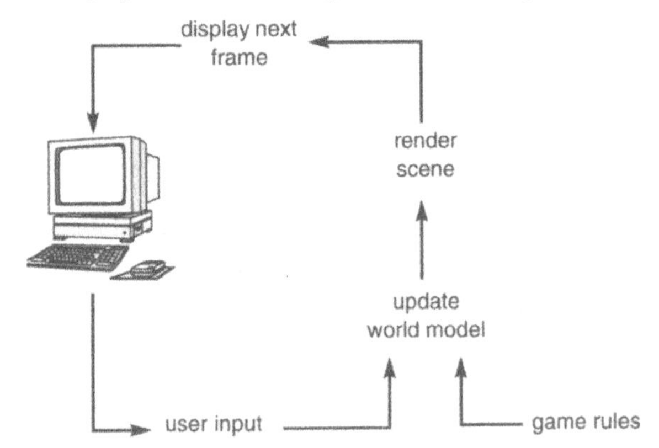

Fig. 15.1 Single player game loop.

In a multiplayer version, the situation is greatly complicated by the need to ensure that each player appears to be interacting with the same environment (i.e. the same internal model). While each player will have a unique viewpoint on the environment, they must all experience the same changes to the environment in the same order.

The simplest means to achieve this is to centralize the model of the world and sequence each player's input into this model, while distributing the relevant changes to all players. This is the approach taken by MUDs, where a single central process sequences text input from each player in turn, calculates the changes to the world, and transmits a textual description of the new world state to the relevant players (see Fig. 15.2).

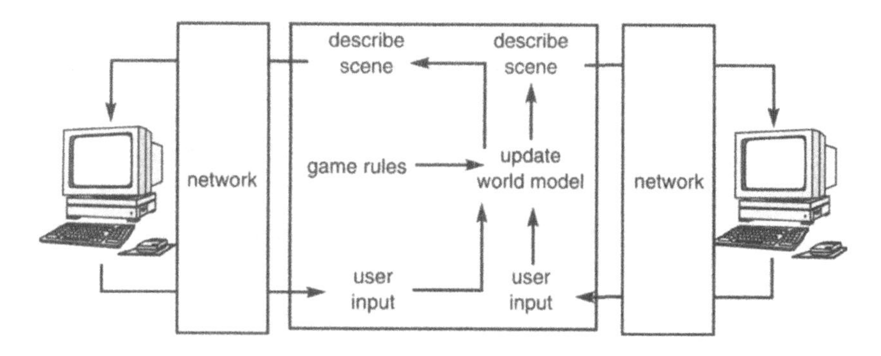

Fig. 15.2 MUD game loop.

While this ensures a consistent view of the environment for each player (by maintaining only a single model of the world), it is inappropriate for more complex graphical environments, as the link between the world model and the visual output consumes far more bandwidth than that required by a text MUD. For a complex interactive 3-D environment it is necessary to present (as an approximate minimum) an image of 320×200 pixels, 15 times a second. With 8 bits of colour depth per pixel this translates to approximately 1 Mbit/s of data. As a result, the approach taken by most games is both to allow each player to maintain a local model of the environment, with which he/she interacts, and to ensure that these local copies of the environment remain consistent with each other. This is shown in Fig. 15.3.

15.2.2 Scalability

One possible approach to implementing the architecture shown in Fig. 15.3, is to form peer-to-peer network connections for all clients, and alternate between updating the world and performing local rendering. This gives a simple system which effectively synchronizes the clients around a repeating data exchange (see Fig. 15.4).

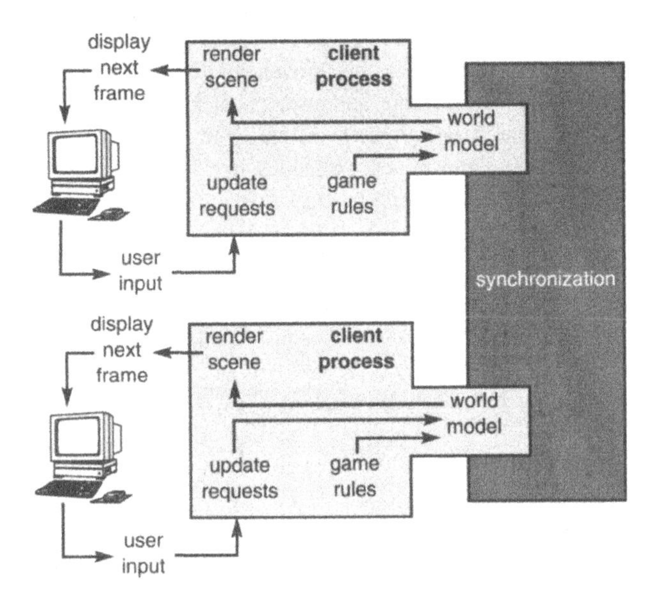

Fig. 15.3 Synchronizing shared environments.

This is the approach used by the popular game Doom, but it has two main drawbacks. Firstly, it links the state synchronization to the rendering loop. This means that the speed of the screen update (a critical factor in producing an interactive environment) is linked to the speed of the slowest client network connection.

This problem can be partially resolved by allowing clients to make local predictions about future world states, and then implementing roll-back to a previously agreed state if the predictions prove invalid. Duke Nukem, a later variation on the Doom genre, implements this approach.

The second and more serious drawback to this approach is that it lacks scalability. As the number of clients grows the amount of traffic each must send and receive grows, quickly overwhelming the available bandwidth to each client.

A possible solution to this problem is the provision of a central server, as shown in Fig. 15.5.

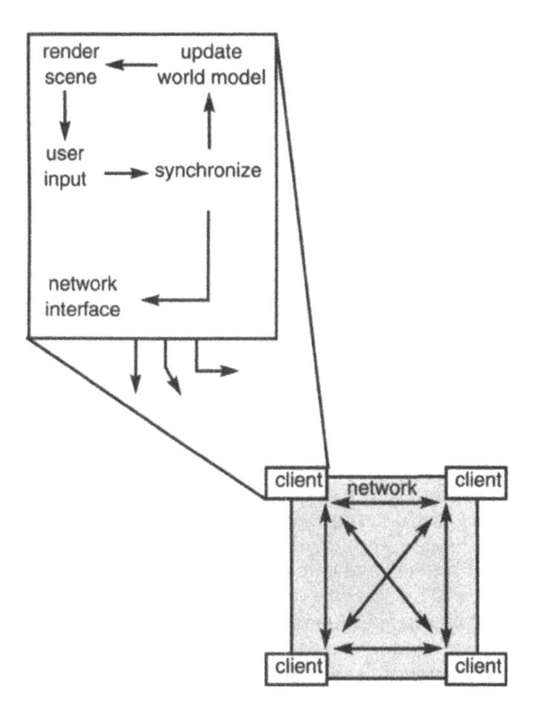

Fig. 15.4 Client-to-client network architecture.

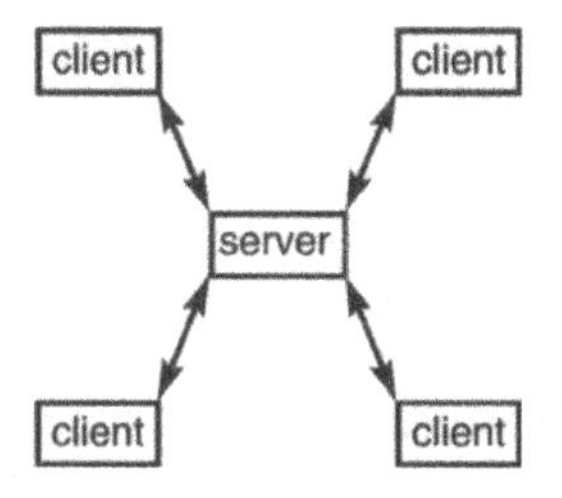

Fig. 15.5 Server-oriented architecture.

In its simplest form the server acts as a broadcaster, sending the updates from each client to all other clients. This reduces the upstream traffic requirement from each client. A more sophisticated solution places a model of the world at the server, which not only acts as the reference description of state, but also validates changes to this state, e.g. ensuring that, when two players want to take an object, only one of them succeeds.

It should be noted that this centralized world model does not replace the client-side models, but instead supplements them with a definitive state description.

Quake, the follow-up to Doom, uses this approach, allowing it to support up to 16 players, rather than the 4 of Doom.

The server can also reduce the downstream traffic to each client by only transmitting the world state changes that are currently relevant to each client's player. For example, a player will have no awareness of objects outside their current visual range. It is therefore unnecessary to keep the client informed of state changes to objects in this category.

However, while the central server approach offers improved scalability over the peer-to-peer client arrangement shown in Fig. 15.4, it still has a potential bottle-neck in the form of the server process. While network traffic to/from each client is reduced, the server process must cope with:

- all environment modification requests from each client;
- validating each request and calculating the new environment state;
- calculating which clients each environment modification affects;
- generating and sending the environment updates to all relevant clients.

With games involving hundreds of clients, moving objects and autonomous game creatures, a single central server will quickly become overwhelmed by both the network traffic and the processing required.

One approach to distributing this network traffic load is to divide the spatial environment into a number of pre-defined areas, and associate a multicast network address with each area. State changes to the environment are broadcast on the multicast address associated with the area in which the change occurred. Clients subscribe to the multicast address of the areas in which they are interested (i.e. the areas in visual range on the player's viewpoint), and hence receive all the state changes they require to maintain a valid model of the local environment. To move the player's avatar the client transmits the new position to the relevant multicast address, which in turn informs any other clients visualizing the relevant area.

This approach, shown in Fig. 15.6, is used in the NPSNET architecture, designed for military simulations [4].

15.2.3 Persistence

The NPSNET approach both helps to reduce the amount of information transmitted to each client and aids scalability by having no single central server. However, in the form shown in Fig. 15.6, it lacks support for persistence.

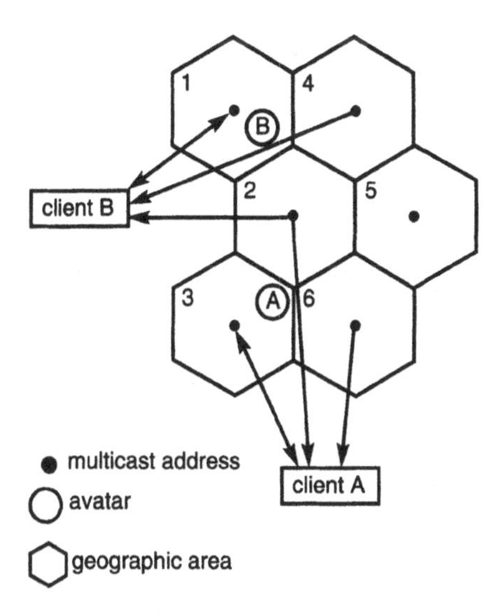

Fig. 15.6 NPSNET approach to distributing network traffic.

Persistence is the ability of a virtual world to retain state over an extended period, so changes made by players persist between their gaming periods. Provision of persistence is a critical difference between simple multi-player games and VE games. It is also a major factor in building an on-line community around a particular game, as it allows individuals to feel part of an evolving environment, rather than a single transitory game.

One approach to adding persistence to the NPSNET architecture is to associate an 'oldest client' with each multicast address. This is usually a process started when the world is first created and is responsible for storing the state of its particular area. As avatars move within range of a particular area, its current state can be requested from the relevant 'oldest client'.

15.2.4 Integrity

It is critical that all players can trust the integrity of the environment in which they are participating. Typical examples of integrity failures might be:

- players able to modify client code to provide them with a gaming advantage, e.g. turning off collision detection for their avatar, allowing them to move through walls;
- events and object behaviour depending on the quality of clients' network connections, e.g. objects disappearing or ceasing movement when a client's network connection fails;

- players able to modify locally stored state information to provide them with a gaming advantage, e.g. altering the effectiveness of weapons in their inventory.

Delivering consistency means ensuring all players perceive the same environment. Delivering integrity means ensuring that the environment behaves as players would expect. Without integrity the sense of immersion needed for a compelling game cannot be created.

This represents one of the key drawbacks to the NPSNET system (see section 15.2.2). With no centralized world management, responsibility for modelling behaviour is distributed to clients. This puts the integrity at risk, as clients are open to modification by the players. By contrast, the use of a central server helps deliver integrity, but at the risk of insufficient scalability.

15.2.5 Balance

Balance in a game is one of the essential factors to its success. A well-balanced game will always provide a challenge to the player, continually opening up new problems and new experiences, matching these to the current skill of the player. For example, in a single player game, the puzzles and hostile creatures within the game are gradually increased in toughness as the game progresses. Hence, as the player's skill level and the equipment available to their game character increases, the problems faced increase proportionally.

By contrast a multi-player game is far more complex to balance, for example:

- a new player may encounter a seasoned veteran within the first few minutes of play;
- the designer has little control over the state of the world encountered by any of the players after the first one has arrived;
- with a currency within the game a miniature economy may develop, and the presence of 'wealthy' players will impact on the behaviour of others;
- a particular weapon may prove too powerful, making the owner too significant an influence in the world.

To achieve balance in a persistent on-line environment, it may be necessary to introduce objects with new behaviour, or modify an existing object's behaviour, or introduce new rules governing the interaction between players; and, since the environment is persistent, these changes must be made without closing the system down and recompiling all the code.

15.2.6 Extensibility

If players are not to become bored with a virtual gaming world it is important to ensure that it can be extended. This means not only extending the volume of the world that can be explored, but also the range of objects within it and the behaviour of existing objects. One key feature of some MUDs is that players are rewarded for reaching a certain level of experience by being allowed to design new areas of the game. This not only acts as a spur to playing, but also ensures that the environment is regularly refreshed and redeveloped.

Having successfully built a community of users, extensibility offers a means to retain them.

15.3 DISTRIBUTED ENTERTAINMENT ENVIRONMENT

DEE (distributed entertainment environment) is an on-line gaming architecture being developed at BT Laboratories, designed to meet the requirements outlined in section 15.2. As such it draws on ideas from academic and military virtual environments, from MUDs, from Internet technologies and from existing games.

Its target network delivery platform is a BT product known as Wireplay.

15.3.1 Wireplay

In trying to deliver a shared gaming environment, the performance of the network is critical, and the most important factor in the performance is latency. This is defined as the time period it takes information to propagate between nodes on the network. Without consistently low latency it is impossible to deliver highly interactive multi-player games.

To meet this particular network requirement BT developed Wireplay. This is designed to work with PCs using standard 28.8 kbit/s modems, and allows multi-player games to be played over guaranteed low-latency links (120 ms being the current client-to-client latency).

While Wireplay was initially targeted at supporting standard multi-player games, it is ideally suited for supporting virtual world gaming. This is because its architecture offers a centralized location suitable for placing servers capable of delivering the consistency, persistence and integrity discussed in section 15.2.

15.3.2 DEE — architecture overview

DEE has two distinguishing ideas. The first is concerned with the distribution of modelling behaviour within the VE, and is termed entity representation. The second is concerned with achieving scalability, and is termed dynamic management.

15.3.2.1 Entity representation

Anything that can be placed within the virtual environment is termed an entity. For example, a player's avatar is an entity, as is any equipment or scenery the player encounters.

While any entity represents a single concept, its behaviour and state can be grouped into three categories:

- visual — entities must be viewable, and have an interface with which players can interact (e.g. doors have handles allowing them to be opened);
- conceptual — entities incorporate game concepts, that relate to how they function within the designer specified rules (e.g. guns can be fired, which generate bullets, switches start machinery, medicine heals people, etc);
- dynamic — entities can move within the environment and collide with other entities.

DEE suggests that each of these categories is represented by a separate object, termed a representative, and that each entity is formed from the grouping of three of these representatives. For example, a sliding door in the environment might be represented by the triplet <DoorConcept, SlidingDoorDynamic, AnimatedVisual>. The DoorConcept representative would understand that doors can be opened by avatars, if they have the correct key. The SlidingDoorDynamic would be capable of calculating the movement as the door opened or closed, together with any associated collisions. The AnimatedVisual would be capable of rendering the correct geometry, and ensuring that as the door moved it was correctly animated.

In terms of placement within a network architecture, the various representatives would be arranged as shown in Fig. 15.7.

Clients communicate dynamic change requests (e.g. move avatar, wave sword) to the dynamics manager (DM), and higher level game-related requests (i.e. open door, pull switch) to the rule manager (RM). The DM validates changes to the environment, updates the environment state and broadcasts the changes to clients. The RM responds to requests by changing the game state and possibly placing requests with the DM (e.g. add velocity x to entity y).

A typical messaging sequence for opening a door, might be:

- client requests action 'open' on its visual representation of a door — this generates a message from the visual representative to the conceptual representative (route 1 on Fig. 15.8);

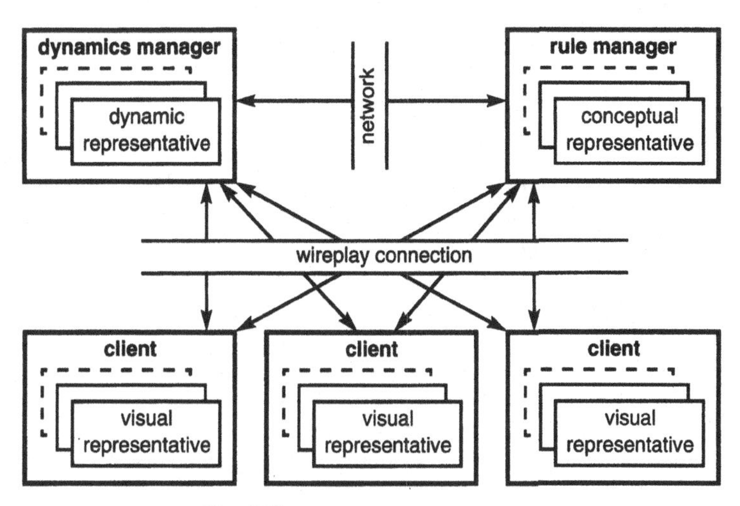

Fig. 15.7 DEE architecture overview.

- the conceptual representation of the door at the RM decides if the player can perform the action (e.g. Is the player strong enough? Is the door unlocked? If it is locked, does the player have the key? etc);
- if the player can open the door, the door's conceptual representative will store the new state, and instruct the dynamic representative to swing the door open, thus generating a message from the conceptual representative

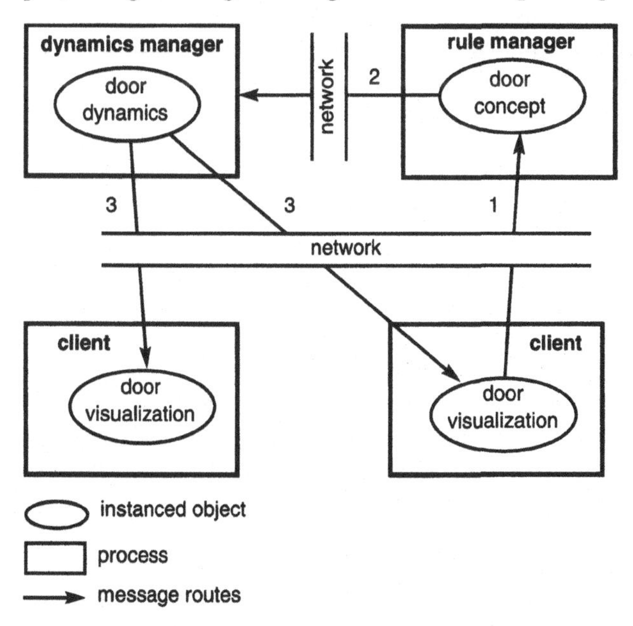

Fig. 15.8 Message flow for opening a door.

to the dynamic representative (route 2 on Fig. 15.8) — a typical message might contain just the request to open, or might also specify angle and speed criteria; the conceptual representative may also provide direct feedback to the client (e.g. 'You unlocked the door with the gold key');

- the dynamic representative of the door, instanced at the DM, will model the door movement, and broadcast changes to the client, thus generating a series of messages from the dynamic representative to the visual representatives (route 3 on Fig. 15.8) — a typical set of messages might be 'start rotating at velocity *x*' followed by 'stop rotation at angle *y*';
- the visual representation of the door at each client will render its new appearance as it rotates into the open position (note that the visual representation should be capable of prediction and roll-back, so as to allow separation of the client rendering loop from the dynamic updates).

15.3.2.2 Advantages of entity representation

There are numerous advantages to dividing the modelling of the VE in the way suggested.

- Separation

  The dynamic/conceptual separation allows the use of different technical implementations for the management of the VE. To facilitate **balance** the conceptual representatives should be implemented in a programmer-friendly, dynamically linkable, object-oriented language like Java [5]. This will allow the rules governing the environment to be easily modified. However, using Java for the dynamic representatives would be far too slow, as dynamic modelling is traditionally computationally expensive. Hence, a compiled language such as C++ is more suitable [6].

- Centralization

  The centralization of the dynamic management guarantees consistency. Effectively, the dynamic representatives form a definitive description of the VE dynamic state, with the visual representatives shadowing this state as closely as possible. Requests to alter the dynamic state must be agreed by the dynamic representative, and since the dynamic state is updated centrally, unreliable clients will not impact on the overall consistency.

- Graphics

  The abstraction of the dynamics from the visual representation facilitates development of the client graphical capabilities. Not only is the client able to concentrate processing power on the visual effects, the separation simplifies redesigning visual representatives to make use of new graphics cards or APIs.

- Integrity

  The centralization of the games rules, in the conceptual representatives, guarantees integrity. The players have no access to the conceptual state, and hence cannot perform local modifications.

- Extensibility

  The dynamic/conceptual separation aids extensibility. Developing conceptual representatives is a game-oriented task as the complexity related to visual/movement management has been removed. Hence, the game designer can focus on the rules and interaction of game concepts, rather than ensuring objects collide and render properly.

- Persistence

  The provision of persistence is possible, as all persistent state is held centrally. Standard persistent object technology can therefore be used to store the state of all dynamic/conceptual representative state, and hence the state of the VE [7].

15.3.2.3 Dynamic management

While the concept of entity representation meets many of the requirements specified in section 15.2, it does not address the problem of scalability. Indeed, with a highly centralized, server-oriented architecture, it has potentially a serious scalability problem. DEE addresses this with a dynamic management system that provides efficient scaling of both the client network load and the server processing load.

It is assumed that the network and processing load associated with the conceptual representatives does not have a scalability problem. This is because players generate a large amount of network traffic related to the dynamics (moving, throwing objects, etc), but comparatively little in terms of game level actions (pulling switches, opening doors). This focuses the problem of scalability on to the dynamic manager.

DEE utilizes an architecture similar to NPSNET, whereby the spatial environment is considered to be broken into predefined areas, termed zones. However, unlike NPSNET, responsibility for modelling state within these zones is not held by clients, but by processes termed zone managers. Each zone manager instances the dynamic representatives that fall within its zone, and is responsible for providing them with processing time to perform state updates. As dynamic representatives move they can be passed between zone managers, hence moving processing load and network traffic with the dynamic representative.

Each client determines the zones that it wishes to display locally and connects to the relevant zone manager. Each zone manager maintains a list of all connected clients, and, as dynamic representatives update state (e.g. move, change velocity, etc), they inform all clients connected to their zone manager. In this way clients can constrain the network traffic to the areas of the environment in which they have an interest, while the dynamic representatives automatically ensure the relevant clients are kept informed of their state.

With potentially thousands of zones within any one VE it is not feasible to create a zone manager for every zone, as this would create too high a process management overhead. Instead when the state of a zone becomes dynamically deterministic (i.e. completely predictable over time), its state is written back into a persistent object store and the relevant zone manager closed. If new activity occurs within the zone (e.g. an avatar arrives, somebody throws an object into it) a new zone manager is started and the previous zone state fetched. This will result in clusters of active zones (i.e. those with assigned zone managers) around the avatars in the environment, while most zones are dormant and held in the persistent store. Figure 15.9 shows an overall schematic for the system.

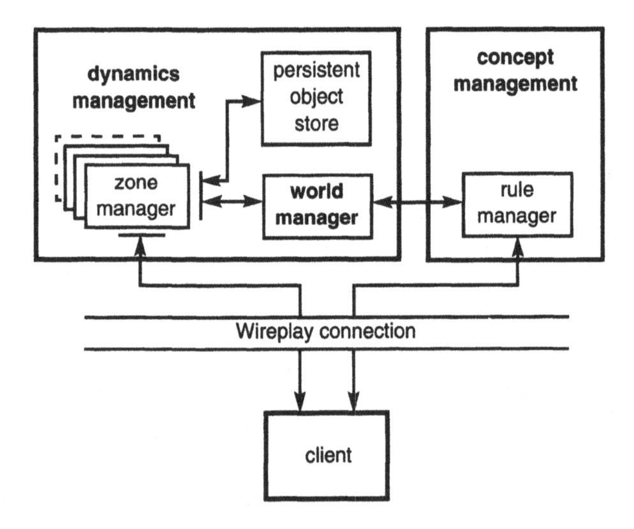

Fig. 15.9 DEE process architecture.

Figure 15.9 has introduced a new process — the world manager. Its responsibility is the routeing of messages between the rule manager and the zone managers (i.e. from conceptual representatives to dynamic representatives), the creation of zone managers, and load management. Creation of a zone manager occurs when either a request is placed from an existing zone (e.g. dynamic representative crossing zone boundaries) or when a conceptual representative tries to send a message to a dynamic representative in a dormant zone. In this situation the world manager creates a new zone manager process, and allocates the relevant key for it to extract the zone's current state from the persistent store. Coarse load management can be achieved by ensuring that the creation of the zone manager occurs on the least-loaded processor available.

It should be noted that the persistent object store shown in Fig. 15.9 should be designed for highly optimized, rapid storage and retrieval of dynamic state only. Conceptual state resides in the rule manager, which should run for the life of the environment, with modifications made for balance and extensibility performed 'on the fly'.

If it were necessary to support closure and re-starting of the rule manager, and hence storage of conceptual representative state, a separate persistent object store should be provided. Since this is not a standard part of normal game operation (unlike zone state storage), speed is not an issue, and the implementation can be based on issues of simplicity and standardization.

15.4 FUTURE WORK

The DEE architecture described here appears to offer a number of important ideas relevant to both VEMUDs and more general, shared virtual environments. It addresses all the major requirements outlined in section 15.2, while avoiding strict enforcement of a particular implementation technology.

An initial prototype is currently being developed at BT Laboratories, and work is under way on developing the test metrics and test harness that will allow an assessment of its potential. If the architecture proves its worth in testing, it is expected that game developers will be able to utilize it in delivering complex and scalable graphical environments over Wireplay.

REFERENCES

1. Bartle R: 'Interactive multi-user computer games', ftp:// ftp.parc.xerox.com/pub/ MOO/papers/ mudreport.txt (December 1990).

2. Lea R, Honda Y, Matsuda K, Rekimoto J: 'Technical issues in the design of a scalable shared virtual world', Sony Computer Science Lab, Tokyo, Japan (1990).

3. Morningstarr C and Farmer F: 'The lessons of Lucasfilm's Habitat', in 'Cyberspace: First Steps', MIT Press, Cambridge, Mass (1990).

4. Macedonia M R, Zyda M J, Michael J P, David R, Barham P T and Zeswitz S: 'NPSNET — a network software architecture for large scale virtual environments', Presence, 3, No 4 (Fall 1994).

5. Gosling J and McGilton H: 'The Java language environment. A White Paper', Sun Microsystems Computer Company, 2550 Garcia Avenue, MoutainView, CA (October 1995).

6. Stroustrup B: 'The C++ Programming Language', Addison-Wesley Publishing Company (June 1994).

7. Saljoughy A: 'Object Persistence and Java', JavaWorld on-line, http://www.javaworld.com/javaworld/jw-05-1997/jw-05-persistence.html (May 1997).

16

THE APPLICATION OF TELEPRESENCE IN MEDICINE

P Garner, M Collins, S M Webster and D A D Rose

16.1 INTRODUCTION

Telemedicine has been defined most clearly as 'the use of electronic information and communications technologies to provide and support health care when distance separates the participants' [1]. Applications of telemedicine in remote consultation fall into two broad categories — store and forward of electronic patient-related information for remote diagnosis, and real-time interactive remote consultation between geographically separate clinicians or patient and clinician. Store-and-forward applications are the fastest growing area of commercial telemedical services, for example, the managed radiological case-reporting service offered by Worldcare Ltd [2]. Patient-related images or case reports are transmitted to a remote expert clinician for report back within 6 to 48 hours. Store-and-forward applications are not explored in this chapter since the remote expert is not telepresent during a consultation. This chapter details the BT trials and pilot schemes which have helped to establish the credentials for real-time medical telepresence. It highlights technical solutions to clinical concerns which can change real-time teleconsultation from an interesting technology for enthusiasts to a potentially important means of making more effective use of limited human resources, leading to improved patient care at acceptable costs.

16.2 REMOTE FETAL SCANNING

The remote fetal ultrasound trial [3] was designed to assess the clinical validity of the transmission of real-time fetal ultrasound images over BT's dial-up ISDN30 service. The system was implemented to provide an expert consultant opinion from a central specialist referral centre to the local clinicians in a hospital. Expectant mothers are routinely scanned at 14-18 weeks of pregnancy at their local antenatal clinic. Up to 5% of scans in the UK have an ambiguity in the image which would normally result in the mother being referred to one of the UK specialist fetal care centres for a second opinion. During this trial, high-quality ultrasound images generated in the antenatal clinic at St Mary's Hospital on the Isle of Wight were transmitted over the BT ISDN network to Queen Charlotte's Hospital in London, where for the purposes of referral they were viewed by a consulting expert with virtually no delay or degradation in quality. The technical and geographic configuration is shown in Fig. 16.1.

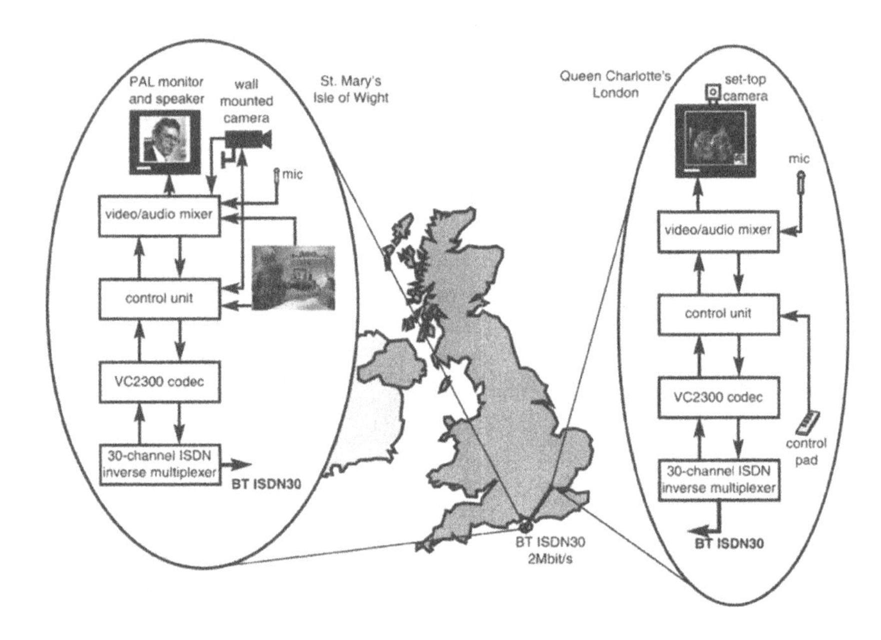

Fig. 16.1 Schematic of remote scanning equipment.

The equipment in the antenatal clinic at St Mary's consisted of a stand-alone cabinet containing a BT VC2300 videocodec, an ISDN channel inverse multiplexer, video and audio mixers, plus a small floor-mounted mobile control console for use next to the scanner. There was also a wall-mounted low-light camera within the scanning room which captured a general view of the patient and sonographer during the scan. This view allowed the remote consultant to correlate the ultrasound image on view with the position and orientation of the ultrasonic transducer held by the sonographer. The control console allowed the sonographer to determine which images were sent to the consultant or how they should be mixed together. The console had a small LCD monitor which displayed the image being transmitted to the consultant so that the sonographer could be confident the configuration was correct. Figure 16.2 indicates the view most often seen by the remote consultants at Queen Charlotte's.

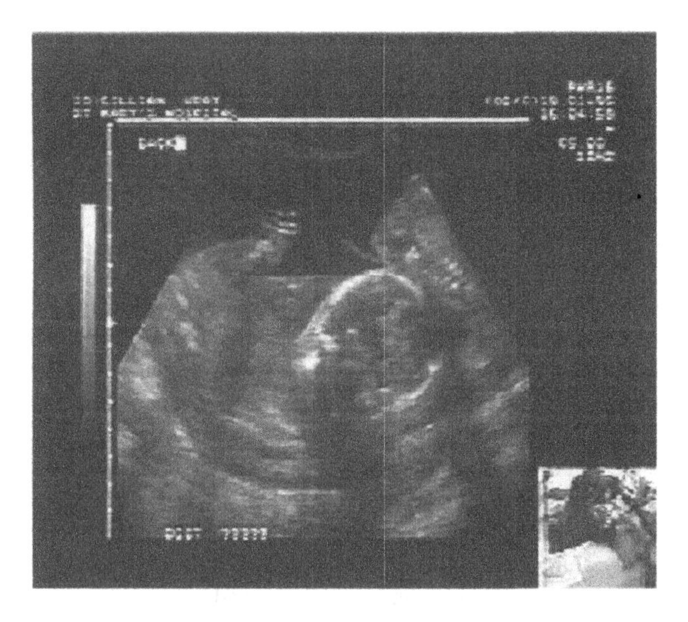

Fig. 16.2 Typical ultrasound image with picture in picture.

A number of human interface issues became apparent after installation. At the scanning end of the link, on the Isle of Wight, the audio system had been configured to allow the remote consultant's voice to be broadcast into the scanning room for all present to hear or for private listening by the sonographer. In practice, the local consultant or other interested clinicians often wished to be present in the room with the sonographer. They required private access to the voice of the remote consultant during their preliminary discussions in order to avoid possible alarm to the expectant mother. This was resolved by the use of cordless infra-red audio headsets within the scanning room. The infra-red headsets allowed any number of interested parties to hear the remote consultant's comments without introducing variable loading on the electrical audio circuits, and gave the additional benefit of allowing complete freedom of movement for the users.

The remote fetal scanning system was clinically evaluated [4] for a period of 6 months during the pilot scheme and produced the following summary results:

- 39 consultations carried out;

- 86% definitive diagnoses were made over the link;

- 76% abnormalities were detected;
- four mothers had to physically attend Queen Charlotte's;
- 80% of mothers felt less anxious using the system.

The four physical referrals were required for more complex tests, not available at St Mary's. Anecdotally the consultants have said that they were as confident in making a primary diagnosis over the system as they were when actually present in the scanning room. The improved quality of life for the mothers should not be overlooked. Mothers, already in a state of anxiety, felt at ease with the system and were happier in the familiar surroundings of their local hospital. At the end of the diagnostic phase of the referral, the sonographer could switch the video input to display the face of the remote consultant, who was able to counsel the mother regarding the outcome of the referral in the identical way as if they had been in the same room. Clinician's time was saved at Queen Charlotte's because the evaluation and management advice was provided without having to provide the ancillary support normally required for a referral where the person physically attends.

The remote fetal scanning trial has shown that videoconferencing technology can be packaged to meet the clinical requirements for antenatal telereferral and to be easy and intuitive to use by the clinicians involved. The use of such a system has enabled referrals to be carried out more quickly by enabling the expert consultant in London to be telepresent in the scanning room of a hospital on the Isle of Wight, which has improved the quality of care provided to expectant mothers during the trial.

It has also been suggested that a remote referral system will reduce the number of births of severely handicapped babies and thus reduce the long-term medical and social costs to the state. The higher abnormality detection rate enabled by the remote referral system would reduce the number of medico-legal claims against a hospital trust. Each successful case against a hospital can cost up to £2M. Nationwide costs for fetal medicine legal cases are approximately £60M-£100M per annum [5].

Further developments of the system will focus on lower cost terminal equipment and video distribution techniques by which a high-quality videoconferencing gateway could be provided as a shared facility for hospitals, allowing other clinical departments access to their preferred specialist centres for remote consultation. This would increase the system utilization and therefore make the purchase of the videoconferencing equipment more justifiable for the hospitals.

16.3 REMOTE SKILL ACQUISITION IN SURGERY

The remote fetal scanning trial has shown that videoconferencing systems with standard two-dimensional (2-D) displays can be used to construct telepresence systems suitable for remote medical referrals. However, when telepresence is applied to distance learning where physical manipulative skills are taught, such as in surgery or dentistry, the lack of the third dimension (3-D) or 'depth' in the transmitted images can inhibit the teaching process. To overcome this limitation a 3-D videoconferencing system was developed and tested in surgical teaching. The videoconferencing element of the system was a BT VC2300 and ancillary networking equipment, identical to that used in the remote fetal scanning trial. Two cameras were used instead of one at the image acquisition end of the system and a high-quality 100 Hz video monitor with synchronized LCD shuttered glasses used to display 3-D images at the viewing end of the system. The video signal from each of the cameras was allocated approximately 50% of bandwidth on the 2 Mbit/s ISDN30 circuit linking the two ends of the system.

In the first experiment [6], the system was used to allow a 'trainee' surgeon to be advised by a consultant surgeon who was physically remote from the operation. The trainee surgeon wore a lightweight headset on which two miniature cameras were mounted in order to capture his left and right eye views, as shown in Fig. 16.3.

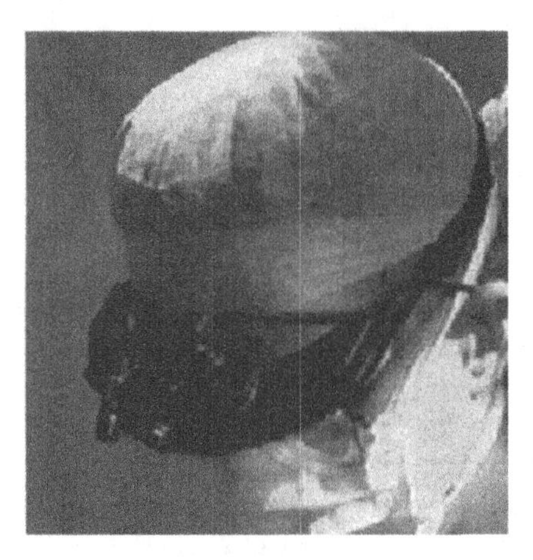

Fig. 16.3 Twin camera headset.

The cameras were angled relative to each other so that their fields of view converged at the normal point of focus of the surgeon's eyes during the operation. A microphone was integrated within the headset so that the surgeon could speak to the remote consultant. The video signals from the cameras were transmitted, with the audio from the surgeon, through the videoconferencing system, to the 3-D viewing station where the consultant surgeon watched the images on the screen through LCD glasses (see Fig. 16.4).

The 3-D system allowed the remote consultant to accurately see, and so comment on, the depth of incisions and the angle at which the surgical tools were being used. The remote consultant surgeon was therefore able to interact with the operating surgeon to give guidance on technique. The 3-D system created a strong sense of shared presence between the surgeons to the extent that the consultant felt he was present at the side of the operating surgeon.

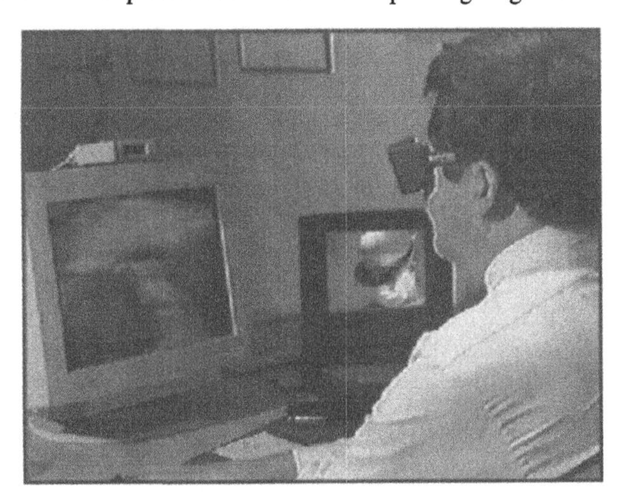

Fig. 16.4 3-D consultant viewing station.

The remote consultant experiment was technically and clinically successful and so a second experiment was designed to test the 3-D system in surgical teaching to larger audiences of medical students. Standard teaching techniques for medical students include watching truncated video sequences of operations recorded by a video camera positioned to one side of the surgeon in the theatre. The usefulness of the 2-D view presented in these videos is limited in the sense that it is not the view seen by the surgeon. To quantify the teaching benefits of real-time 3-D images to large audiences, medical students were invited to view a total hip replacement operation in both 2-D and 3-D video.

The video from the surgeon's headset cameras was transmitted to two distant lecture theatres and presented to a group of 20 students — 2-D video displayed in one theatre, and 3-D video displayed in the other. In both theatres the video was displayed by projection on to a large screen approximately 3 m × 3 m in size. The 3-D video was displayed using a polarizing projector, which circularly polarized the left and right eye views, by splitting the views between adjacent horizontal video scan lines. This 3-D video projection technique produced a 3-D image with 50% of the horizontal resolution of the 2-D video image. The students observed the video through polarizing glasses. The students were allowed to view the 2-D and 3-D transmissions at will, and were asked to rate their opinions of the quality of each in a structured questionnaire. The operation took two hours during which time an experienced surgeon joined the students to moderate questions to the operating surgeon. The two surgeons discussed working methods and specific details about the operative process over the videoconferencing system.

The results of the questionnaires indicated that the compromise in resolution required for practical large screen 3-D displays impaired the fidelity of 3-D video to the extent that there was no perceptible benefit over standard-resolution 2-D video. However, developments in large screen immersive 3-D displays such as the VisionDome [7], in which large 3-D video images can be viewed without compromising the video resolution, are likely to make 3-D remote surgical training effective in the future. It will then be possible for large groups of medical students to be simultaneously telepresent with the operating surgeon.

16.4 TELEPRESENCE IN ACCIDENT AND EMERGENCY

Accident and emergency (A&E) medical care is usually delivered by paramedics at the scene of an incident, e.g. a road traffic accident. A telemedicine system has been investigated which enabled other expert clinicians to be telepresent at the accident to support the work of the paramedic, or to plan subsequent medical care. A primary requirement for a telemedical solution for paramedics is that the system be transportable and not impede the free movement of the paramedic around the accident site. CamNet [8], a system previously designed for engineering applications, was tested as a mobile telemedical solution. CamNet is an audiovisual headset system which had been developed to allow the user to interact with a distant expert using both audio, video and graphical information, in order to transport the expert's skills to the user in the field. In addition to carrying a conventional microphone and earphone, the CamNet headset has a small CCD camera and a miniature view screen. The screen is viewed with the right eye at a distance of a few centimetres. The optics inside the view screen create a display which appears equivalent to that of a standard PC monitor viewed at a distance of 1 m from the eye. The image produced in the display is an intense monochrome red-on-black high-resolution dot matrix with a contrast ratio of 70:1, making it ideal for use in both bright and dark ambient light conditions.

The CamNet headset has been used on trials within the medical facility on the Fortes Bravo North Sea oil rig to provide expert medical diagnosis from the mainland to paramedics on the rig. The previous level of support offered to the rig paramedic was via a structured fax questionnaire from the Aberdeen Royal Infirmary. The questionnaire was completed by the paramedic and faxed back to the mainland where a consultant evaluated the information and called the rig to give his opinion. The patient was then either treated on board the oil rig or air lifted as a medivac (medical evacuation). The deployment of CamNet allowed clinicians on the mainland to see, for the first time, the physical condition of the remote patient which enabled the clinicians to make a more informed decision about the management of the patient. In one example of its use, CamNet enabled clinicians on the mainland to give instruction to a paramedic on how to stabilize a patient who had suffered a head injury.

The successful trials of the CamNet system on the oil rig were followed by a one-day event in which the crew of an East Anglian ambulance were equipped with CamNet to evaluate its use in a simulated road traffic accident [9]. The CamNet system was deployed with the ambulance using a portable Inmarsat 64 kbit/s dial-up satellite communications unit to provide mobile access to the BT ISDN2 network. The connection was made to a BT VC7000 desktop videoconferencing system in the Accident and Emergency Department of Ipswich Hospital. The paramedic wore the CamNet headset and belt pack, which communicated via a commercial digital radio local area network to the Inmarsat terminal in the ambulance vehicle (Fig. 16.5).

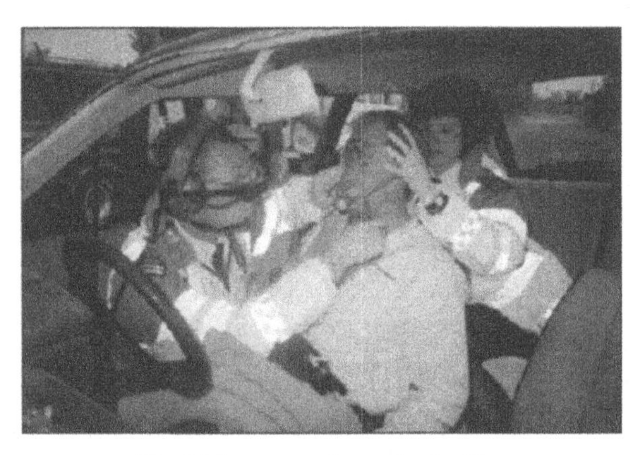

Fig. 16.5 Paramedic wearing CamNet.

The information transmitted to the consultant was found to be useful in establishing a fast and accurate diagnosis of the casualty's injuries. By viewing the received video from the CamNet headset on the VC7000 videoconferencing monitor, and by transmitting graphics of the structure of the human chest to the paramedic, the A&E consultant at Ipswich hospital was able to guide the paramedic through the simulated procedure of inserting a lung drain into the casualty. The trial of CamNet showed that telepresence technology can be successfully used to speed the delivery of A&E care by linking the roadside paramedic and the hospital-based consultant, but the present cost of CamNet makes it appropriate for use only in major incidents.

Capitalizing on the experience gained from the CamNet trial, BT Laboratories, in collaboration with Apple Computer, developed and tested

TraumaLink [10]. TraumaLink is a low-cost system that combines digital still imaging and GSM cellular network technologies to transfer still images from paramedics at the scene of an incident to an A&E department (see Fig. 16.6). The images enable the A&E staff to be forewarned of the nature of the accident and of the likely treatments the casualty will require on arrival at the hospital.

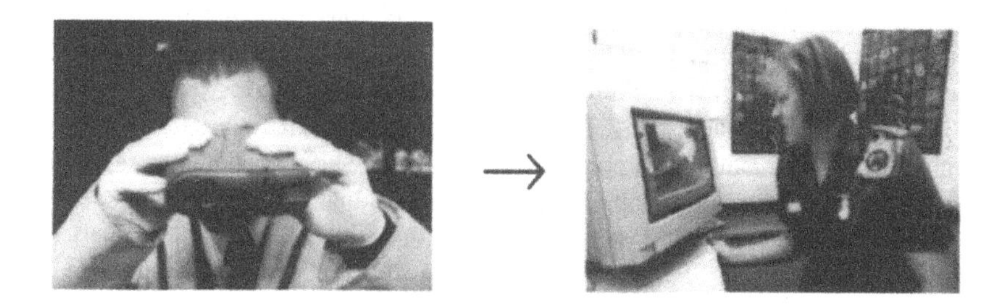

Fig. 16.6 TraumaLink image capture and display.

TraumaLink consists of a small fixed-focus digital still-image camera with integrated flash unit, a dash mounted cradle for the camera, a notebook PC and a GSM cellular phone. The camera is used by the ambulance technician while the paramedic carries out normal patient care procedures. After capturing pictures of the incident and the position of the casualty, the technician places the camera into the cradle. The technician has no other operations to perform to operate the system. The images are automatically compressed and downloaded into the on-board PC. The PC initiates a GSM data call at 9.6 kbit/s from the ambulance to the receiving PC in the hospital A&E department. Each full-colour image requires approximately 30 seconds transmission time. Clinical A&E staff in the hospital view the received images on a PC workstation and make colour prints for reference in the treatment bay. The TraumaLink system, which has been designed to be as simple and intuitive to use as possible, is being evaluated by the Norfolk and Norwich hospital.

16.5 CONCLUSIONS

Telepresence has proved to be clinically effective in specific aspects of medical care, particularly in the provision of remote specialist consultation where real-time interaction is required between the consultant and the clinician attending a patient. The acceptability of telepresence in the delivery of health care is highly dependent on the simplicity of the user interface and on ensuring that the technology is properly integrated with other clinical systems. It should also be recognized that clinicians will not change working practices unless clear benefits can be demonstrated in terms of improved patient care. In addition, health care managers expect demonstrable efficiency gains from new systems, particularly where new technology requires widespread changes in working practices and hence investment in change-management training and support.

Many successful pilots of the use of telepresence in medicine are ongoing in the UK in both urban [11] and rural environments. One approach to implementing telemedicine is multilayered applications [12], using store-and-forward technology initially with real-time telepresence options being reserved for difficult cases. In truly remote areas telepresence is growing in acceptance [13] as a means of facilitating the delivery of health care. As hardware and communications technologies continue to reduce delivery costs, health services will be revolutionized by telemedicine applications that use telepresence. This leads to the prospect of world-class health care that is available to all, irrespective of their location or ability to travel.

REFERENCES

1. Field M J (Ed): 'Telemedicine — a guide to assessing telecommunications in Health Care', Institute of Medicine, US National Academy Press (1996).

2. Richardson R J, Goldberg M A, Sharif H S and Matthew D: 'Implementing Global Telemedicine: experience with 1,097 cases from the Middle East to the USA', Proc Int Conf on Telemedicine and Telecare, Telemed '95, London, pp 147-149 (November 1995).

3. Fisk N M, Bower S, Sepulveda W, Garner P, Cameron K H, Matthews M, Ridley D, Drysdale D and Wootton R: 'Fetal telemedicine: interactive transfer of real-time ultrasound and video via ISDN for Remote Consultation', J of Telemedicine and Telecare, 1, No 1, pp 38-44 (1995).

4. Fisk N M, Bower S, Sepulveda W, Garner P, Ridley D, Drysdale D and Wootton R: 'Fetal Telemedicine: six month pilot of real time ultrasound and video consultation between the Isle of Wight and London', British Journal of Obstetrics and Gynaecology, 103, pp 1092-1095 (1996).

5. Fisk N M: Personal communication (1996).

6. Webster S M, Garner P, Cameron K H and O'Brien T: 'Real Time 3-D Remote Medical Skills Acquisition', Proc Int Conf on Telemedicine and Telecare, Telemed '95, London, pp 78-81 (November 1995).

7. Walker G R, Traill D, Hinds M, Coe A and Polaine M: 'VisionDome: a collaborative virtual environment', British Telecommunications Eng J, 15, Part 3, pp 217-223 (October 1996).

8. Matthews M R, Cameron K H, Heatley D J and Garner P: 'Telepresence and the CamNet remote expert system', Proc of Primary Health Care Specialist Group, British Comp Soc, pp 12-15, Cambridge (September 1993).

9. TV simulation for BBC 'Watchdog Healthcheck', broadcast on BBC1 (July 1995).

10. TV simulation of TraumaLink for BBC 'Look East', broadcast on BBC1 (June 1997).

11. Harrison R, Clayton W and Wallace P: 'Can telemedicine be used to improve the communication between primary and secondary care?', British Medical Journal, 313, pp 1377-1380 (November 1996).

12. Freeman K, Wynn-Jones J, Groves-Phillips S and Lewis L: 'Teleconsulting: a practical account of pitfalls, problems and promise, experience from the TEAM project group', Proc Int Conf on Telemedicine and Telecare, Telemed '95, London, pp 11-13 (November 1995).

13. Armstrong I J and Haston W S: 'Medical decision support for general practitioners using telemedicine', J of Telemedicine and Telecare, 3, No 1, pp 27-34 (1997).

17

APPLYING TELEPRESENCE TO EDUCATION

C J H Fowler and J T Mayes

17.1 INTRODUCTION

A Victorian time traveller visiting us today would be amazed at the use of advanced technology in today's hospitals, homes, offices and factories. However, a glance at our schools and colleges would suggest very little has changed, at least in our delivery methods. Education is under increasing pressure. It is being accused of failing to deliver quality or to embrace efficiencies. Our classrooms and lectures halls are becoming overcrowded, the numbers are increasing but resources are remaining stable or indeed in some sectors reducing. Against this rather depressing backcloth many are seeing technology as the new saviour [1, 2]. However, too often, technology is advocated solely for increasing efficiency, whereas, it is argued here that its real power lies not just in efficiency savings but also in improving the quality of education.

The concept of presence and therefore telepresence is not simply an issue of 'being there'. There are circumstances, particularly in education, where the learner is present physically but is mentally miles away! Psychological and physical distance are not the same thing — many campus-based students will have little or no sense of 'presence'. For learning situations we need to stretch the concept of presence to include 'psychological' presence which could cover such concepts as 'engagement', 'relationships' and 'audience'. For learning, it is critical for the individual to engage in the learning process — this ensures motivation and assumes that the learner is grounded in the concepts being taught. If a learner cannot relate to the new concepts then they will always remain distant and probably misunderstood. Also, presence is often associated with active participation — being there is not sufficient, you also have to contribute. Yet it is well known that 'mere presence' or the presence of a passive audience can motivate people, and simply knowing others are there may be as important or even more important than the individual being there.

Relationships are important because of their role in determining expectations during conversations or communications. For example, they help to determine the interaction patterns or communication structures. In Carey's five models of information flow [3], the structure of the interaction pattern is determined by the different roles and relationships of the participants (e.g. in the 'classroom' model, the moderator role assumes authority that is manifested in a controlled one-to-many interaction pattern). Relationship can also provide insight into the purpose of the interaction. People rarely communicate without some purpose or topic of communication. This does not rule out the casual or informal meetings which often have the purpose to explore or update understanding [4]. Relationship may also suggest the form and content of the interaction. Artefacts, either conceptual (e.g. an idea) or real (e.g. an object), are important for supporting or 'scaffolding' the conversation. The ability to easily access and share such artefacts could be critical in learning situations, e.g. through the use of 'whiteboards', or document cameras. The nature of the relationship may also influence the formality or emotional tone of the communication, e.g. mother to child or boss to employee relationship. More critically, relationships also determine the nature of task interactions and the subsequent dialogues around those tasks, e.g. learner/teacher, designer/implementor relationships.

This chapter presents a vision of education based on sound principles of learning and educational psychology — a vision that essentially sees learning and teaching as a social and collaborative process, and a process ripe for the application of telepresence principles. The focus will be on educational technology in general and not just to support distance learning. The chapter will both cover work being undertaken at BT Laboratories to test these ideas by building robust prototypes and by using them in real educational settings, e.g. schools and colleges, and also provide a glimpse of the future where we will be less inhibited by bandwidth and processing constraints. Finally, there is a discussion on what has been learnt from these examples and how they have changed people's perceptions of the importance of telepresence to the learning context.

17.2 UNDERSTANDING EDUCATION

There are, at least, two competing paradigms or approaches found in the education literature [5]. One approach, the didactic or instructionist approach, advocates that the teachers possess the knowledge and understanding and have the difficult job of 'transferring' that knowledge from their heads into the heads of their students. In contrast, a constructionist approach argues that effective learning is best achieved when the student is in control of their own learning. Figure 17.1 summarizes the main differences between the two paradigms.

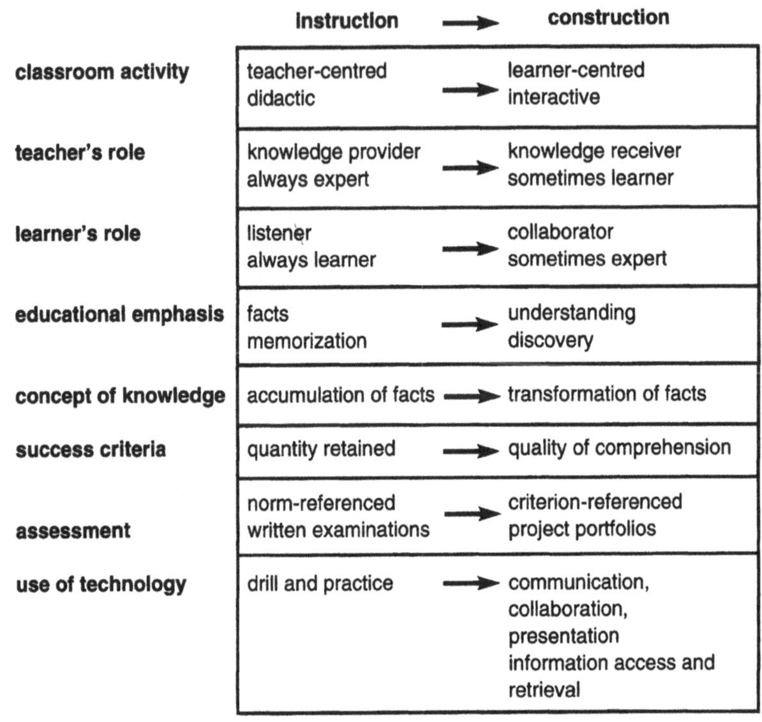

	instruction	→	construction
classroom activity	teacher-centred didactic	→	learner-centred interactive
teacher's role	knowledge provider always expert	→	knowledge receiver sometimes learner
learner's role	listener always learner	→	collaborator sometimes expert
educational emphasis	facts memorization	→	understanding discovery
concept of knowledge	accumulation of facts	→	transformation of facts
success criteria	quantity retained	→	quality of comprehension
assessment	norm-referenced written examinations	→	criterion-referenced project portfolios
use of technology	drill and practice	→	communication, collaboration, presentation information access and retrieval

Fig. 17.1 Paradigm shifts — from instruction to construction (adapted from Dwyer [6]).

However, a new paradigm is not sufficient. There is a need for a way of linking our understanding of learning with appropriate technologies. A project was commissioned to create a conceptual framework to help us bridge the pedagogy with the technology [7]. The framework assumes that:

- knowledge is situated — it is situated in action, and the meaning of the action is itself determined by task and social environments; we use knowledge, and without that use it has no or limited value;
- knowledge is relative — understanding is relative to what you already know;
- knowledge acquisition is a dynamic and iterative process — it involves a conceptualization cycle where new ideas are continually tested and refined, and the refinement process changes our new knowledge state, thus producing a reconceptualization;
- knowledge can be viewed as a set of interrelated concepts.

From these basic assumptions three top-level learning stages are offered:

- conceptualization — the coming into contact with other people's concepts;
- construction — the building and testing of one's knowledge through the performance of meaningful tasks;
- dialogue — the debate and discussion that results in the creation of new concepts.

It is important to note that conceptualization is about 'other people's concepts', that construction is about 'building knowledge' from combining your own and other people's concepts into something meaningful, and that dialogue refers back to creation of 'new concepts' (rather than knowledge) which then triggers another cycle of the conceptualization process. However, the framework does not just assume that dialogue only takes place at the dialogue stage. There will be dialogue for clarification at the conceptualization stage, and dialogue of collaboration at the construction stage.

17.3 APPLYING TELEPRESENCE IN THE SCHOOL SECTOR

In some earlier studies with primary school children using videoconferencing technology (the PC-based VC8000 over ISDN2), it became clear that the constructionist approach is extremely effective in increasing children's levels of motivation, their general attitude to learning, and their levels of self-esteem. It also became clear that within this particular context the videotelephony element had a powerful social role to play in the children's interaction (the 'hi' and the 'bye'), but added little to the successful completion of their particular tasks. This is not an uncommon finding [8], and increasing the level of presence is unlikely to change the finding. For the particular collaborative tasks undertaken by these children it was the dataconferencing facilities (e.g. whiteboards, application sharing, and the use of file transfer protocol (FTP)) in conjunction with high-quality audiotelephony that were critical to their efficient and effective completion of their tasks.

The children also became experts in specialist areas (e.g. in the use of spreadsheets) and soon were 'teaching' other children and their teachers. For these children to be seen and valued as experts was an important contribution to the high levels of self-esteem and subsequent high motivation levels. It was also found that children liked the fact that there was a wider audience, beyond the teacher and their parents, who could see their work. Again this sense of audience was a powerful motivator to produce quality materials of which the children could feel justly proud and wish to share. The effects of audiences on performance has a long psychology pedigree and can be traced back to the pioneering work of Allport [9]. A fuller description of the videoconferencing work can be found in Fowler et al [5].

In the subsequent years it was decided to increase the number of schools to include some secondary schools and to explore the application of these concepts to an Internet-based service. In line with the constructionist approach, children were asked to spend more time authoring the Web rather than surfing the Web. To help achieve this a Netscape-based framework was set up, a range of authoring tools (e.g. AOL press, MacroMedia Director), and some data collection tools (e.g. DAT recorders and Digital cameras) were provided, together with Web-based support in terms of 'hints and tips' for developing your own Web site, as well as an example of good practice that the children and teachers could follow.

The constructionist approach emphasizes the importance of process as well as the products of learning. To reflect this, the Web site was divided into two areas for the content. One area was for the completed products. The children placed their material in this part of the site when they felt happy enough to share it, and thus encourage other schools to enter into a collaborative partnership with them. One school, for example, has set up an international rivers project with schools participating from all over the world. The idea is for the schools to compare and contrast their local rivers. The other area is for work in progress. This area also will be public and other schools will be encouraged to make constructive comments at this formative stage of the creation process.

The level of communication, and indeed by implication the degree of 'presence' required, was thought to be best provided through a standard groupware product (FirstClass) which provides e-mail, text conferencing and 'discussion groups' facilities. These were chosen because in this particular context the children are being encouraged to be more reflective and thoughtful, and an asynchronous solution seemed best to support this level and type of communication.

One problem encountered fairly early on was the difficulty some teachers had in finding a focus that had curriculum relevance and that could be meaningfully shared across schools both in the UK and abroad. To help stimulate the project, the team, in conjunction with the local newspaper (the Ipswich Evening Star), suggested a common theme of '1000 years on the Orwell'. The Orwell is a local river in Ipswich near which many of the trial schools were located. Schools were encouraged to create projects addressing all areas of the curriculum (for example, pollution on the Orwell, a collaborative novel — 'Murder on the Orwell', the history of Landguard Fort at the mouth of the Orwell, etc).

The richness of the context, and the fact that the children 'owned' and were ultimately responsible for their own work, changes the nature of any on-line or off-line communications. The off-line communication is rich in the dialogue of collaboration and negotiation. The children work in small teams and have to plan as well as do the work. The same richness and sense of ownership attenuated the need for a high level of presence for the on-line communications, making fairly low fidelity communications systems with their associated low levels of 'presence' perfectly acceptable. This was possible because of the commonality and shared understanding of the task context. Rivers do, after all, exist nearly everywhere in the world, and children are children regardless of their background, colour or creed.

17.4 APPLYING TELEPRESENCE IN FURTHER AND HIGHER EDUCATION

When the role of technology in supporting distance learning in further or higher education (FE and HE respectively) came to be considered, it became clear that the social aspects were not only important for supporting collaboration, but also because people like a sense of group identity (i.e. the student cohort).

Putting people, who are essentially strangers, into contact with each other is always going to be difficult. Conversation usually occurs for a purpose, and often to be successful the communicators must have a shared understanding of their subject matter. Indeed, Vygotsky [10] would argue that lovers, as an extreme case, are so close that communication between them becomes sparse and telegraphic as they already know what each other is thinking. A system was therefore needed that allowed students to quickly provide and share information on each other that could then be used to 'seed' further conversation, until at least the course itself provided enough common experiences for the students to share, and the cohort to form.

A further factor to be considered was the importance of real-time interactions in effective communications. The question yet again came down to quality or fidelity of the communications medium. The ability to appropriately sense and react to non-verbal cues is clearly important for mediating and controlling conversations. In particular the use of eye contact to control turn-taking is well understood [11]. In contrast Bales [12] stresses the importance of the social and emotional activities that surround effective task-based group interactions. These can be either positive (e.g. shows solidarity, shows tension release) or negative (e.g. disagrees, shows tension), but from our perspective what is important about Bales' categories is that they are communicated equally effectively by spoken rather than visual cues.

To support these social needs, an Internet-based system for teaching English as a foreign language (EFL), RISE/MERLIN, was developed and tested with Hull University. The screen shot in Fig. 17.2 shows the main screen. The important features to note from a telepresence perspective is the 'People' and 'Meeting Place' facilities.

Activating the People link allowed students to access brief biographical information about their tutors and other members of the course. It was hoped that this would provide some information to help 'break the ice' in subsequent real-time interactions. These real-time interactions take place in the Meeting Place (see Fig. 17.3).

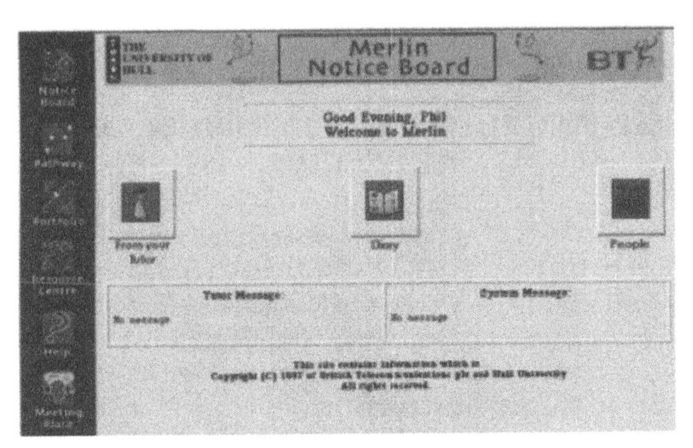

Fig. 17.2 The Merlin Notice Board.

Fig. 17.3 Merlin's Meeting Place.

The Meeting Place allows the setting up of multipoint audioconferences from the Web browser. All students who are logged on to the system are listed under 'on-line now'. If an audioconference is already taking place the participants are placed in a 'room'. Others can be invited to join the conference or even denied entry into the room. There was a need to ensure the use of high-quality speech both for teaching purposes and for maintaining acceptable feelings of presence. It was felt that the current versions of the Internet phone could not consistently deliver the required quality, and so the PSTN facilities were used. This does have the implication that students and teachers need two 'lines' — one for data and one for voice. This could be implemented by two PSTN lines, one for the telephone and the other for a modem (minimum 33 kbit/s), or an ISDN-2, or LAN access and a telephone.

One of the outcomes of the first trial was that people were reluctant to engage in chance meetings. They preferred to know in advance who was on-line and when. In the second version of RISE/MERLIN a diary was added (see Fig. 17.4) so that people could schedule their meetings.

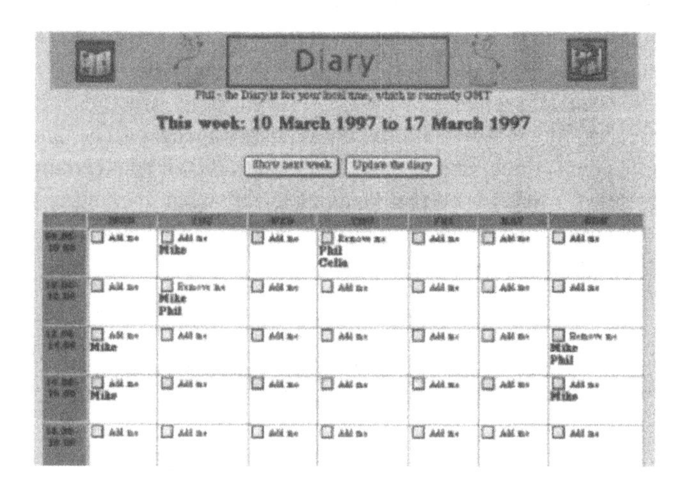

Fig. 17.4 The Diary.

RISE/MERLIN also provides more traditional learning support (see Fig. 17.2). The 'pathways' contain most of the teaching content provided by Hull University. Hull also provided materials for 'resources' which could be material specific to EFL (e.g. dictionaries) or interesting links to other Web sites which could encourage conversation and debate. Finally RISE/MERLIN's 'portfolio' provides facilities for students to store their exercises either as personal documents or as shared (group) documents. The 'documents' can be a mix of text and/or audio files. The audio files could be, for example, a tutorial recorded in the meeting place.

The two technically unique features of the RISE/MERLIN prototype were the use of a dynamic webserver which allowed access to data specific to a student and a course. To ensure this, the students have to log on and provide a password. The system can then access material relevant for their course and files personal to them. The second unique feature is the ability to set up an audioconference from a Web page using high-quality PSTN services rather than relying on the more variable quality of the Internet phone.

RISE/MERLIN was tested on 35 students from Western Europe, Central Europe, the Far East, USA and South America. Of these students who undertook the 15-week intermediary EFL course, ten accessed MERLIN from their homes, sixteen at their colleges, and six from their work. Other students had to make special arrangements to access an appropriate machine. Early results from the evaluation study suggest that the twelve triallists who completed the studies enjoyed doing the course, were satisfied with it, had received considerable help with their English, and would definitely want to do another MERLIN course. The findings suggested that the students who dropped out either suffered unacceptable technical problems or tended to favour more structured classroom-based approaches, found it difficult to talk to other students on-line, and felt the need for more conventional tutor contact time. Overall, and with certain caveats, the findings suggest that for certain kinds of learners and learning situations a carefully designed and interactive on-line course can be as effective as undertaking a traditional course.

17.5 BEYOND RISE — ADVANCED LEARNING ENVIRONMENTS

The work described above in schools and colleges uses today's technology (predominantly PCs over ISDN) and was constrained by that technology. The conceptual framework, by definition, is technology independent, so the question was asked what an advanced learning environment (ALE) would look like if there were no bandwidth, processing or hardware constraints. In such an ideal world a high-quality or fidelity telepresence solution would become available.

The questions then are: when do we need such fidelity and how does it help us learn? The project is new and to date there are only visualizations of what a high-bandwidth (100 Mbit/s) desktop version of RISE might look like.

At the conceptualization stage in Mayes' framework, we imagined that small groups of learners (about three) would like to share the 'lecture' experience.

In Fig. 17.5 a mock-up of what the user interface may look like for the conceptualization stage can be seen. There is a central space or 'learning canvas', and below that there is a 'film strip' from which a lecture can be chosen. The boxes to the left are communication windows, and in this version they would be using high-quality multipoint videoconferencing.

In Fig. 17.6, the lecture is in progress and is being watched by three students. The students are able (with mutual agreement) to fast forward, pause, rewind, etc, the lecture which is 'streamed' from a large central video server. They are also able to make notes while the lecture is in progress and to attach the notes to the relevant part of the lecture material. Notes can be private or shared.

Fig. 17.5 The user interface (UI) for ALE's conceptualization stage.

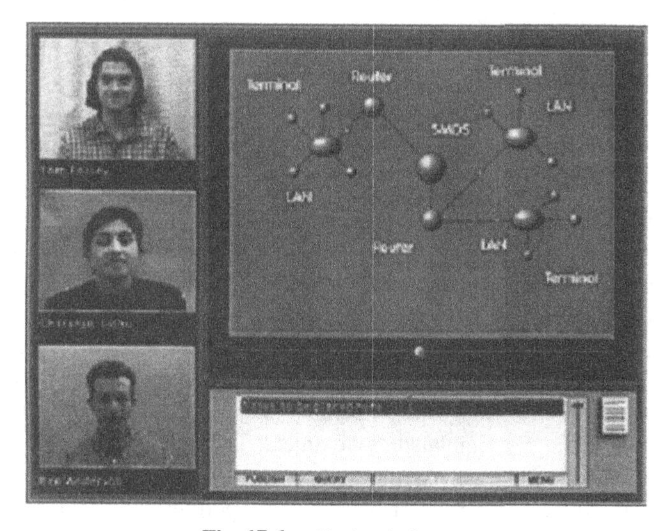

Fig. 17.6 Sharing the lecture.

If at any time a student wishes to ask a question for clarification purposes, they can stop the lecture, type in and submit their question, and a clip from a video server of the lecturer answering the question then appears (see Fig. 17.7). If in the unlikely event no suitable answer is available, the lecturer will suggest e-mailing a tutor or expert for an answer. The answer will, at a later date, be added to the video database. Although technically it is possible (but difficult) for the student to speak the question, the use of a written question seemed more appropriate as it supports a more reflective and thoughtful approach. This is similar to the traditional situation where learners will often spend time composing their question either in their minds or on paper to ask the lecturer at the end of a session.

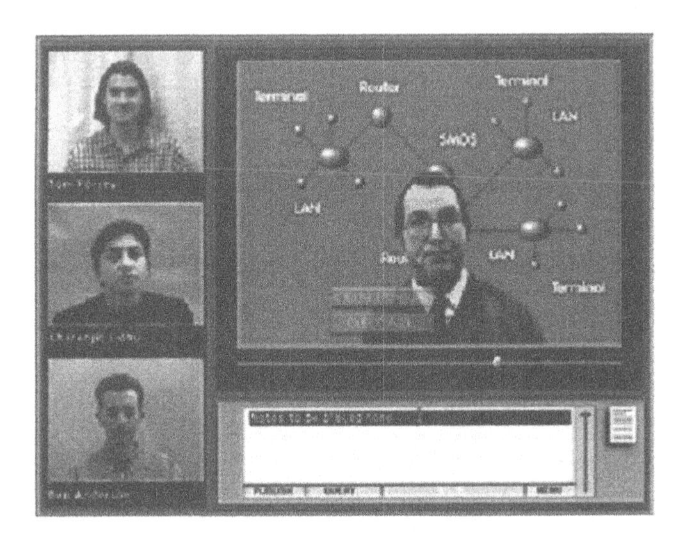

Fig. 17.7 Asking questions at the conceptualization stage.

Although learning in a group is important, it is most probable that 'protocols' for social learning will have to be developed. For example, a simple protocol may be that no student should interrupt on the first viewing of the lecture material. Only on subsequent viewing would the students be encouraged to question each other's understanding and to use the VCR controls to access relevant parts of the lecture to illustrate or support any point being made.

At the construction stage, the learning canvas now supports a highly interactive, modelling or simulated environment. This stage could involve more traditional data-conferencing facilities to support collaborative writing or creation of multimedia materials. It could also support virtual worlds where the learners will be able to touch and manipulate objects and have the means of sharing their experiences in a collaborative way. The collaboration can be achieved in two ways — through the conventional videoconferencing windows for the small group, and in virtual worlds where others are invited from outside that group who can have virtual representations through the use of avatars.

Finally, for the dialogue stage we return to a traditional videoconferencing set-up. Initially the tutor appears on the learning canvas and assumes the role of chair person (see Fig. 17.8), but if a tutee wishes to take control this can be signalled and if the chair person permits they can move into the central canvas and take control of the dialogue.

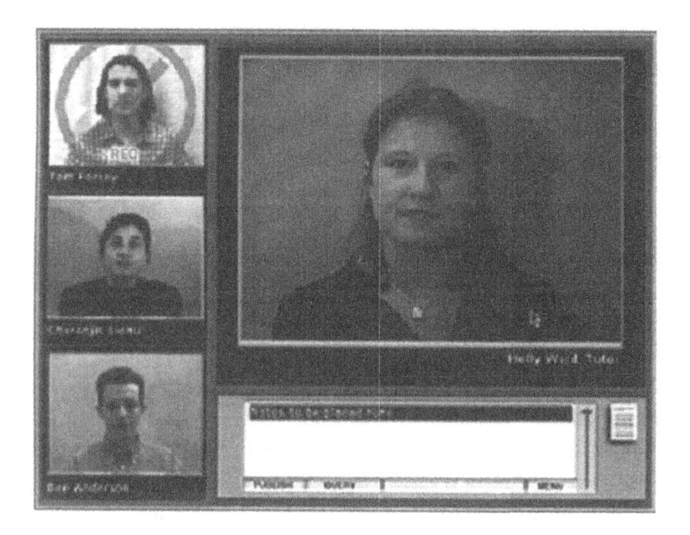

Fig. 17.8 The UI for ALE's dialogue stage.

Notes can also be taken and shared. The videoconference can be recorded and reused either for revision purposes, or as source material for subsequent lectures.

It is also the intention to specify and build a microlearning centre. The design of such a centre will still be based on the basic principles described in Mayes' Learning Framework, but will try to escape the constraints of the traditional learning 'furniture'. In current distance-learning technology, that furniture is represented by the desk-top computer. The microlearning centre will build a fully interactive virtual environment (IVE).

The technical challenges will be to understand the interfaces between many different types of technology, from a simple telephone, through to a desktop computer, to fully emersive virtual worlds involving high-definition wrap-round projections.

The educational challenges will be to design such centres to be educationally valid and to prove them to be so. To achieve this, the same content currently used for one of the BT MSc modules will be used. The content will then be converted and presented in the two new technical learning environments — the desktop and the microlearning centre. However, the effectiveness will then be measured across the three environments — the traditional, the desktop and the mircolearning centre.

17.6 DISCUSSION

Our original thoughts on the relationship between presence and learning were driven by such concepts as engagement, relationships and audience. These undoubtedly proved important. On reflection the key and superordinate concept seems to be the notion of relationships. The relationship concept embodies notions of task and purpose, and also the type of relationship sets up expectations according to the roles of the participants. The nature of the relationship could also determine whether or not it is appropriate to include a wider audience as part of the communication and the degree of formality of the interaction. In formal situations a full range of cues are required and a shared understanding cannot be assumed between the communicators. In such a situation, high-fidelity systems would be advantageous. In less formal situations, less cues are required and lower fidelity systems would be acceptable.

The importance of the learning task cannot be underestimated in designing effective systems. There appear to be two important concepts. Firstly, there is the main learning 'content' which in ALE and RISE systems takes place in the learning canvas. In terms of Mayes' conceptual framework, the content of the learning, both in form and type, will vary according to the learner's stage in the framework (i.e. conceptualization, construction and dialogue). The other concept appears to address the dialogue or conversations that 'go around' the learning content. In a similar fashion to content, the nature of the dialogue (i.e. clarification and collaboration) will vary according to Mayes' first two stages.

Clearly the two variables are related or bound together by the learning framework. However, in terms of applying telepresence there seems to be important differences. When dealing with the conceptualization stage the critical issues appear to be fidelity or realism of the material. The conceptualization stage needs to be rich and flexible, allowing learners with different backgrounds and abilities to quickly grasp new concepts. The issue is often to build upon what the learners already know and to use concrete illustrations to push the learner into the 'conceptual' unknown. Often at this stage the narrative or story line needs to be explicit and in the control of the teacher. This, in association with relative passivity required of the learner, it is argued, demands high-quality well-structured, balanced and engaging material. The necessity for engaging content suggests parallels with a well-made TV documentary, and indeed traditional producers of broadcast material may have an important role to play in educating teachers in the production of engaging learning materials.

At the construction stage the emphasis is more on 'doing'. Photorealism can be sacrificed for higher levels of interactivity with objects represented in the 'constructive' world. Realism is still required but the emphasis is now on the quality of the behaviour of the object rather than the quality of its representation. In other words, the learners should be able to interact with objects in their virtual world and the behaviour of those objects should be both internally and ecologically valid. Internal validity refers to the appropriateness of the behaviour within the virtual world, whereas ecological validity refers to the correlation between the behaviour of virtual objects with their real world counterparts (where they exist). Furthermore, in the construction stage the control of the narrative needs to move from the teachers to the learners. The role of the teacher is one of facilitation or guidance in ensuring that the narrative is made explicit in order that the learning lessons can be abstracted and generalized to new situations.

At the dialogue stage there is a movement from the concrete to the abstract. Such a movement assumes a change in knowledge state from the novice to the more expert. Dialogue between experts is often carried out in real time, has little predetermined structure (i.e. the arguments 'flow' from point to point), and often requires access to shared whiteboards and notepads to graphically illustrate points or arguments.

The essential element of discourse in such situations demands flexible or dynamic communications facilities. For example, to regulate and support a highly dialectic situation requires high-fidelity videotelephony where participants can easily sense critical non-verbal cues that will regulate the argument and indicate the emotional tone of the discussion. Equally there will be situations when the discussants will want to access whiteboards or show artefacts in order to support their positions. In such situations high-quality audio combined with data-conferencing is required.

The dialogue aspects of the framework also make differential demands on telepresence principles. Dialogue for clarification only requires a real-time presence of the questioner rather than the respondent. This is based on the principle of 'most frequently asked questions' which assumes in the vast majority of cases the questions asked are not original and a database of answers can be created and accessed. Moreover, the formulation of the question often requires reflection and so the use of text, albeit in real time, with the ability to edit before submitting, would meet the 'reflection' requirement. The reflection requirement also demands a greater temporal separation between the content that provoked the question and the question itself. It is often only after reflection that a failure to understand a concept becomes apparent. Indeed the design of the conceptualization stage needs to support such dissociation through the ability to replay, edit and/or tag 'conceptual' material.

Dialogue for collaboration would also need to be in real time in order to map on to the current 'constructive' process, and would be symmetrical in requiring the real-time presence of all participants. The need to map in real time between the conversation and the content requires the ability for participants to easily identify (e.g. pointing), manipulate (e.g. gripping) or observe realistic changes in behaviour (i.e. good speed or latency of response is required). These behavioural elements combined with real-time verbal abilities (either voice or text) are critical to this stage. There is no requirement for the ability to record or play back collaborations other than to support clarification dialogues (e.g. 'why did you do that?') or as input (i.e. as content) into the conceptualization stage.

The marriage of the dialogues with the learning stage is also important in suggesting future research directions. It helps direct our research effort away from the syntax of the dialogue (i.e. the rules that govern the utterances) which are becoming well understood [13], at least in video-mediated communications, to the semantics of the dialogue. Meaning or semantics can be best understood within the situation or context of the utterance, which itself can be captured through the concept of relationships.

17.7 CONCLUSIONS

In the learning context at least, the continual push for higher and higher quality or more and more realistic telepresence systems is not always necessary, and indeed in some situations may distract rather than benefit the learner. The notable exception to this may be in cases where a learning task requires a high-quality and realistic visual content to be valid at the construction stage (e.g. in teaching surgery, or engineering), or again at the construction stage where in the 'doing' it is important to detect emotional changes in peoples' interactions (e.g. in teaching counselling or interviewing skills).

A more general conclusion arising from this work, is that basic telepresence principles like 'presence' or 'engagement' are best understood within applied contexts. Indeed it may be more useful for designers to have a set of heuristics (or rules) rather than principles which can then inform design decisions. More research needs to be undertaken to understand the rules governing the different types and quality of representations and communication required in different learning situations. Only then will we be in a better position to truly answer the question of whether or not 'being there' is important.

REFERENCES

1. MacFarlane A G J: 'A report of a working party of the committee of Scottish University Principals on Teaching and Learning in an Expanding Higher Education Systems', CSUP (1992).

2. Dearing R: 'Higher education in the learning society', Published by The National Committee of Inquiry into Higher Education (1993).

3. Carey J: 'Interaction patterns in audio conferencing', Telecommunications Policy, pp 304-313 (December 1981).

4. Fisher E: 'Characteristics of Children's talk at the computer and its relationship to computer software', Language and Education, 7, pp 187-215 (1992).

5. Fowler C J H, Mayes J T and Bowles B A: 'Education for changing times', British Telecommunications Eng J, 15, pp 32-38 (1996).

6. Dwyer D: 'Apple classroom of tomorrow: What we've learned', Educational Leadership, Apple Incorporation (1994).

7. Mayes J T: 'Learning Technology and Groundhog Day', in Strang W, Simpson V B, Slater D (Eds): 'Hypermedia at Work: Practice and Theory in Higher Education', University of Kent Press, Canterbury (1995).

8. Williams E: 'Experimental comparisons of face-to-face and video mediated communication', Psychological Bulletin, 84, pp 963-976 (1977).

9. Allport F H: 'Social psychology', Houghton Mifflin, Boston (1924).

10. Vygotsky L S: 'Thought and language', MIT Press (1962).

11. Argyle M: 'The psychology of interpersonal behaviour', Penguin, London (1967).

12. Bales R F: 'Interaction process analysis', Addison-Wesley, Cambridge, Mass (1950).

13. Anderson A H, O'Malley C, Doherty-Sneddon G, Langton S, Newlands A, Mullin J, Fleming A M and Van der Velden J: 'The impact of VMC on collaborative problem solving: an analysis of task performance, communicative process, and user satisfaction', in Finn K, Selleu A J and Wilbur S B (Eds): 'Video-Mediated Communications', Lawrence Erlbaum Associates, New Jersey (1996).

18

CONCEPT 2010 — BT'S NEW GENERATION DEALING DESK

D J T Heatley, I B Cockburn, F T Lyne, A K Williamson, K J Fisher, I Neild and N Haque

18.1 INTRODUCTION

Today's wholesale finance sector relies heavily on state-of-the-art telecommunications to conduct business in an efficient and competitive manner. This is particularly evident in the trading community where deals are won or lost on the strength of the information that traders can access electronically across the globe, and the speed and effectiveness with which they can act on that information.

For many years BT has been a major player in this sector, and now commands a 55% share in the UK market for dealing room technology. With some 50 000 dealing positions installed at over 1400 sites in 45 countries, BT's market share world wide is 25%. In order to retain and indeed increase this coverage into the next millennium it is essential that BT keeps its customers in the finance sector abreast of the very latest in technologies and working practices. Furthermore, this enlightening process must be continuous and driven by the customers and the markets in which they operate. This can be achieved through the use of an evolving conceptual platform on which these technologies and working practices can be demonstrated and evaluated. BT Laboratories (BTL) [1] and Syntegra [2], the company's systems integration arm, have jointly developed such a platform for the dealing room market and it is called Concept 2010.

This chapter describes the key features of Concept 2010, paying particular attention to those novel aspects that will bring real benefits to traders. Also presented is a vision of how this platform will evolve, with discussions of some of the new features currently under development.

18.2 THE BT ADVANTAGE

Everything about Concept 2010 is radical. The latest sound, video and data technologies have been integrated to enable traders to operate in the most flexible and productive manner. Flat panel displays are used throughout to make the most of screen real estate while at the same time reducing the overall size of the desk. Powerful data-handling and visualization techniques minimize the quantity of information presented to the dealer and maximize its content. Multipoint videoconferencing and a spatial sound system ensure that the dealer gets the most out of link-ups to remote locations. Speech platforms can be accommodated to allow the dealer to interact with and control the desk by spoken commands. The latest broadband wireless technologies do away with all cable communication links to the desk, thus eliminating the high cost incurred when moving desks, and allowing the latest structured wiring techniques to be exploited. Finally, the 'furniture' aspect of the desk comprises the latest modular design which can be configured to any requirement.

Concept 2010 was unveiled at Innovation'97, a showcase exhibition held at BT Laboratories in April 1997. Since then the desk has been demonstrated in many of the world's dealing centres. Such has been the impact at these events that prospective customers now include some of the world's largest finance institutions.

18.3 DEALING PLATFORM

All dealing operations on the Concept 2010 desk are performed using the open trading system (OTS) market data platform, developed and marketed by Syntegra. Such is the performance and flexibility of this platform that it has become an industry standard. OTS-4 is the latest version and is designed to work openly with all major market data systems. It gives the dealer unprecedented control over the presentation and manipulation of on-screen data, allowing use of the display space to its best advantage (Fig. 18.1). By incorporating a state-of-the-art object-oriented user interface, OTS-4 gives intuitive access to data and enables linking to other dealing applications in real time, thus ensuring that the dealer receives the most power from the desk and the best overall competitive advantage.

The use of a voice/data integration (VDI) platform within the Concept 2010 desk means that a variety of time-critical functions can be performed automatically to assist the dealer. For example, when the dealer answers an incoming telephone call from a corporate client, that client's trading history, portfolio valuation and exposure will automatically appear on the displays. Similarly, if a telephone call is received from another dealer, the desk will

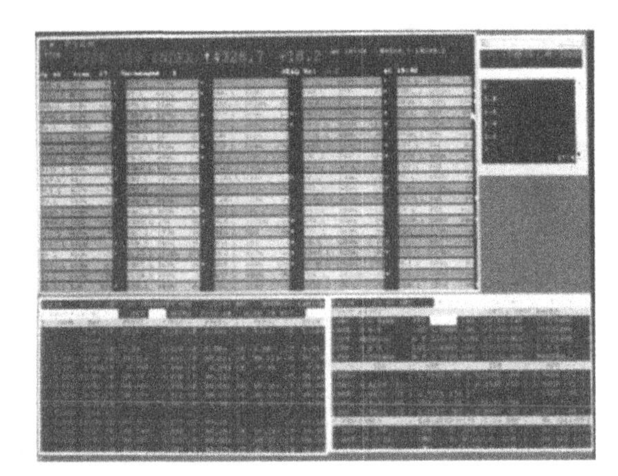

Fig. 18.1 Typical OTS dealing data display.

automatically display the source of the call and the counter party information to make deal capture and settlement quicker and easier.

All of these features, including others described in later sections, are placed under the control of a Syntegra-developed Desk Manager. This provides the integration framework that 'glues' the various applications together into a seamless whole, and provides dealers with an easy visual interface through which they can choose and manage the applications they require.

A crucial feature of the dealing platform within Concept 2010 is its compatibility and upgradability. As newer versions and different types of dealing software emerge, they can be readily installed on the desk. By ensuring that the desk can incorporate developments as they emerge, the desk itself becomes future proofed and dealers are assured of a continued competitive advantage.

18.4 DISPLAYS

One of the most striking features of the Concept 2010 desk is its use of the latest flat panel displays. Gone is the need to accommodate deep, heavy, cathode ray tube (CRT) displays with their heat dissipation problems. Flat panel displays afford the minimum of depth, weight, heat, and overall power consumption, while at the same time delivering a picture clarity and resolution that nowadays can easily match that of the best CRT displays. In addition, flat panel displays do not flicker as CRTs do and so afford comfortable viewing for much longer periods.

Concept 2010 utilizes four 13.3-inch colour flat panel displays (Fig. 18.2), manufactured by Hitachi and packaged to BT's specific requirements by Microvitec [3]. Each screen is configurable to XGA resolution (1024 × 768 pixels), offering a picture real estate equivalent to that of a 17-inch CRT. By exploiting the latest super-TFT (thin film transistor) fabrication techniques these displays give a much brighter and sharper image than previous flat panel technologies, and a much wider viewing angle (140° horizontal and vertical). Three displays are mounted vertically on the desk and effectively form a single panoramic viewing surface. The dealer may choose to display different information on each display, or instead may choose to 'spread' the same information across all three.

Fig. 18.2 Flat panel displays and touch-screen.

The fourth display is mounted horizontally and is used as an active desk top. This display incorporates a touch-sensitive panel so that the dealer can set up and control many functions of the desk by touch and it also provides the telephony interface for the desk. Figure 18.3 shows a typical layout that enables the dealer to control all telecommunications functions.

Fig. 18.3 Typical touch-screen layout.

18.5 VIDEOCONFERENCING

Modern day dealing is increasingly making use of videoconferencing and TV distribution, and the Concept 2010 desk embraces this to the fullest extent. It can incorporate a multipoint videoconferencing system, such as the FreeVision system developed by C-C-C [4], which then enables the dealer to establish videoconference links with a number of colleagues simultaneously, and if necessary share their working data. Any of the vertical screens can be configured to display a videoconference image as an overlay, while still displaying dealing and market data in the background. Figure 18.4 shows the C-C-C system in operation on a four-way videoconference.

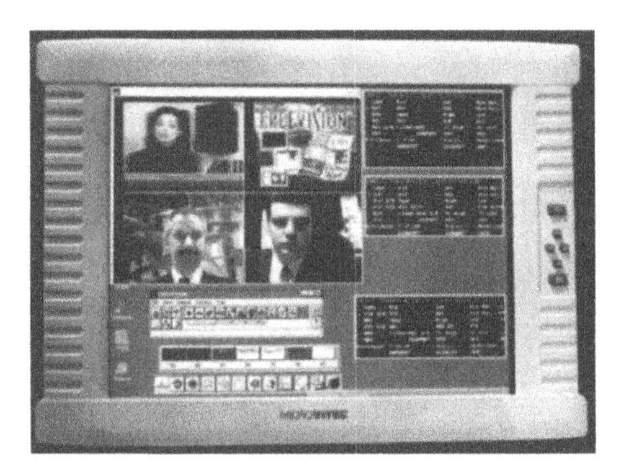

Fig. 18.4 Four-way videoconferencing.

By placing this system under the control of the Desk Manager described earlier, conference links can be set up manually by the dealer or automatically by the desk in response to, for example, a change in the dealer's position which requires him to consult urgently with clients or colleagues.

In order to maximize the overall convenience of videoconferencing, the Concept 2010 desk uses a video camera that tracks the movement of the dealer. This ensures that the dealer is always 'in shot' from the point of view of the remote participants. This camera can be seen above the middle screen in Fig. 18.2.

In addition to videoconferencing, dealers are increasingly using TV distributed to their desks to assist in their decision-making process. This is incorporated into the Concept 2010 desk with the audio being delivered to the handset or the spatial speaker system described later in this chapter.

18.6 DATA VISUALIZATION

Identifying market trends quickly and acting upon them in the correct manner is central to successful dealing; however, the volume of information with which dealers are confronted can be overwhelming. Concept 2010 utilizes the latest in BTL-developed visualization platforms to give the dealer complete control over how best to display data and market trends, and to highlight changes the instant they occur in a simple, intuitive manner.

One of the many visualization options that Concept 2010 can offer dealers is 'virtual worlds' in which objects and surfaces can be made to represent either individual items of data or whole assemblages of data. The colour or 'emotional behaviour' of these objects and surfaces can be programmed to signify the state of the data to which they relate. Figure 18.5 shows a typical example as viewed on one of the desk's displays.

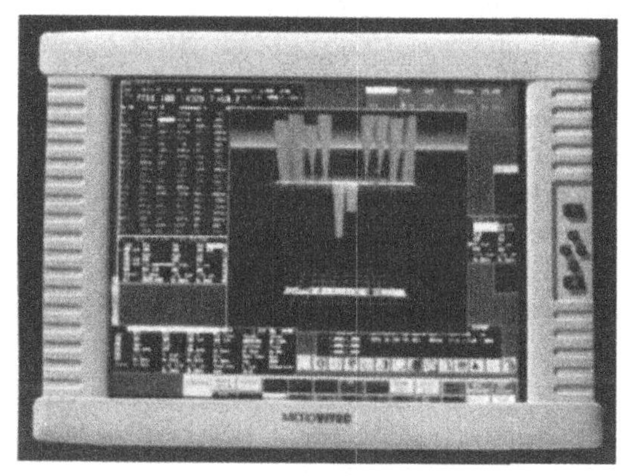

Fig. 18.5 Data visualization in a virtual world.

Individual stocks, bonds or currencies are represented in Fig. 18.5 by vertical columns. The colour of each column signifies whether the value of that particular stock has risen or fallen from its opening value, and the length defines the size of the shift in value. The shape of the columns indicates whether the value is rising (upward pointing triangle) at that precise moment or falling (downward pointing triangle). If the rate of this change in value exceeds a threshold (defined by the dealers themselves), these columns then also begin to wobble like jelly, i.e. they look worried. This signifies to the dealers that urgent action is required.

Future developments in this field will see the use of 'intelligent agents' to perform search and retrieve functions and to build a profile of the dealer's pattern of usage so that the displayed information can be prioritized and formatted as preferred. As virtual reality and 3-D visualization techniques improve, much larger multi-dimensional databases will be able to be visualized at the desktop with only a small overhead on processing power.

18.7 SPATIAL SOUND SYSTEM

Voice telecommunications is a vital part of modern-day dealing, so much so that conventional dealing desks often incorporate an array of loudspeakers, each conveying a single audio link from, for example, an information supplier, a counter party, or fellow dealers at other finance centres, and so on. These loudspeakers unavoidably create clutter at the desk. The Concept 2010 desk removes this clutter by first mixing the various audio feeds and then delivering them to the dealer via a single pair of stereo loudspeakers. However, this on its own would have created new problems because all the simultaneous incoming voices would be effectively sitting on top of one another, thereby making it difficult or even impossible for the dealer to distinguish individual voices. To avoid this, the desk incorporates a spatial sound system, developed specifically for this application by Edinburgh Communications Ltd.

This system accepts audio signals from up to eight distinct sources and distributes them spatially around the desk in a 3-dimensional (3-D) configuration chosen by the dealer (Fig. 18.6). For example, the audio channel from New York could appear to emanate from the left of the desk, with London in the middle and Tokyo on the right, with other locations distributed in between. For dealing environments that require more than eight channels, the system is readily scalable.

Fig. 18.6 Spatial sound system.

The spatial sound system also improves the effectiveness of the videoconferencing facilities described earlier by creating the impression that the voices of the remote participants emanate from the particular display screen on which they appear.

18.8 FURNITURE

The furniture constituents of the desk are the very latest in modular designs from Specialised Banking Furniture International Ltd (SBFI). This furniture can be configured exactly to the operating requirements of the dealers and their working environment.

The double-sided configuration shown in Fig. 18.7 is particularly appropriate to Concept 2010 as it exploits the significant space-saving features of the flat panel displays. The distance between the vertical viewing surfaces on each side of the desk is only about 40 cm, which compares very favourably with the 150 cm or more for conventional desks that are populated by CRTs.

Fig. 18.7 Concept 2010 dealing desk.

With Concept 2010 it is therefore possible to accommodate two dealing positions within a floor area that one conventional position would occupy. In a large dealing room this equates to an enormous saving in space and a reduction in operating costs.

A further important benefit afforded by the furniture is the unimpeded view that the dealers have of each other while seated at their desks. Since the flat panel displays have the same resolution and information area as the 17-inch CRTs used by many dealers today but are themselves only 13.3 inch, it is significantly easier to ensure that no equipment protrudes above the topmost element of the desk, thus ensuring full visual contact throughout the room.

18.9 INFRA-RED WIRELESS COMMUNICATIONS

In dealing institutions it is vital that desks can be moved to different locations on the floor to respond to changes in the business requirements of dealing teams. Such moves might be frequent (e.g. monthly or even weekly), involving a great many desks and costing several thousand pounds per desk. The high cost is due in large part to the difficulty in recovering and relocating the underfloor communications cabling to the desk. Indeed, to minimize this cost the old cabling is often left in place and new cabling is overlaid to the new location. Clearly a point will be reached where the underfloor cabling cavities within the building are congested or floor loading becomes a concern, beyond which further moves are impossible without a major clear-out and re-cabling programme.

Concept 2010 addresses this problem by incorporating state-of-the-art wireless communications. The wireless channel is delivered to/from the desk via an infra-red communications system that is being developed jointly by BTL and Microvitec.

The 155 Mbit/s capacity of this system ensures that it is significantly future proofed and ATM compatible. Furthermore, the use of low-power lasers operating at a wavelength of 1550 nm ensures that the system has a good operating margin and is eye-safe at all times.

The infra-red wireless system creates an optical 'cell' around the desk from a transceiver located in the ceiling (Fig. 18.8). In so doing, the system does away with the need for underfloor cabling and instead utilizes the latest structured wiring techniques in the ceiling cavity. However, as the Concept 2010 desk is fully compatible with underfloor cabling, the dealing institutions can adopt the communications option that best suits their circumstances.

All voice, data and video to/from the desk are conveyed within the infra-red cell. The use of infra-red rather than radio as the wireless medium offers a unique evolutionary path because of the significantly greater bandwidth that is available. Infra-red also offers an element of security since the communications cell is confined to a single desk and the edge of the cell is very sharply defined.

With a sufficient number of ceiling stations distributed throughout the room (Fig. 18.9), dealing desks can be moved to any other location without the need to recover and relocate cabling.

18.10 MODULARITY

A crucial feature of the Concept 2010 desk is modularity. All aspects of the desk, from the furniture through to the various technologies, can be configured to the exact requirements of individual customers. For example, the furniture can be free standing in a single-sided or double-sided form (see Fig. 18.7 for the latter), or arranged end-to-end in rows or islands.

.Fig 18.8 A representation of infra-red wireless communications to Concept 2010.

Fig. 18.9 Infra-red wireless in a dealing room.

The mix of technologies at each dealing position is similarly flexible. The front position shown in Figs 18.2 and 18.7 makes full use of the latest technologies, with a vertical viewing surface that nearly spans the full width of the desk. In contrast, the rear position (Fig. 18.10) uses a blend of new and existing technologies. For example, the conventional dealer's console on the right of the desktop is interchangeable with the touch-sensitive panel in use on the other position (Figs 18.2 and 18.7). The rear position also makes use of powerful data compression and visualization techniques (Fig. 18.5) to reduce the required vertical viewing surface to just one display. This configuration might therefore be particularly suited to dealing environments in which space is at a premium and the width of the desks is less than that shown in Fig. 18.7.

Fig. 18.10 View of the rear dealer position.

18.11 FEATURES UNDER DEVELOPMENT

As relevant new technologies and services become available, they will be downstreamed to the Concept 2010 desk. For example, CORBA, voice over IP, and information agents are currently at an advanced stage of development and will be ready for deployment during 1999. The following sections describe a selection of these and other areas that will be explored over the coming one or two years.

18.11.1 Speech recognition

At a conventional dealing desk the dealer will often have to look away from critical information on the displays in order to instigate a course of action. This may be as simple as a telephone call, or it could involve the setting up of a complex multiway conference. In either case the dealer's attention is being distracted and hence their effectiveness is impaired to some extent. Speech recognition [5] offers a solution since it would allow dealers to speak commands to the desk without their attention being diverted.

The Concept 2010 desk can accommodate a variety of speech recognition systems, each with its own pros and cons in relation to the particular operating environment. These systems can be broadly categorized as speaker independent and speaker dependent. A speaker-independent system is generally the more robust of the two in the sense that it will respond to any speaker and will minimize the number of instances when a spoken command is incorrectly interpreted. However, the vocabulary of such systems may be limited to a few hundred words and phrases. Conversely, a speaker-dependent system can accommodate a vocabulary spanning thousands of words and phrases, and also affords a degree of security since it will only respond to one dealer. However, misinterpretations and the need to repeat a spoken command might occur relatively more frequently.

Speech recognition could have a profound impact on dealing practice. For example, 'deal tickets' which carry all the details of a particular transaction could be completed in real time by the dealer largely, or even entirely, by voice command. Not only is this likely to speed up the process, but the data carried by the ticket could also include authentication of the dealer's voice, in addition to the usual time and date stamps, at which desk the transaction was conducted, etc.

Dealers might also wish to use speech recognition to create documents by oral dictation. Such documents could, for example, give details on a transaction and could be sent to colleagues by e-mail for them to read at a convenient time.

18.11.2 Speech synthesis

When dealers are faced with visual information spread across multiple displays it is inevitable that critical events may occur on one of the displays at which the dealers are not presently looking. The conventional solution to this is for the desk to sound a simple audio prompt, such as a beep, and to overlay an accompanying message window on the relevant display. Although this approach is perfectly adequate, it represents another distraction of the kind alluded to in the previous section. Once again, speech can be used to remove this distraction, but the roles are reversed as it is now the desk that speaks to the dealer by the use of speech synthesis.

The Concept 2010 desk can accommodate a variety of speech synthesis systems such as BTL's own Laureate system [5]. This system produces a very realistic voice which it then uses to speak commands and phrases from a library that is tailored to the particular application. The gender and timbre of the voice can also be selected to suit the application. With such a system integrated into the

Concept 2010 desk, not only can all prompts to the dealer be vocal, but each prompt will orate their exact nature, for example, 'the FTSE index has just risen by X points', or, 'you have an urgent call coming in from John in New York'. All of this represents a significant reduction in distraction and a corresponding increase in the effectiveness of the dealer.

Speech synthesis can overcome other distractions that commonly occur at dealing desks. For example, a conventional desk will display streamed news from agencies such as Reuters as text on a display. To access this information dealers must focus their attention on the relevant display. However, speech synthesis can be used to 'read aloud' streamed news to dealers, which allows them to attend to other displays while listening to the news.

18.11.3 Talking head

Speech synthesis systems such as Laureate can be linked to a virtual 'talking head' on one of the displays whose lip movements are synchronized to the synthetic words (Fig. 18.11).

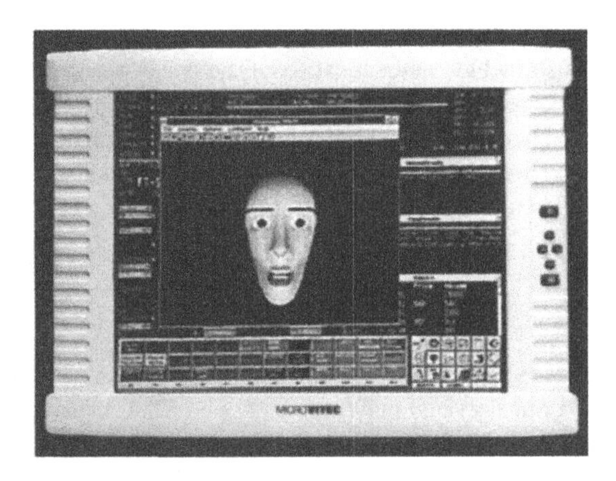

Fig. 18.11 Virtual talking head.

This feature is being developed at BTL as a way of increasing the 'presence' of prompts and spoken information to the dealer. The head has the effect of bringing oral information to the foreground and so is a very powerful way of controlling the balance of importance between oral and visual information. For example, the desk might convey low importance prompts and audio messages in voice only, whereas critical prompts and messages might also be conveyed via the talking head. In parallel with this, the timbre of the voice could be selected to accentuate the level of urgency.

18.11.4 Text summarizer

In addition to the 'streamed news' alluded to in the previous section, dealers may also wish to obtain information from various financial newspapers and journals. Nowadays, this can be done with ease via Web browsers. However, whereas streamed news is highly abridged by the source agencies to maximize content within the fewest number of lines, news obtained directly from Web sites is largely unabridged. Consequently, dealers spend significantly more time absorbing the latter forms of information, which in turn represents yet another distraction.

With applications such as this in mind, BTL has developed and BT now markets a text summarizer called Netsum [6] which abridges streamed text or whole documents that have been downloaded. With text summation incorporated into the Concept 2010 desk, dealers can then receive abridged information from a wide range of news sources. Similarly, dealers could use text summation to reduce the length of a document that they wish to transmit to colleagues.

Text summarizers such as Netsum allow the extent and 'flavour' of the abridgement to be selected to suit the content of the original text. For example, a dealer might wish to abridge two multi-paged news articles, one concerning market trends, the other concerning a natural disaster somewhere in the world. Although both sets of information are important to the dealer because they may influence their subsequent decisions, their flavours are nevertheless very different. In order to produce abridgements that correctly encapsulate the overall meaning of the original texts, different settings would be used.

Abridgements such as these can be delivered to the dealer as text on a display, or as a spoken message using the speech synthesis systems described earlier.

18.11.5 Information agents

New techniques are becoming available that enable dealers to be supplied with information from Web sites and other sources which matches preferences that are set manually by each dealer or are 'learned' automatically by the desk over time. This data is 'found' by using information agents and would include some data that dealers might otherwise miss during their normal routine. Information agents are software utilities that run independently and report back their findings to the dealer so that informed decisions can be made. When, where and how these agents are used is entirely flexible. For example, they could be embedded in each client's portfolio, then when launched the agents will seek out relevant data, such as cross references, similarly performing stocks or bonds, etc, and feed this back to the dealer for action. This process, being iterative, improves with time. Furthermore, by linking these agents to text summation as described above, the collected data may be presented to the dealer in abridged form, thus enhancing its overall value.

It is clear that information agents represent a powerful new tool for dealers that will help to raise their overall efficiency, particularly when linked to other information platforms such as text summation.

18.11.6 Active desktop

The coming few years will see rapid strides in the development and widespread use of flat panel displays. For example, as this book goes to press, large size (e.g. 20-inch) flat panel displays are becoming available which offer the same wide viewing angle and picture clarity as the 13.3-inch models used on the Concept 2010 desk and described earlier in this chaper, but with a SXGA resolution (1600 × 1200 pixels) rather than XGA (1024 × 768 pixels). Aside from the obvious advantage of these displays in giving the dealer a larger vertical viewing surface within which to distribute the various data windows illustrated earlier, these displays could also be embedded horizontally into the desktop fabric, flush with the top surface, whereupon the desktop surface becomes one large, customizable viewing surface. Then add a touch-sensitive panel and the dealer has a desktop sized, customizable touch-screen. This feature will introduce valuable opportunities for new working styles. For example, the active desktop could incorporate the current touch-screen illustrated in Fig. 18.3 in addition to 'virtual' representations of a keyboard, tracker ball, telephone keypad, notepad, etc, all shown as life size graphics on the desktop and all movable and configurable with the wave of a hand. Similarly, data windows on the vertical flat panel displays could be moved to the active desktop whereupon touch-driven interaction becomes available. Such a desktop would deliver the ultimate in flexible, clutter-free work spaces.

18.11.7 Security

It is vital that dealers, and the institutions within which they operate, have absolute confidence in the security of their dealing desks because of the extreme sensitivity of the data that they can access. This is particularly true if individual desks are time shared between several dealers. In the present realization of the Concept 2010 desk, security is effected by means of conventional user IDs and entry passwords when the dealer logs on. Although this level of security is entirely adequate, new developments at BTL in the fields of iris recognition [7] and voice authentication [5] will raise this to an unprecedented level, thus maximizing overall confidence. With these technologies integrated into the Concept 2010 desk a typical logging-on might then comprise the following:

- the dealer sits at the desk;
- the iris is scanned and the dealer is recognized to a very high certainty;
- the dealer then speaks a specific word or phrase, or enters a PIN;
- the desk recognizes the voice and the specific password;
- the overall certainty that the dealer is bona fide is now absolute;
- the desk then activates and auto-configures to the last known work pattern for that particular dealer.

A unique advantage that security systems like iris authentication and voice authentication have over conventional user IDs and passwords is that they can be in continuous operation. While the dealer is seated at the desk his voice is authenticated every time he speaks a command to the desk and his iris is periodically scanned. These features could be integral to a future voice-operated deal-capture system.

A further level of security might include active badges. During the logging on procedure the desk interrogates the dealer's badge via radio or infra-red, depending on the particular badge technology in use. Such badges could also be part of a building-wide security and tracking system.

18.12 CONCLUSIONS

This chapter has shown that Concept 2010 represents a truly radical approach to dealing desk design. The use throughout of state-of-the-art technologies and the latest in software platforms ensures that dealers have the best possible working environment and competitive edge. Furthermore, the modular design of the desk allows the mix of technologies, services and furniture to be tailored to the exact requirements of individual dealers and institutions. Indeed, this modularity also means that the desk can be configured for a wide range of other key applications, for example, command and control centres, multimedia work stations.

As the majority of the desk features described in this chapter are available today as off-the-shelf items, dealers whose requirements are satisfied by that particular technology mix can be accommodated today. The other features will be introduced when the underlying technologies become sufficiently robust. Indeed, Concept 2010 is a demonstration platform that will continue to evolve. For example, the use of virtual worlds in dealing is a new paradigm being pioneered by BT. The value it adds to a dealer's effectiveness is still being evaluated, but as the 'Sega' generation start to appear on the dealing room floor, the benefits will become more quickly realized. Another example is the use of speech platforms. The dealing environment is arguably one of the most challenging for speech applications, which means that Concept 2010 is the ideal platform to demonstrate and evaluate these new ideas.

Through the facilities afforded by Concept 2010, BT now has a powerful demonstration and evaluation platform designed to show strategic customers in the finance sector the latest technologies and applications in a relevant environment. The feedback and validation that this will generate will help BT to take those customers into the next millennium, and to continue to offer solutions that give an assured competitive edge.

REFERENCES

1. http://www.labs.bt.com/
2. http://www.syntegra.com
3. http://www.microvitecplc.com/
4. http://www.cccgroup.co.uk/
5. Westall F A, Johnston R D and Lewis A V (Eds): 'Speech technology for telecommunications', Chapman & Hall (1997).
6. Preston K R: 'From books to bytes — managing information in the information age', British Telecommunications Eng J, 15, Part 1, pp 72-77 (April 1996).
7. http://www.labs.bt.com/library/papers/PAMIpaper/PAMIpaper.html

Appendix

List of acronyms

A&E	accident and emergency
ADSL	asymmetric digital subscriber loop
ALE	advanced learning environment
ALIVE	Artificial Life Interactive Video Environment
ANOVA	analysis of variance
ARC	Alternate Realities Corporation
AVC	Audio-Video Control
BAP	body animation parameter
BB	bulletin board
BDP	body definition parameter
BGP	border gateway protocol
BTL	BT Laboratories
CAD	computer aided design
CCI	Conference Call 'Instant'
CDR	customer data record
CIF	common intermediate format
CN	customer number
CPU	central processor unit
CRT	cathode ray tube
CSCW	computer-supported co-operative work
CTI	computer/telephony integration
CVE	collaborative virtual environment
DCT	discrete cosine transform
DEE	distributed entertainment environment
DLIC	digital line interface card
DM	dynamics manager
DNS	domain name service
DSP	digital signal processor (*or* processing)
EFL	English as a foreign language

FAP	facial animation parameter
FAQ	frequently asked questions
FDP	facial definition parameters
FTP	file transfer protocol
GAT	Generic Application Template
GCC	Generic Conference Control
GUI	graphical user interface
HCI	human/computer interaction
HDTV	high definition television
HF	human factors
HMD	head mounted display
HRTF	head-related transfer function
HTTP	hypertext transfer protocol
IETF	Internet Engineering Task Force
IGMP	Internet group management protocol
IMTC	International Multimedia Teleconferencing Consortium
IP	Internet protocol
IP	information place
IRTF	Internet Research Task Force
ISDN	integrated services digital network
ISP	Internet service provider
ITU	International Telecommunication Union
IVE	interactive virtual environment
LAN	local area network
MC	multipoint controller
MCS	Multipoint Communication Service
MCT	Millennium-CT
MCU	multipoint control unit
medivac	medical evacuation
MIDI	Musical Instrument Digital Interface
MITG	media integration of text and graphics
MMUSIC	multi-party multimedia session control
MUD	multi-user dungeon/multi-user dimension
NSAS	network spatial audio server
PBX	private branch exchange
PCL	personal conference list
PDU	protocol data unit
PIN	personal identification number
PSTN	public switched telephone network
QoS	quality of service
RevRooms	Reverb Rooms
RGB	red, green, blue
RM	rule manager

RPC	remote procedure call
RSVP	resource reservation protocol
RTCP	real-time control protocol
RTP	real-time transport protocol
RTSP	real-time streaming protocol
SAP	session announcement protocol
SDP	session description protocol
SGi	Silicon Graphics
SIP	session initiation protocol
SIS	sensory interface simulator
SNHC	synthetic/natural hybrid coding
SRM	scalable reliable multicast
SURVIVE	Simulated Urban Recreational Violence IVE
TCP	transport control protocol
TLS	transport layer security
URL	uniform resource locator
UDP	user datagram protocol
VCR	video cassette recorder
VDI	voice/data integration
VE	virtual environment
VEMUD	virtual environment MUD
VLSI	very large scale integration
VM	verification model
VOP	video object plane
VQ	vector quantization
VR	virtual reality
WWW	World Wide Web

Index